Lecture Notes in Economics and Mathematical Systems

477

Springer
Berlin
Heidelberg
New York
Barcelona
Hong Kong
London
Milan
Paris
Singapore
Tokyo

Michel Théra Rainer Tichatschke (Eds.)

Ill-posed Variational Problems and Regularization Techniques

Proceedings of the "Workshop on Ill-Posed Variational Problems and Regulation Techniques" held at the University of Trier, September 3–5, 1998

Springer

Editors

Prof. Dr. Michel Théra
University of Limoges
LACO, URA 1586
123 Avenue Albert Thomas
87060 Limoges Cedex, France

Prof. Dr. Rainer Tichatschke
University of Trier
Fachbereich IV – Mathematik
54286 Trier, Germany

Library of Congress Cataloging-in-Publication Data

Ill-posed variational problems and regularization techniques / [edited by] Michel Théra, Rainer Tichatschke.
p. cm. -- (Lecture notes in economics and mathematical systems, ISSN 0075-8442 ; 477)
Includes bibliographical references.
ISBN 978-3-540-66323-2 (softcover : alk. paper)
1. Variational inequalities (Mathematics)--Improperly posed problems. I. Théra, M. A. (Michel A.), 1946- . II. Tichatschke, Rainer, 1946- . III. Workshop on Ill-Posed Variational Problems and Regularization Techniques (1998 : Trier, Germany) IV. Series.
QA316.I45 1999
515'.64--dc21 99-40871
CIP

ISBN-13: 978-3-540-66323-2 e-ISBN-13: 978-3-642-45780-7
DOI: 10.1007/978-3-642-45780-7

Typesetting: Camera ready by authors/editors
SPIN: 10699801 42/3143-543210 - Printed on acid-free paper

Preface

This volume is a collection of papers presented at the *Workshop on Ill-Posed Variational Problems and Regularization Techniques* held at the University of Trier (Germany) in September 1998.

From September 3 through September 5, 1998, about 50 scientists met at Trier to discuss recent developments in the field of ill-posed variational problems and variational inequalities. 17 lectures were delivered, covering a large range of theoretical, numerical and practical aspects. The topics, as well as the invited speakers, were selected by the organizers. The main topics discussed were

- Regularization methods for equilibrium problems
- Variational inequalities and complementarity problems and their regularization
- Regularization of fixed point problems
- Links between approximation, penalization and regularization
- Bundle methods, nonsmooth optimization and regularization
- Error bounds for regularized optimization problems

The organizers are gratful to all participants for their contribution to the success of this workshop. We also wish to express our cordial thanks for the financial support granted by the *Deutsche Forschungsgemeinschaft*, Bonn and the *University of Trier.*

We are indebted to the referees for their helpful comments and suggestions and to our colleagues of the University Trier for their assistance in preparing this volume.

M. Théra, Université de Limoges (France)

R. Tichatschke, University of Trier (Germany)

Contents

Regularization Method for Equilibrium Programming Problem with Inaccurate Initial Data

Anatoly Antipin * Fedor Vasil'ev **

Abstract

Regularization methods for solving general problems of equilibrium programming are applied to canonical forms where the objective functions given are inaccurate. The regularized penalty-function method is offered as a solution for equilibrium problems in which the admissible set is given inaccurately. The convergence of the proposed methods is investigated.

Keywords: equilibrium program, regularization method, penalty-function method, convergence

AMS subject classification:65K10,90C31

1 Statement of Problem

Let us consider the problem of computing a fixed point of the extreme constrained mapping [1],[2]

$$\textit{find} \quad v^* \in \Omega_0 \quad \textit{such that} \quad v^* \in \text{Argmin}\{\Phi(v^*, w) \mid \ g(w) \leq 0, \ w \in \Omega_0\}. \quad (1)$$

Here the function $\Phi(v, w)$ is defined on the product space $R^n \times R^n$ and $\Omega_0 \subset R^n$ is a convex closed set. It is supposed that $\Phi(v, w)$ is convex in $w \in \Omega_0$ for every $v \in \Omega_0$. The vector-valued function $g(w)$ has the dimension m. Every component of this function is convex. The variable $v \in \Omega_0$ is a parameter and $w \in \Omega_0$ is the optimization variable. We also assume that the extreme (marginal) mapping $w(v) \equiv \text{Argmin}\{\Phi(v, w) \mid \ g(w) \leq 0, \ w \in \Omega_0\}$ is defined for all $v \in \Omega_0$ and the solution set $\Omega^* = \{v^* \in \Omega \mid \ v^* \in w(v^*)\} \subset \Omega_0$ of the initial problem is nonempty. According to Kakutani's fixed point theorem the latter assertion follows from the continuity of $\Phi(v, w)$ and the convexity of $\Phi(v, w)$ in w for any $v \in \Omega_0$, where Ω_0 is compact. In this case $w(v)$ is an upper

*Computing Center Russian Academy of Sciences, 40 Vavilov str., 117967 Moscow. Russia
**Moscow State University, Vorob'evy Gory, 119899 GSP, VMC Faculty, Moscow. Russia

semicontinuous mapping that maps each point of the convex, compact set Ω_0 into a closed convex subset of Ω_0 [9].
According to (1), any fixed point satisfies the inequality

$$\Phi(v^*, v^*) \leq \Phi(v^*, w) \quad \forall w \in \Omega, \tag{2}$$

where $\Omega = \{w | \ g(w) \leq 0, \ \ w \in \Omega_0\}$ is an admissible set of (1). Let us introduce the function $\Psi(v,w) = \Phi(v,w) - \Phi(v,v)$ and using it present (2) as

$$\Psi(v^*, w) \geq 0 \quad \forall w \in \Omega. \tag{3}$$

This inequality is a consequence of (1). But if this inequality is considered as primary then it is known as Ky Fan's inequality [9] since it is proved in [12] that there exists the solution of (3), that is vector v^*. In this case the existence of the fixed point of (1) results from (3).
The problem of (1) can be regarded as a scalar convolution of diverse game problems, which describe the situation of the reconciliation of contradictory interests and factors for many agents. For instance, we show, that a $m-$ person game with Nash equilibrium [13] can be scalarized and presented as a equilibrium programming problem (1). Indeed, let M be the number of agents in a system. Each agent i of the system has certain variables or actions it controls $x_i = (x_{i_1}, ..., x_{i_{n_i}})$, $x_i \in R^{i}$, $i = 1, ..., M$. Let $x = (x_1, ..., x_M)$ describe the variables of all M agents, where $x \in R^N$ for $N = \sum_{i=1}^{M} n_i$. The variables of agent i, x_i depends upon the variables of the other agents. To describe this, let x_{-i} be the vector consisting of all x except x_i, where $x_{-i} \in R^{N-n_i}$, that is $x_{-i} = (x_1, ..., x_{i-1}, x_{i+1}, ..., x_M)$. Using the introduced notation we can write $x = (x_i, x_{-i})$ Each agent of the system tries to minimize his payoff function $f_i(x_i, x_{-i})$ only over the x_i. There are constraints $g_i(x_i) \leq 0$, $x_i \in X_i$ that depend upon variable x_i only. Both functions $f_i(x_i, x_{-i})$ and $g_i(x_i)$ are convex in $x_i \in X_i$. The equilibrium solution of $m-$ person game is the solution of the system of extreme inclusions

$$x_i^* \in \text{Argmin}\{f_i(x_i, x_{-i}^*) \mid \ g_i(x_i) \leq 0, \ x_i \in X_i\}. \tag{4}$$

In particular, if $i = 1, 2$, and $f_1(x_1, x_2) + f_2(x_1, x_2) = 0$, we receive a saddle point programming problem [3].
Let us enter a normalized function of a kind [16]

$$\Phi(v,w) = \sum_{i=1}^{m} f_i(x_i, x_{-i}),$$

where $w = (x_1, ..., x_M)$, $v = (x_{-1}, ..., x_{-M})$, $g(w) = (g_i(x_i))$, $i = 1, 2, ..., M$ and $\Omega = X_1 \times X_2 \times ... \times X_n$, With the help of this convolution the problem (4) can be shown to be of the form (1).

Many inverse optimization problems [4] can be introduced also alike (1). As an example, we shall consider a inverse convex programming problem in the form

$$x^* \in \text{Argmin}\{\langle \lambda^*, f(x)\rangle \mid g(x) \le 0,\ x \in Q\}, \quad G(x^*) \le d. \tag{5}$$

In this problem, one must choose nonnegative coefficients of linear convolutions $\lambda = \lambda^*$ so that the optimal solution $x = x^*$ corresponding to these weights belongs to the preassigned convex set. In particular, this set may contain only one point. It is supposed, that all functions in this problem are convex.
The system (5) can be recorded as a two-person game with the Nash equilibrium [1]

$$\begin{aligned} x^* &\in \text{Argmin}\{\langle \lambda^*, f(x)\rangle \mid G(x) \le 0, \quad x \in Q\}, \\ \lambda^* &\in \text{Argmin}\{\langle \lambda, G(x^*) - d\rangle \mid \lambda \ge 0\}. \end{aligned} \tag{6}$$

The problem (6) in turn, as it was shown above, can be written in the form of a fixed point problem for an extreme map (1). Other examples of problems, which can be considered as (1), see in [11].
The adduced examples show that the areas for the possible applications of equilibrium problems are extensive, therefore the development of methods for solving equilibrium problems will generate sizeable interest. In the present paper we consider and justify the regularization and regularized penalty-function methods for the solution of equilibrium problems. This approach is based on the idea of an expansion of the objective function as a sum of functions from two special subclasses. Each function from the first subclass we shall call pseudo-symmetric, and those from the second subclass shall be called skew-symmetric. The types of problems are determined by their objective functions, which it is natural to describe as potential and skew-symmetric equilibrium problems. In the following sections we shall consider sequentially their properties.

2 Pseudo-Symmetric Functions

We consider a class of pseudo-symmetrical functions.

Definition 1 *A differentiable function $\Phi(v, w)$ from $\mathbb{R}^n \times \mathbb{R}^n$ in $\mathbb{R}^1$ is called pseudo-symmetric on $\Omega_0 \times \Omega_0$, if there exists a differentiable function $P(v)$ such that*

$$\nabla P(v) = 2\nabla_w \Phi(v, w)|_{w=v} \quad \forall v \in \Omega_0, \tag{7}$$

where $\nabla P(v)$ is the gradient of $P(v)$ and $\nabla_w \Phi(v, w)$ is the partial gradient of the function $\Phi(v, w)$ in w.

The function $P(v)$ is called the potential of the operator $\nabla_w \Phi(v, w)\,|_{w=v}$.

If the function $P(v)$ is twice continuously differentiable, then the Lagrange formula follows from (7)

$$P(v+h) = P(v) + 2\int_0^1 \langle \nabla_w \Phi(v+th, v+th), h\rangle dt. \tag{8}$$

On the contrary, if the Jacobi matrix $\nabla F(v)$ for the operator $F(v) = \nabla_w \Phi(v,v)$ is symmetric for all $v \in \Omega_0$, then (8) holds and in this case operator $\nabla_w \Phi(v,v)$ is potential [17].
So, if the objective function in (1) satisfies (7) or (8), then the equilibrium problem is called potential.
The set of all pseudo-symmetric functions makes by itself a linear space.
The pseudo-symmetric functions include all symmetric functions. These functions are possible to define with the help of following conditions

$$\Phi(w,v) = \Phi(v,w) \quad \forall w \in \Omega_0, \quad \forall v \in \Omega_0. \tag{9}$$

If $\Phi(v,w)$ is a differentiable function, then we shall obtain by differentiating identity (9) in w,

$$\nabla_v \Phi(w,v) = \nabla_w \Phi(v,w) \quad \forall w \in \Omega_0, \quad \forall v \in \Omega_0. \tag{10}$$

Let's assume $w = v$ in (10), then we have

$$\nabla_v \Phi(v,v) = \nabla_w \Phi(v,v) \quad \forall v \in \Omega_0. \tag{11}$$

Thus we can formulate the following

Property 1 *The contractions of partial derivatives of symmetric functions onto the diagonal of the square $\Omega_0 \times \Omega_0$ are identical.*

By the definition of the differentiability of the function $\Phi(v,w)$, we have [21]

$$\begin{aligned}\Phi(v+h, w+k) = \Phi(v,w) + \langle \nabla_v \Phi(v,w), h\rangle + \\ \langle \nabla_w \Phi(v,w), k\rangle + \omega(v,w,h,k),\end{aligned} \tag{12}$$

where $\omega(v,w,h,k)/(|h|^2+|k|^2)^{1/2} \to 0$ as $|h|^2+|k|^2 \to 0$. Let $w = v$ and $h = k$, then with regard to (11) we get from (12)

$$\Phi(v+h, v+h) = \Phi(v,v) + 2\langle \nabla_w \Phi(v,v), h\rangle + \omega(v,h), \tag{13}$$

where $\omega(v,h)/|h| \to 0$ as $|h| \to 0$. Since the formula (13) is a particular case of (12) it means that the contraction of the gradient $\nabla_w \Phi(v,w)$ onto the diagonal of the square $\Omega_0 \times \Omega_0$ is the gradient $\nabla \Phi(v,v)$ of function $\Phi(v,v)$, i.e.

$$2\nabla_w \Phi(v,w)|_{v=w} = \nabla \Phi(v,v) \quad \forall v \in \Omega_0. \tag{14}$$

Thus, we can prove

Property 2 *If $\Phi(v,w)$ is a symmetric function, then the operator $\nabla_w\Phi(v,v)$ is potential and coincides with the contraction of the gradient for the function $\Phi(v,w)$ onto the diagonal of square, i.e. $2\nabla_w\Phi(v,w)|_{w=v} = \nabla\Phi(v,v) = \nabla P(v)$.*

The concept of potentiality has been considered in the scientific literature rather for a long time. Apparently, one of the first articles where the potential was used for the analysis of asymptotic stability for a gradient method to solve the m-person game, was the publication [19]. In one of recent papers [15], on an example of m-person game the concept of potential game was introduced as follows. If for the m-person game with Nash equilibrium

$$x_i^* \in \text{Argmin}\{f_i(x_i, x_{-i}^*) \quad | \quad x_i \in X_i\},$$

where $f_i(x_i, x_{-i})$ is a payoff function of the i-th player, $x_i, i = 1, ..., M$, $x_{-i} = (x_1, ..., x_{i-1}, x_{i+1}, ..., x_M)$, $x_i \in X_i$, there is a function $P(x_1, x_2, ..., x_n)$ such that

$$\frac{\partial P(x_1, x_2, ... x_m}{\partial x_i} = 2\frac{\partial f_i(x_1, x_2, ..., x_m)}{\partial x_i}, \tag{15}$$

then the game is called potential. In other words, partial derivatives of payoff functions in own variables of the players will make the gradient of some function $P(x_1, x_2, ..., x_m)$ which is called a potential. We shall present the right-hand side of (15) in the form

$$\frac{\partial f_i(x_1, x_2, ... x_m)}{\partial x_i} = \frac{\partial f_i(x_i, x_{-i})}{\partial x_i}, \tag{16}$$

and then we shall introduce a normalized function for the considered game

$$\Phi(v,w) = \sum_{i=1}^{n} f_i(x_i, x_{-i}),$$

where $w = (x_1, ..., x_M)$, $v = (x_{-1}, ..., x_{-M})$ $g(w) = (g_i(x_i))$, $i = 1, 2, ..., M$ and $\Omega = X_1 \times X_2 \times ... \times X_m$. Since the function $\Phi(v,w)$ is separable in w and the set X has a block structure, then

$$\nabla_w\Phi(v,w)\mid_{v=w} \;\; = \left(\frac{\partial f_i(x_i, x_{-i})}{\partial x_i}\right). \tag{17}$$

Comparing (15), (16) and (17), we have

$$\nabla P(v) = 2\nabla_w\Phi(v,w)\mid_{v=w}.$$

Thus, potential games in the sense of (15) are potential games and in the sense of (7).

If the objective function of problem (1) is subordinate to the condition (7), in this case, an equilibrium problem can be considered as an optimization problem. Really, from (2) we have

$$\langle\nabla_w\Phi(v^*, v^*), w - v^*\rangle \geq 0 \;\; \forall w \in \Omega. \tag{18}$$

By virtue of (7) from (18) we get

$$\langle \nabla P(v^*), w - v^* \rangle \geq 0 \quad \forall w \in \Omega. \tag{19}$$

If the operator $\nabla P(v)$ is monotone, then $P(v)$ is a convex function over Ω and $v^* \in \Omega^*$ is its optimal solution. In this case an equilibrium potential problem (1) can be replaced by a optimization problem of the function $P(v)$ over Ω. However we shall mark that the function $\Phi(v^*, w)$, generally speaking, is not convex and the condition (18) is only a necessary one. But, if $\Phi(v^*, w)$ is the convex function in $w \in \Omega_0$, then v^* is an equilibrium solution of (1).

3 Skew-Symmetric Functions

We pass to the consideration of the second class of objective functions for problem (1).

Definition 2 *([5]) A function $\Phi(v, w)$ is called skew-symmetric onto $\Omega_0 \times \Omega_0$ if it satisfies*

$$\Phi(w, w) - \Phi(w, v) - \Phi(v, w) + \Phi(v, v) \geq 0 \quad \forall w \in \Omega_0, \quad \forall v \in \Omega_0. \tag{20}$$

If the inequality

$$\Phi(w, w) - \Phi(w, v^*) - \Phi(v^*, w) + \Phi(v^*, v^*) \geq 0 \quad \forall w \in \Omega_0, \tag{21}$$

holds, where $v^* \in \Omega^*$, then the function $\Phi(v, w)$ is called skew-symmetric concerning an equilibrium point.

The class of skew-symmetric functions is non-empty, as it includes in itself all antisymmetric functions which can be determined with the help of the following relations

$$\Phi(w, v) = -\Phi(v, w) \quad \forall w \in \Omega_0, \quad \forall v \in \Omega_0. \tag{22}$$

Put $v = w$ in (22), then $\Phi(v, v) + \Phi(v, v) = 0$, that is the antisymmetric function is identically equal to zero on the diagonal of the square $\Omega_0 \times \Omega_0$. If the antisymmetric function is convex in w, then it follows from (22), that it is concave in v, that is in this case $\Phi(v, w)$ is a saddle point function. To illustrate the condition (20), consider the normalized function $\Phi(v, w)$ of the saddle-point problem, which can be obtained from (4) in $i = 1, 2$. In this case, as pointed out in [5], Φ satisfies

$$\Phi(v, v) = 0, \quad \Phi(v, w) + \Phi(w, v) = 0 \quad \forall w \in \Omega_0, v \in \Omega_0.$$

Note that the authors of [10] earlier attempted to extend these conditions to non-saddle-point problems.
The skew-symmetric functions have properties, which can be considered as analogues of monotonicity conditions of a gradient for convex functions.

Property 3 *If the function $\Phi(v,w)$ is differentiable, skew-symmetric and convex in w, then the contraction of the partial gradient $\nabla\Phi_w(v,v)$ onto the diagonal of the square $\Omega_0 \times \Omega_0$, is monotone i.e.*

$$\langle \nabla_w\Phi(v+h,v+h) - \nabla_w\Phi(v,v), h\rangle \geq 0 \quad \forall v \in \Omega_0,\ \forall v+h \in \Omega_0. \tag{23}$$

This inequality follows from (20), if we apply the condition of convexity for $\Phi_v(w) := \Phi(v,w)$

$$\langle \nabla\Phi_v(w), h\rangle \leq \Phi_v(w+h) - \Phi_v(w) \leq \langle \nabla\Phi_v(w+h), h\rangle \tag{24}$$

for all w and $w+h$ from some set.
In particular, if $\Phi(v,w)$ is a normalized function of the saddle-point problem, it follows from (24), that $(-\nabla_x L(x,y),\ \nabla_y L(x,y))^T$ is a monotone operator, where $L(x,y) = f_1(x,y) = -f_2(x,y)$ from (4). This fact was established in [18]. From the above it follows that the skew-symmetric equilibrium problems largely inherit properties of saddle-point problems.
We introduce a useful inequality which allows us to estimate the growth rate of a function $\Phi(v,w)$ in a neighbourhood of a point $v, w \in \Omega_0 \times \Omega_0$.

$$|\{\Phi(w+h,v+k) - \Phi(w+h,v)\} - \{\Phi(w,v+k) - \Phi(w,v)\}| \leq L\,|h|\,|k| \tag{25}$$

for all $w,\ w+h \in \Omega_0,\ \ v,\ v+k \in \Omega_0$, where L is constant. The class of functions satisfying condition (25) is non-empty [5].
It was established earlier that the symmetrical functions have the potential properties. However some of them have also skew-symmetric ones. Indeed, consider a subset of functions subject to the condition: $\Phi(v,w) \leq \sqrt{\Phi(w,w)\Phi(v,v)}\ \forall v,w \in \Omega_0 \times \Omega_0$. Let us write out an expression similar to the left-hand side of (20). Using (9) and the condition introduced, we rewrite this expression to obtain $\Phi(w,w) - \Phi(w,v) - \Phi(v,w) + \Phi(v,v) = \Phi(w,w) - 2\Phi(w,v) + \Phi(v,v) \geq \Phi(w,w) - 2\sqrt{\Phi(w,w)\Phi(v,v)} + \Phi(v,v) = (\sqrt{\Phi(w,w)} - \sqrt{\Phi(v,v)})^2 \geq 0\ \forall v,w \in \Omega_0$. From here and from (23) it follows that if $\Phi(v,w)$ is convex in w for any $v \in \Omega_0$, then $v \to \nabla_w\Phi(v,v)$ is monotone.
Now we shall consider the properties of the equilibrium solution of (1) in the case when the objective function is skew-symmetric. Let $v = v^* \in \Omega^*$, then taking into account (2) we get from (21)

$$\Phi(w,w) \geq \Phi(w,v^*) \quad \forall w \in \Omega. \tag{26}$$

Thus, any equilibrium solution of a skew-symmetric problem (1) satisfies the condition (26).
Inequalities (2) and (26) are of fundamental importance for proving the convergence of methods to equilibrium solutions of skew-symmetric problems [1],[2]. Using a function $\Psi(v,w)$ from (3) we present inequalities (2) and (26) as the following system

$$\Psi(w,v^*) \leq \Psi(v^*,v^*) \leq \Psi(v^*,w) \quad \forall w \in \Omega. \tag{27}$$

Here the right-hand inequality is known as Ky Fan's inequality (3) [9]. It follows from (27) that v^*, v^* is a saddle point of $\Psi(v,w)$ onto $\Omega \times \Omega$, and $\Psi(v^*,v^*) = 0$. Note that the function $\Psi(v,w)$ in general is never convex-concave, even if $\Phi(v,w)$ is convex-concave. Inequality (27) shows us that the skew-symmetric equilibrium problems inherit properties of saddle-point problems.

4 Splitting of Functions

In the previous sections we considered classes of potential and skew-symmetric equilibrium problems. There is still the question of how to proceed in the case where the objective function of an equilibrium problem does not belong to either of those classes. We shall show that in general, the objective function can be decomposed into a sum of two functions such that one is pseudo-symmetric and the other is skew-symmetric.

Remember that the pair of points with coordinates w, v and v, w is located symmetrically concerning a diagonal of square $\Omega_0 \times \Omega_0$, i.e. relatively to the linear manifold $v = w$. This gives us a capability to introduce the concept of a transposed function [6]. If we assign the value of the function $\Phi(v,w)$ calculated at the point v, w to the point with coordinates w, v, we obtain the transposed function $\Phi^T(v,w) = \Phi(w,v)$. In terms of this function, condition (9) and (22) look like

$$\Phi(v,w) = \Phi^T(v,w), \quad \Phi(v,w) = -\Phi^T(v,w).$$

Using the obvious relations $\Phi(v,w) = (\Phi^T(v,w))^T$, $(\Phi_1(v,w) + \Phi_2(v,w))^T = \Phi_1^T(v,w) + \Phi_2^T(v,w)$, we can readily verify that any real function $\Phi(v,w)$ can be represented as the sum

$$\Phi(v,w) = S(v,w) + K(v,w), \tag{28}$$

where $S(v,w)$ and $K(v,w)$ are symmetric and antisymmetric functions, respectively. This expansion is unique, and

$$S(v,w) = \frac{1}{2}\left(\Phi(v,w) + \Phi^T(v,w)\right), \quad K(v,w) = \frac{1}{2}\left(\Phi(v,w) - \Phi^T(v,w)\right). \tag{29}$$

As the conditions of pseudo-symmetry and skew-symmetry are generalized concepts of symmetry and antisymmetry it is possible to regard always that any objective function of a problem (1) can be splitted into a sum of pseudo-symmetrical and skew-symmetric functions, which can be done certainly not uniquely. For instance, any sum of two latter functions always can be considered as an example of such an expansion.

So, expansion (28), where the functions $S(v,w)$ and $K(v,w)$ are pseudo-symmetric and skew-symmetric is justified. From(28) we have

$$\nabla_w \Phi(v,v) = \nabla_w S(v,v) + \nabla_w K(v,v) \tag{30}$$

Using the condition of pseudo-symmetry (7), it is possible rewrite the equality (30) in the form

$$\nabla_w \Phi(v,v) = \frac{1}{2}\nabla P(v) + \nabla_w K(v,v) \tag{31}$$

We recall that the fixed point of (1) satisfies inequality (2). If $K(v,w)$ is differentiable in $w \in \Omega_0$ for any $v \in \Omega_0$, from (2) and (31) the necessary condition follows

$$\langle \frac{1}{2}\nabla P(v^*) + \nabla_w K(v^*,v^*), w - v^* \rangle \geq 0 \quad \forall w \in \Omega. \tag{32}$$

Note that this condition is necessary (and sufficient in the case of convexity of function $\frac{1}{2}P(w) + \Phi(v^*,w)$) for problem

$$v^* \in \text{Argmin}\{R(v^*,w) = \frac{1}{2}P(w) + K(v^*,w) \mid \quad g(w) \leq 0, \quad w \in \Omega_0\}. \tag{33}$$

We call the representation of the objective function of problem (1) in the form $R(v,w) = \frac{1}{2}P(w) + K(v,w)$ **canonical.** Here $P(w)$ is any real function and $K(v,w)$ is a skew-symmetric one. In particular, if $S(v,w)$ is a symmetrical function, then $P(v) = S(v,v)$. It is important to underline that the **canonical** representation always exists for any objective function $\Phi(v,w)$ of problem (1). The function $R(v,w)$ is in turn skew-symmetric and consequently, primarily, $\Phi(v,w)$ would be possible to consider as skew-symmetric, but in case, when $P(w)$ is potential of the operator $\nabla_w S(v,v)$ the condition (30) will be only necessary and it needs convexity $S(v,w)$ in w to be the point v^* of the equilibrium solution for (1).
In consequent sections we shall assume that the objective function of problem (1) is given in the canonical form. Note that if in this representation one of the components in $K(v,w)$ is not given, then we deal with a nonlinear programming problem.

5 Regularization Method

In this section we consider problem (33) in the case, where its functional constraints are not present.

$$v^* \in \text{Argmin}\{R(v^*,w) \mid \quad w \in \Omega_0\}. \tag{34}$$

We discuss now the practicability of applying the regularization method to solve this equilibrium problem. Let us show that we have an ill-posed problem. To this end consider the following example.

The Example. Consider the problem with objective function of kind $\Phi(v,w) = vw$, where $|v| \leq 1$, $|w| \leq 1$. The marginal map $w(v) = \text{argmin}\{vw \mid w \in$

$\Omega_0\}$, $\Omega_0 = [-1, 1]$ of this problem looks like

$$w(v) = \begin{cases} -1, & v > 0, \\ [-1, +1], & v = 0, \\ +1, & v < 0. \end{cases}$$

The graph of this map intersects with the line $v = w$ in the point $v^* = 0$ which is a fixed point of this map.
Alongside with the function $\Phi(v, w)$ we consider $\Phi_\delta(v, w) = vw - \delta w^2$, where $\delta > 0$, which we shall treat as perturbed with respect to $\Phi(v, w)$. The size of a disturbance is uniform onto $\Omega_0 \times \Omega_0$ and does not exceed δ. Indeed, $|\Phi_\delta(v, w) - \Phi(v, w)| = \delta w^2 \leq \delta \ \forall w, v \in \Omega_0, \ \forall \delta > 0$.
Let us present $\Phi_\delta(v, w)$ in the form $\Phi_\delta(v, w) = -\delta w \left(w - \frac{v}{\delta}\right)$. It is easy to see that the marginal map of problem $w_\delta(v) = \text{argmin}\{-\delta w(w - \frac{v}{\delta}) \mid w \in \Omega_0\}$, looks like

$$w_\delta(v) = \begin{cases} -1, & v > 0, \\ \pm 1, & v = 0, \\ +1, & v < 0. \end{cases}$$

The graph of this map does not intersect the line $w = v$ and, therefore, the perturbed map has no fixed points for all $\delta > 0$. In other words, the problem has no equilibrium solution for the perturbed function $\Phi_\delta(v, w)$ over the compact set Ω_0 and with arbitrarily small $\delta > 0$.
Moreover, this problem also does not have ϵ-approximate equilibrium solutions. Let us offer evidence of that. At first let parameter $v > 0$ be given as positive at function $\Phi_\delta(v, w) = -\delta w \left(w - \frac{v}{\delta}\right)$, then we obtain a parabola which is concave and intersected with the X-axis in points 0 and $\frac{v}{\delta}$. On the segment $[-1, +1]$ this function will reach the minimal value which equals $\Phi(v, -1) = -\delta - v$ at the point $w = -1$. All points from interval $[-1, -1 + \epsilon]$ get values $-\delta - v + (2\delta + v)\epsilon - \delta\epsilon^2$. If $\epsilon \to +0$, then $-\delta - v + (2\delta + v)\epsilon - \delta\epsilon^2 \to -\delta - v$. Thus a set of the approximate solutions of our parabola on an interval $[-1, +1]$ is a segment as $[-1, -1 + \epsilon]$ for all $v > 0$. Similar reasoning is correct for values of parameter $v < 0$. From these methods the general structure of the ϵ-marginal map for the perturbed problem clears up.

$$w_{\delta,\epsilon}(v) = \begin{cases} [-1, -1 + \epsilon], & v > 0, \\ [-1, -1 + \epsilon], [+1, +1 - \epsilon], & v = 0, \\ [+1, +1 - \epsilon], & v < 0. \end{cases}$$

Thus this map represents two bars: that is the "negative" bar for positive v and the "positive" bar for negative v and, still, this map does not intersect the line $v = w$, i.e. the ϵ-marginal map of a perturbed problem onto a compact set does not have any fixed points.
Remember that in the optimization field the perturbed problems on compact sets always have solutions. Therefore the adduced example has basic characteristics. This shows that the solution of the perturbed equilibrium problem is hardly possible without preliminary regularization.

We also formulate a regularization method with reference to the solution of equilibrium problems and we discuss its use at first in the absence of perturbations (for easier understanding of the logic of proof) and then they are given. To simplify the character of reasoning all calculations are accomplished in the finite-dimensional Euclidean space. Actually all theorems proved here are correct in any infinite-dimensional Hilbert space.

1. Convergence in the absence of perturbations. The regularization method for an equilibrium problem consists in the replacement of the initial unstable problem (34) with the family of stable equilibrium problems [22]

$$v_\alpha = \operatorname{argmin}\{R(v_\alpha, w) + \alpha\,|w|^2 \ \mid\ w \in \Omega_0\}, \tag{35}$$

where v_α is the regularized equilibrium solution. If the set Ω_0 is a convex compact set and the continuous function $R(v, w)$ is convex in w for any v, then the regularized problem (35) always has a fixed point. Methods for computing a fixed point of the extreme map (35) are considered in [5], [7].
By definition any solution of problem (35) satisfies the inequality

$$R\,(v_\alpha, v_\alpha) + \alpha\,|v_\alpha|^2 \le R\,(v_\alpha, w) + \alpha\,|w|^2 \quad \forall w \in \Omega_0. \tag{36}$$

It is possible to expect that the family of regularized solutions v_α will converge to the solution of the initial problem v^* if the effect of the term $\alpha|w|^2$ to the function $R(v, w)$ will be reduced as $\alpha \to 0$. The following basic lemma confirms this supposition.

Lemma 1 *Suppose that the objective function of problem (1) is presented in the canonical form $R(v, w) = \frac{1}{2}P(w) + K(v, w)$, where $K(v, w)$ is skew-symmetric function. Suppose also that $R(v, w)$ is a convex function in $w \in \Omega_0$ for any $v \in \Omega_0$, where $\Omega_0 \in R^n$ is a convex closed set.*
Then the family v_α induced by the regularization method (35) converges to the set of equilibrium solutions Ω^* of problem (34) with the minimal norm, that is

$$\rho(v_\alpha, \Omega^*) = \inf_{v^* \in \Omega^*} |v_\alpha - v^*| \to 0 \text{ as } \alpha \to 0$$

and also $\lim_{\alpha \to 0} |v_\alpha|^2 = \inf_{v^* \in \Omega^*} |v^*|^2$.

P r o o f. Take any $v^* \in \Omega^*$ and put $w = v^*$ in (36). Use condition (26), then we get

$$R\,(v_\alpha, v_\alpha) + \alpha\,|v_\alpha|^2 \le R\,(v_\alpha, v^*) + \alpha\,|v^*|^2 \le R\,(v_\alpha, v_\alpha) + \alpha\,|v^*|^2. \tag{37}$$

From here we have

$$|v_\alpha|^2 \le |v^*|^2 \quad \forall v^* \in \Omega^*. \tag{38}$$

The inequality (38) means that the family v_α is bounded. Therefore, there is a sequence v_{α_i}, such that $v_{\alpha_i} \to v^\infty \in \Omega_0$ as $\alpha_i \to 0$. We regard the family of inequalities (36) for each v_{α_i}

$$R(v_{\alpha_i}, v_{\alpha_i}) + \alpha_i |v_{\alpha_i}|^2 \le R(v_{\alpha_i}, w) + \alpha_i |w|^2 \quad \forall w \in \Omega_0. \tag{39}$$

Passing to the limit as $\alpha_i \to 0$ for each vector $w \in \Omega_0$, we obtain a limiting inequality

$$R(v^\infty, v^\infty) \le R(v^\infty, w) \quad \forall w \in \Omega_0. \tag{40}$$

Thus, any limit point v^∞ of the family v_α is the solution of problem (34). That means that $\rho(v_\alpha, \Omega^*) \to 0$, as $\alpha \to 0$, where $\rho(v, \Omega^*) = \inf_{v^* \in \Omega^*} |v - v^*|$. I fact, if it is not true, then there exists v_{α_i} so that $\rho(v_{\alpha_i}, \Omega^*) \ge 0$. But that conflicts to the fact that every limit point of this sequence is the solution of the problem. Note that from (38) we have: $|v^\infty|^2 \le |v^*|^2 \;\; \forall v^* \in \Omega^*$. From here it follows that $|v^\infty|^2 = \inf_{v^* \in \Omega^*} |v^*|^2$. i.e. v^* is a solution of problem (34) with the minimal norm.

If in addition to the conditions, listed in the theorem, function $P(w)$ is a potential for the operator $\nabla_w S(v, v)$ from expansion (28), and the function $\Phi(v, w)$ is convex in w for all $v \in \Omega_0$, then by virtue of conditions (24) the point v^* will be the solution of problem (2), if $\Omega = \Omega_0$. □

2. Convergence in the presence of convex perturbations. Let us consider the convergence of a regularization method in the case when the objective function $R(v, w)$ is given with some inaccuracy and the auxiliary regularized problem is solved approximately. The inaccuracy of the determination of $R(v, w)$ is simulated by a family of functions $R_\delta(v, w)$ satisfying the condition introduced in [23]

$$|R_\delta(v, w) - R(v, w)| \le \delta(1 + |w|^2) \quad \forall v \in \Omega_0, \quad \forall w \in \Omega_0, \tag{41}$$

where $\delta > 0$. This condition means the uniform proximity of $R(v, w)$ and $R_\delta(v, w)$ on a bounded set $\{w \mid |w|^2 \le C, \; w \in \Omega_0\}$. Hereby we assume that each function of family $R_\delta(v, w)$ is convex in $w \in \Omega_0$ for any $v \in \Omega_0$ and any $\delta > 0$.

The regularization method in this situation receives the form

$$v_{\alpha\delta} = \text{argmin}\{R_\delta(v_{\alpha\delta}, w) + \alpha |w|^2 \mid w \in \Omega_0\}. \tag{42}$$

Here the solution $v_{\alpha\delta}$ depends on both the regularization parameter α and the approximation parameter δ of the objective function. By virtue of the strong convexity of Tikhanov's function $T_{\alpha\delta}(v, w) = R_\delta(v, w) + \alpha|w|^2$, the map

$$M_{\alpha\delta}(v) = \text{argmin}\{R_\delta(v, w) + \alpha |w|^2 \mid w \in \Omega_0\}$$

is a single-valued operator for any $\alpha > 0$ and $\delta > 0$. It is possible to find a fixed point $v_{\alpha\delta}$ from (42) as the solution of the equation

$$v_{\alpha\delta} = M_{\alpha\delta}(v_{\alpha\delta}),$$

which is equivalent to (42).
By definition the problem (42) is equivalent to solving the following inequality

$$R_\delta(v_{\alpha\delta}, v_{\alpha\delta}) + \alpha\,|v_{\alpha\delta}|^2 \le R_\delta(v_{\alpha\delta}, w) + \alpha\,|w|^2 \quad \forall w \in \Omega_0. \tag{43}$$

The sense of this inequality is very simple: Here one needs to choose the value of the parametric variable $v = v_{\alpha\delta}$ such that this point will be the minimum $R(v_{\alpha\delta}, w)$ in w. It is very natural to solve this problem approximately, with inaccuracy to $\varepsilon > 0$ for the functional. In this case we come to the concept of an $\varepsilon-$approximate equilibrium solution or an $\varepsilon-$approximate fixed point. More precisely the $\varepsilon-$ fixed point is determined as the solution of the following inequality

$$R_\delta(v_{\alpha\delta\varepsilon}, v_{\alpha\delta\varepsilon}) + \alpha\,|v_{\alpha\delta\varepsilon}|^2 \le R_\delta(v_{\alpha\delta\varepsilon}, w) + \alpha\,|w|^2 + \varepsilon \quad \forall w \in \Omega_0\,. \tag{44}$$

This inequality generalizes (43) and means that one needs to determine the value of the parametric variable $v = v_{\alpha\delta\varepsilon}$ so that this point will be $\varepsilon-$optimal to the functional $R_\delta(v_{\alpha\delta\varepsilon}, w)$. It is easy to see that $M_{\alpha\delta}(v_{\alpha\delta}) \subset M_{\alpha\delta\varepsilon}(v_{\alpha\delta\varepsilon})$ and, consequently, the solution of (44) there exists.
Assume that the parameters $\alpha, \delta, \varepsilon$ tend to zero in a special coordinated way, namly: $\alpha = \alpha(\delta) \to 0$, $\varepsilon = \varepsilon(\delta) \to 0$ as $\delta \to 0$. Hereby it is proposed that $v_{\alpha\delta\varepsilon} = v_{\alpha(\delta)\varepsilon(\delta)} = v_\delta$.
Note that a family of approximate solutions v_δ converges to v^* provided that parameters $\alpha(\delta) \to 0$ and $\varepsilon(\delta) \to 0$ as $\delta \to 0$.

Theorem 1 *Suppose that the objective function of problem (1) is presented in the canonical form* $R(v, w) = \frac{1}{2}P(w) + K(v, w)$, *where* $K(v, w)$ *is a skew-symmetric function. Suppose also that* $R(v, w)$ *and* $R_\delta(v, w)$ *are convex functions in* $w \in \Omega_0$ *for any* $v \in \Omega_0$, *the perturbed functions* $R_\delta(v, w)$ *satisfy condition (41),* $\Omega_0 \in R^n$ *is a convex compact set, the parameters* $\alpha = \alpha(\delta)$, $\varepsilon = \varepsilon(\delta)$ *of method (44) are such that* $\alpha(\delta) \to 0$, $\varepsilon(\delta) \to 0$ *as* $\delta \to 0$ *and* $\sup_{\delta>0} \delta/\alpha(\delta) < 1$, $\sup_{\delta>0} \varepsilon(\delta)/\alpha(\delta) \le C$.
Then the family v_δ, *induced by a regularization method (44) converges to the set of equilibrium solutions* Ω^* *of problem (34), that is*

$$\rho(v_\delta, \Omega^*) = \inf_{v^* \in \Omega^*} |v_\delta - v^*| \to 0 \text{ as } \delta \to 0.$$

If in addition $\delta/\alpha(\delta) \to 0$, $\varepsilon(\delta)/\alpha(\delta) \to 0$, *then the solution has the minimal norm among all elements of* Ω^*.

P r o o f. We evaluate with the help of (41) the size for $R_\delta(v_\delta, w)$ from (44)

$$R_\delta(v_\delta, v_\delta) + \alpha\,|v_\delta|^2 \le R_\delta(v_\delta, w) + \alpha\,|w|^2 + \varepsilon \le$$
$$\le R(v_\delta, w) + \delta(1 + |w|^2) + \alpha\,|w|^2 + \varepsilon.$$

In the inequality obtained let's put $w = v^* \in \Omega^*$ and apply condition (26) and use (41) once again

$$R_\delta(v_\delta, v_\delta) + \alpha\,|v_\delta|^2 \le R(v_\delta, v^*) + \delta(1 + |v^*|^2) + \alpha\,|v^*|^2 + \varepsilon \le$$
$$\le R(v_\delta, v_\delta) + \delta(1 + |v^*|^2) + \alpha\,|v^*|^2 + \varepsilon \le R_\delta(v_\delta, v_\delta) + \delta(1 + |v_\delta|^2) +$$
$$+\delta(1 + |v^*|^2) + \alpha\,|v^*|^2 + \varepsilon. \tag{45}$$

Comparing the first and latter inequalities in system (45), we obtain

$$(\alpha - \delta)\,|v_\delta|^2 \le 2\delta + \varepsilon + (\alpha + \delta)\,|v^*|^2$$

or

$$|v_\delta|^2 \le \frac{\alpha + \delta}{\alpha - \delta}\,|v^*|^2 + \frac{2\delta + \varepsilon}{\alpha - \delta} \quad \forall v^* \in \Omega^*. \tag{46}$$

Due to the conditions $\sup\limits_{\delta > 0} \delta/\alpha(\delta) < 1, \quad \sup\limits_{\delta > 0} \varepsilon(\delta)/\alpha(\delta) \le C$,
(46) implies the boundedness of the family v_δ as $\delta \to 0$. Therefore, there is a sequence v_{δ_i}, such that $v_{\delta_i} \to v^\infty \in \Omega_0$ as $\delta_i \to 0$.
We consider the inequality (44) for all points of the sequence v_{δ_i}

$$R_{\delta_i}(v_{\delta_i}, v_{\delta_i}) + \alpha_i\,|v_{\delta_i}|^2 \le R_{\delta_i}(v_{\delta_i}, w) + \alpha_i\,|w|^2 + \varepsilon_i \quad \forall w \in \Omega_0. \tag{47}$$

Passing to the limit as $\delta_i \to 0$ for each $w \in \Omega_0$ and taking into account the following evaluations from (41)

$$R(v_{\delta_i}, v_{\delta_i}) - \delta_i(1 + |v_{\delta_i}|^2) \le R_{\delta_i}(v_{\delta_i}, v_{\delta_i}),$$
$$R_{\delta_i}(v_{\delta_i}, w) \le R(v_{\delta_i}, w) + \delta_i(1 + |w|^2),$$

we get a limit inequality

$$R\,(v^\infty, v^\infty) \le R\,(v^\infty, w) \quad \forall w \in \Omega_0. \tag{48}$$

Thus, any limit point v^∞ of the family v_δ is a solution of problem (34). That means that $\rho(v_\delta, \Omega^*) \to 0$, as $\delta \to 0$. Indeed, if it is not true, then there exists v_{δ_i} so that $\rho(v_{\delta_i}, \Omega^*) \ge 0$. But that conflicts to the fact that every limit point of this sequence is the solution of the problem.
Let $\delta/\alpha(\delta) \to 0, \ \ \varepsilon(\delta)/\alpha(\delta) \to 0$ be as $\delta \to 0$, then from (46) as $\delta \to 0$ we get $|v^\infty|^2 = \inf\limits_{v^* \in \Omega^*} |v^*|^2$, i.e. v^∞ is the solution of (34) with the minimal norm.
If in addition to the conditions listed in the theorem the function $P(w)$ is a potential for the operator $\nabla_w S(v, v)$ from expansion (28) and $\Phi(v, w)$ is convex in w for any $v \in \Omega_0$, then by virtue of the conditions (24) the point obtained v^* will be the solution of problem (1).

6 Regularized Penalty-Function Method

Let us study the case where both the objective function and the admissible set of an equilibrium problem are given with some inaccuracy. The regularized penalty-function method for the present takes the form

$$v^* \in \operatorname{Argmin}\{R(v^*, w) \mid \quad g(w) \leq 0, \quad w \in \Omega_0\}, \tag{49}$$

where $R(v, w) = P(v) + K(v, w)$. In this approach with the help of the penalty concept we dispose of functional constraints, i.e. instead of problem (1) we consider the parametric family of problems of the kind

$$v_\beta \in \operatorname{Argmin}\{\left|g^+(w)\right|^2 + \beta R(v_\beta, w) \mid \quad w \in \Omega_0\}, \quad v_\beta \in \Omega_0, \tag{50}$$

where $g_i^+(w) = \max\{0, g_i(w)\}, i = 1, 2, ..., m, \quad \beta > 0$, and $\Omega_0 \in R^n$ is a convex compact set. This family of equilibrium problems determines a family of solutions v_β, which represents itself by a set-valued map for all $\beta > 0$. Drawing the analogy to optimization problems it is possible to assume that the family of the solutions v_β will be aimed (in some sense) to the solution of an initial equilibrium problem (49) as $\beta \to 0$.

1. Convergence in the absence of perturbations. To avoid dealing with set-valued maps and simultaneously to improve properties of convergence and stability for the family v_β for various perturbed factors we consider the regularized penalty-function method instead of (50)

$$v_{\alpha\beta} = \operatorname{argmin}\{\left|g^+(w)\right|^2 + \beta(R(v_{\alpha\beta}, w) + \alpha\left|w\right|^2) \mid \quad w \in \Omega_0\}, \quad v_{\alpha\beta} \in \Omega_0. \tag{51}$$

This problem generates a two-parameter family of solutions $v_{\alpha\beta}$. Rewrite the extreme inclusion (51) in the form of the equivalent inequality

$$\left|g^+(v_{\alpha\beta})\right|^2 + \beta(R(v_{\alpha\beta}, v_{\alpha\beta}) + \alpha\left|v_{\alpha\beta}\right|^2) \leq$$

$$\leq \left|g^+(w)\right|^2 + \beta(R(v_{\alpha\beta}, w) + \alpha\left|w\right|^2) \quad \forall w \in \Omega_0. \tag{52}$$

The regularized penalty-function method for convex programming problems was explored earlier in [23]-[25]. The regularized proximal method for the same problems was investigated in [26],[14].
We show that method (51) converges to the solution of the initial equilibrium problem when the term $|w|^2$ becomes small. The proof of convergence, still, is based on inequality (26). However this inequality for problem (49) is true on its admissible set Ω only and, generally speaking, not correct on set Ω_0. As points of the set $v_{\alpha\beta}$ belong to Ω_0, it is necessary to expand the area of validity of inequality (26). On purpose to do this we consider the Lagrange function to problem (49): $L(v^*, w, p) = R(v^*, w) + \langle p, g(w)\rangle$, where $w \in \Omega_0$, $p \geq 0$.

Let v^*, p^* be the saddle point of this function, then the following system of inequalities hods

$$R(v^*,v^*)+\langle p, g(v^*)\rangle \leq R(v^*,v^*)+\langle p^*, g(v^*)\rangle \leq R(v^*,w)+\langle p^*, g(w)\rangle \;\; \forall w \in \Omega_0,\; p \geq 0. \tag{53}$$

From the last inequality we have

$$R(v^*,w) - R(v^*,v^*) \geq -\langle p^*, g(w)\rangle, \quad \forall w \in \Omega_0. \tag{54}$$

Comparing (54) and (21), we obtain

$$R(w,w) - R(w,v^*) \geq -\langle p^*, g(w)\rangle, \quad \forall w \in \Omega_0. \tag{55}$$

Inequality (55) can be examined as generalization of (26).

Lemma 2 *Suppose that the objective function of problem (1) is presented in the canonical form $R(v,w) = \frac{1}{2}P(w) + K(v,w)$, where $K(v,w)$ is a skew-symmetric function. Suppose also that $R(v,w)$ and $g(w)$ are continuous functions, convex in $w \in \Omega_0$, and that for any $v \in \Omega_0$, Slater's condition for Ω is valid, $\Omega_0 \in R^n$ is a convex compact set, the parameters satisfy the condition $\beta/\alpha \leq C$.*
Then the family $v_{\alpha,\beta}$ generated by method (51) converges to the set of equilibrium solutions Ω^* of problem (49), that is

$$\rho(v_{\alpha,\beta}, \Omega^*) = \inf_{v^* \in \Omega^*} |v_{\alpha,\beta} - v^*| \to 0 \text{ as } \alpha \to 0 \;\; \beta \to 0.$$

If in addition $\beta/\alpha \to 0$, then this solution has the minimal norm among all elements of Ω^*.

P r o o f. Let us put $w = v^* \in \Omega^*$ in (52) and (55). Hence

$$\begin{gathered} \left|g^+(v_{\alpha\beta})\right|^2 + \beta(R(v_{\alpha\beta}, v_{\alpha\beta}) + \alpha\left|v_{\alpha\beta}\right|^2) \leq \\ \leq \left|g^+(v^*)\right|^2 + \beta(R(v_{\alpha\beta}, v^*) + \alpha\left|v^*\right|^2) \leq \\ \leq \left|g^+(v^*)\right|^2 + \beta(R(v_{\alpha\beta}, v_{\alpha\beta}) + \langle p^*, g(v_{\alpha\beta})\rangle + \alpha\left|v^*\right|^2). \end{gathered} \tag{56}$$

From here we have

$$\left|g^+(v_{\alpha\beta})\right|^2 + \alpha\beta\left|v_{\alpha\beta}\right|^2 \leq \beta\langle p^*, g(v_{\alpha\beta})\rangle + \alpha\beta\left|v^*\right|^2.$$

Whereas $\langle p^*, g(v_{\alpha\beta})\rangle \leq \langle p^*, g^+(v_{\alpha\beta})\rangle$,

$$\frac{1}{\alpha\beta}\left|g^+(v_{\alpha\beta})\right|^2 - \frac{1}{\alpha}\langle p^*, g^+(v_{\alpha\beta})\rangle + \left|v_{\alpha\beta}\right|^2 \leq \left|v^*\right|^2.$$

We isolate the quadratic term from the last inequality

$$\left|\frac{1}{\sqrt{\alpha\beta}}g^+(v_{\alpha\beta}) - \frac{\sqrt{\alpha\beta}}{2\alpha}p^*\right|^2 - \frac{\alpha\beta}{4\alpha^2}|p^*|^2 + \left|v_{\alpha\beta}\right|^2 \leq \left|v^*\right|^2.$$

Finally

$$|v_{\alpha\beta}|^2 \leq |v^*|^2 + \frac{1}{4}\frac{\beta}{\alpha}|p^*|^2 \leq |v^*|^2 + C|p^*|^2. \tag{57}$$

The inequality (57) means that the family $v_{\alpha\beta}$ is bounded. Therefore, there is a sequence $v_{\alpha_i\beta_i}$ such that $v_{\alpha_i\beta_i} \to v^\infty \in \Omega_0$ as $\alpha_i \to 0,\ \beta_i \to 0$. We consider the set of inequalities (52) for any $v_{\alpha_i\beta_i}$ and obtain

$$\left|g^+(v_{\alpha_i\beta_i})\right|^2 + \beta_i(R(v_{\alpha_i\beta_i}, v_{\alpha_i\beta_i}) + \alpha_i\,|v_{\alpha_i\beta_i}|^2) \leq$$

$$\leq \left|g^+(w)\right|^2 + \beta_i(R(v_{\alpha_i\beta_i}, w) + \alpha_i\,|w|^2) \quad \forall w \in \Omega_0. \tag{58}$$

Passing to the limit for any $w \in \Omega_0$ as $\alpha_i \to 0,\ \beta_i \to 0$, we get a limit inequality

$$\left|g^+(v^\infty)\right|^2 \leq \left|g^+(w)\right|^2 \quad \forall w \in \Omega_0. \tag{59}$$

Since the inequality (59) is correct for all $w \in \Omega_0$, hence, in particular, it is true for all w from the admissible set $\Omega = \{w \mid g(w) \leq 0, \quad w \in \Omega_0\}$. This means that (59) takes a form on this set: $\left|g^+(v^\infty)\right|^2 \leq 0$, i.e. $g(v^\infty) \leq 0$. In other words, any limit point of the family $v_{\alpha\beta}$ is an element of the admissible set Ω. Since all the limit points of family $v_{\alpha\beta}$ belong to the admissible area, it is meaningful to consider a system of inequalities (58) in points of this area only. In this case, the system of inequalities (58) becomes essentially simpler and receives the shape

$$R(v_{\alpha_i\beta_i}, v_{\alpha_i\beta_i}) + \alpha_i\,|v_{\alpha_i\beta_i}|^2 \leq R(v_{\alpha_i\beta_i}, w) + \alpha_i\,|w|^2 \quad \forall w \in \Omega. \tag{60}$$

This inequality means that $v_{\alpha_i\beta_i}$ is a fixed point of the extreme map of the form $\text{argmin}\{R(v, w) + \alpha_i\,|w|^2 \quad | \quad w \in \Omega\}$ over the admissible set.
Passing to the limit for any w from the permissible area Ω as $\alpha_i \to 0,\ \beta_i \to 0$, we receive a limit inequality

$$R(v^\infty, v^\infty) \leq R(v^\infty, w) \quad \forall w \in \Omega. \tag{61}$$

Thus, any limit point v^∞ of the family $v_{\alpha\beta}$ is the solution of problem (49). That means that $\rho(v_{\alpha\beta}, \Omega^*) \to 0$, as $\alpha \to 0,\ \beta \to 0$. Indeed, if it is not true, then there exists $v_{\alpha_i\beta_i}$ so that $\rho(v_{\alpha_i\beta_i}, \Omega^*) \geq 0$. But that conflicts to the fact that every limit point of this sequence is the solution of the problem.
We note that from (57) we have: $|v^\infty|^2 \leq |v^*|^2 \ \ \forall v^* \in \Omega^*$, if $\beta/\alpha \to 0$. Hence it follows that $|v^\infty|^2 = \inf_{v^* \in \Omega^*} |v^*|^2$, i.e. v^* is the solution of (49) with the minimal norm.
If in addition to the conditions listed in the theorem the function $P(w)$ is a potential of the operator $\nabla_w S(v, v)$ from expansion (28) and $\Phi(v, w)$ is convex in w for any $v \in \Omega_0$, then by virtue of (24) the point v^* is the solution of (1). □

2. Convergence in the presence of convex perturbations. In this section we consider the family of approximations

$$v_\delta \in \mathrm{Argmin}\{R_\delta(v_\delta, w) \mid \ g_\delta(w) \leq 0, \quad w \in \Omega_0\}. \tag{62}$$

instead of problem (49). It is supposed that the approximate data of problem (62) satisfy the conditions

$$|R_\delta(v,w) - R(v,w)| \leq \delta(1 + |w|^2) \ \ \forall v \in \Omega_0, \ \ \forall w \in \Omega_0, \tag{63}$$

$$\big||g_\delta^+(w)|^2 - |g^+(w)|^2\big| \leq \delta(1 + |w|^2) \ \ \forall v \in \Omega_0, \ \ \forall w \in \Omega_0. \tag{64}$$

Moreover any function of the family $R_\delta(v,w)$, $g_\delta(w)$ is convex in $w \in \Omega_0$ for any $v \in \Omega_0$ and any $\delta > 0$. Note that the penalty function $|g_\delta^+(w)|^2$ is determined even if the permissible set of problem (62) is empty [23].
To solve problem (62)-(64) we use the perturbed regularized penalty-function method in the form

$$v_{\alpha\beta\delta} = \mathrm{argmin}\{\left|g_\delta^+(w)\right|^2 + \beta(R_\delta(v_{\alpha\beta\delta}, w) + \alpha\,|w|^2) \mid \quad w \in \Omega_0\}, \quad v_{\alpha\beta\delta} \in \Omega_0. \tag{65}$$

For convenience of further reasoning we rewrite (65) as an inequality

$$\left|g_\delta^+(v_{\alpha\beta\delta})\right|^2 + \beta(R_\delta(v_{\alpha\beta\delta}, v_{\alpha\beta\delta}) + \alpha\,|v_{\alpha\beta\delta}|^2) \leq$$

$$\leq \left|g_\delta^+(w)\right|^2 + \beta(R_\delta(v_{\alpha\beta\delta}, w) + \alpha\,|w|^2) \ \ \forall w \in \Omega_0. \tag{66}$$

Problem (66) is convex in w and has a solution. Methods for the solution of such problems are given in [5],[7]. It is not necessary to solve problem (66) precisely. It is enough to find a vector $v \in \Omega$ satisfying the following condition

$$\left|g_\delta^+(v_{\alpha\beta\delta\varepsilon})\right|^2 + \beta(R_\delta(v_{\alpha\beta\delta\varepsilon}, v_{\alpha\beta\delta\varepsilon}) + \alpha\,|v_{\alpha\beta\delta\varepsilon}|^2) \leq$$

$$\leq \left|g_\delta^+(w)\right|^2 + \beta(R_\delta(v_{\alpha\beta\delta\varepsilon}, w) + \alpha\,|w|^2) + \varepsilon \ \ \forall w \in \Omega_0, \ \ \varepsilon > 0. \tag{67}$$

Note that any solution of (66) is a solution of (67) and consequently the solution of (67) exists. We shall further assume that the parameters $\alpha, \beta, \delta, \varepsilon$ tend to zero in a harmonized way, namely: $\alpha = \alpha(\delta) \to 0$, $\beta = \beta(\delta)$, $\quad \varepsilon = \varepsilon(\delta) \to 0$ as $\delta \to 0$. Thus we shall mean that $v_{\alpha\beta\delta\varepsilon} = v_{\alpha(\delta)\beta(\delta)\varepsilon(\delta)} = v_\delta$. We prove now the theorem of convergence.

Theorem 2 *Suppose that the objective function of problem (1) is presented in the canonical form $R(v,w) = \frac{1}{2}P(w) + K(v,w)$, where $K(v,w)$ is a skew-symmetric function. Suppose also that $R(v,w), g(w)$ and $R_\delta(v,w)$ are continuous functions and convex in $w \in \Omega_0$ for any $v \in \Omega_0$, the perturbed functions $R_\delta(v,w), |g_\delta^+(w)|^2$ satisfy the conditions (63), (64) and are convex in w for any $v \in \Omega_0$, Slater's condition for Ω is valid, $\Omega_0 \in R^n$ is a convex compact set,*

the parameters $\alpha = \alpha(\delta)$, $\beta = \beta(\delta)$, $\varepsilon = \varepsilon(\delta)$ *of method (67) are such that* $\alpha(\delta) \to 0$, $\beta(\delta) \to 0$, $\varepsilon(\delta) \to 0$ *as* $\delta \to 0$ *and* $\sup_{\delta>0} \delta/\alpha(\delta) < 1$, $\sup_{\delta>0} \beta(\delta)/\alpha(\delta) < C$, $\sup_{\delta>0} \delta/\alpha(\delta)\beta(\delta) < C$, $\sup_{\delta>0} \varepsilon(\delta)/\alpha(\delta)\beta(\delta) \leq C$.
Then the family v_δ, *induced by the regularization method (67) converges to the set of equilibrium solutions* Ω^* *of problem (49), that is*

$$\rho(v_\delta, \Omega^*) = \inf_{v^* \in \Omega^*} |v_\delta - v^*| \to 0 \text{ as } \delta \to 0.$$

If in addition $\delta/\alpha(\delta) \to 0$, $\varepsilon(\delta)/\alpha(\delta) \to 0$, $\beta(\delta)/\alpha(\delta) \to 0$,
then this solution has the minimal norm among all point Ω^*.

Note that the parameters satisfy the condition of theorem, if $\alpha = \beta = \sqrt{\delta}$, $\varepsilon = \delta$.
P r o o f. We estimate the functions $|g_\delta^+(w)|^2$ and $R_\delta(v_\delta, w)$ from (67) with the help of (63),(64)

$$\left|g_\delta^+(v_\delta)\right|^2 + \beta(R_\delta(v_\delta, v_\delta) + \alpha\left|v_\delta\right|^2) \leq$$

$$\leq \left|g^+(w)\right|^2 + \delta(1 + |w|^2) + \beta(R(v_\delta, w) + \delta(1 + |w|^2) + \alpha\left|w\right|^2) + \varepsilon \leq$$

$$\leq \left|g^+(w)\right|^2 + \beta R(v_\delta, w) + (1 + \beta)\delta(1 + |w|^2) + \alpha\beta\left|w\right|^2 + \varepsilon.$$

Let us put $w = v^* \in \Omega^*$ in the inequality obtained and apply condition (55) and then use (63),(64) once again

$$\left|g_\delta^+(v_\delta)\right|^2 + \beta R_\delta(v_\delta, v_\delta) + \alpha\beta\left|v_\delta\right|^2 \leq$$

$$\leq \left|g^+(v^*)\right|^2 + \beta R(v_\delta, v^*) + (1 + \beta)\delta(1 + |v^*|^2) + \alpha\beta\left|v^*\right|^2 + \varepsilon \leq$$

$$\leq \left|g^+(v^*)\right|^2 + \beta R(v_\delta, v_\delta) + \beta\langle p^*, g(v_\delta)\rangle + (1 + \beta)\delta(1 + |v^*|^2) + \alpha\beta\left|v^*\right|^2 + \varepsilon \leq$$

$$\leq \left|g^+(v^*)\right|^2 + \beta R_\delta(v_\delta, v_\delta) + \beta\delta(1 + |v_\delta|^2) + \beta\langle p^*, g(v_\delta)\rangle +$$
$$+ (1 + \beta)\delta(1 + |v^*|^2) + \alpha\beta\left|v^*\right|^2 + \varepsilon. \tag{68}$$

Comparing the first and latter inequalitis in the system (68) and taking into account also the obvious relation $\langle p^*, g(v_\delta)\rangle \leq \langle p^*, g^+(v_\delta)\rangle$, we obtain

$$\left|g_\delta^+(v_\delta)\right|^2 + \beta(\alpha - \delta)\left|v_\delta\right|^2 \leq$$

$$\leq \beta\delta + (1 + \beta)\delta + (\delta + \beta(\alpha + \delta))|v^*|^2 + \beta\langle p^*, g^+(v_\delta)\rangle + \varepsilon.$$

Therefore

$$\frac{1}{\beta(\alpha - \delta)}\left|g_\delta^+(v_\delta)\right|^2 - \frac{1}{\alpha - \delta}\langle p^*, g^+(v_\delta)\rangle + \left|v_\delta\right|^2 \leq$$

$$\leq \frac{\delta + \beta(\alpha + \delta)}{\beta(\alpha - \delta)}|v^*|^2 + \frac{2\delta}{\alpha - \delta} + \frac{\delta + \varepsilon}{\beta(\alpha - \delta)}.$$

We isolate the full quadratic term from the latter inequality

$$\left|\frac{1}{\sqrt{\beta(\alpha-\delta)}}g_\delta^+(v_\delta)-\frac{\sqrt{\beta(\alpha-\delta)}}{2(\alpha-\delta)}p^*\right|^2-\frac{\beta(\alpha-\delta)}{4(\alpha-\delta)^2}|p^*|^2+|v_\delta|^2\le$$

$$\le\frac{\delta+\beta(\alpha+\delta)}{\beta(\alpha-\delta)}|v^*|^2+\frac{2\delta}{\alpha-\delta}+\frac{\delta+\varepsilon}{\beta(\alpha-\delta)}.$$

Hence

$$|v_\delta|^2\le\frac{\delta+\beta(\alpha+\delta)}{\beta(\alpha-\delta)}|v^*|^2+\frac{\beta}{4(\alpha-\delta)}|p^*|^2+\frac{2\delta}{\alpha-\delta}+\frac{\delta+\varepsilon}{\beta(\alpha-\delta)}. \tag{69}$$

Since the parameters $\alpha,\beta,\delta,\varepsilon$ are subordinate to conditions $\alpha=\alpha(\delta)$, $\beta=\beta(\delta)$, $\varepsilon=\varepsilon(\delta)$ and $\alpha(\delta)\to 0$, $\beta(\delta)\to 0$, $\varepsilon(\delta)\to 0$ as $\delta\to 0$ and $\sup_{\delta>0}\delta/\alpha(\delta)<1$, $\sup_{\delta>0}\beta(\delta)/\alpha(\delta)<C$, $\sup_{\delta>0}\delta/\alpha(\delta)\beta(\delta)<C$, $\sup_{\delta>0}\varepsilon(\delta)/\alpha(\delta)\beta(\delta)\le C$, it is clear that the coefficients at $|v^*|^2$ and $|p^*|^2$ and the free term in the right-hand side of (69) are bounded if the parameters tend to zero. Thus, from the inequality (69) it follows the boundedness of the family $|v_\delta|\le C$. Therefore, there is a sequence v_{δ_i} such that $v_{\delta_i}\to v^\infty\in\Omega_0$ as $\delta_i\to 0$. Consider the family of inequalities (67) for all points of the sequence v_{δ_i}

$$|g_{\delta_i}^+(v_{\delta_i})|^2+\beta_i(R_{\delta_i}(v_{\delta_i},v_{\delta_i})+\alpha_i|v_{\delta_i}|^2)\le$$

$$\le|g_{\delta_i}^+(w)|^2+\beta_i(R_{\delta_i}(v_{\delta_i},w)+\alpha_i|w|^2)+\varepsilon_i\quad\forall w\in\Omega_0. \tag{70}$$

Passing to the limit as $\delta_i\to 0$ for any fixed $w\in\Omega_0$ and taking into account the following estimates from (63) and (64)

$$|g^+(v_{\delta_i})|^2-\delta_i(1+|v_{\delta_i}|^2)\le|g_{\delta_i}^+(v_{\delta_i})|^2,$$

$$|g_{\delta_i}^+(w)|^2\le|g^+(w)|^2+\delta_i(1+|w|^2),$$

$$R(v_{\delta_i},v_{\delta_i})-\delta_i(1+|v_{\delta_i}|^2)\le R_{\delta_i}(v_{\delta_i},v_{\delta_i}), \tag{71}$$

$$R_{\delta_i}(v_{\delta_i},w)\le R(v_{\delta_i},w)+\delta_i(1+|w|^2), \tag{72}$$

we get a limit inequality

$$|g^+(v^\infty)|^2\le|g^+(w)|^2\quad\forall w\in\Omega_0. \tag{73}$$

As inequality (73) is correct for all $w\in\Omega_0$, then in particular, it is valid for all w from an admissible set $\Omega=\{w\mid g(w)\le 0,\quad w\in\Omega_0\}$. This means, that (73) leads to $|g^+(v^\infty)|^2\le 0$, i.e. $g(v^\infty)\le 0$. In other words, any limit point of the family v_δ is an element of the admissible set Ω.

Since all limit points of v_δ belong to the admissible area, we can consider a system of inequalities (70) over this area, and this system takes the form

$$R_{\delta_i}(v_{\delta_i}, v_{\delta_i}) + \alpha_i |v_{\delta_i}|^2 \leq R_{\delta_i}(v_{\delta_i}, w) + \alpha_i |w|^2 + \varepsilon_i \quad \forall w \in \Omega. \tag{74}$$

This inequality means that $v_{\alpha_i \beta_i}$ is the fixed point of the extreme map of the form $\operatorname{argmin}\{R_{\delta_i}(v,w) + \alpha_i |w|^2 \; + \varepsilon_i \mid \; w \in \Omega\}$ over the admissible set.
Passing to the limit as $\delta_i \to 0$ for any w from the admissible set Ω and taking into account (71), (72), we obtain the limit inequality

$$R(v^\infty, v^\infty) \leq R(v^\infty, w) \quad \forall w \in \Omega. \tag{75}$$

Thus, any limit point v^∞ of the family v_δ is the solution of (49). That means that $\rho(v_\delta, \Omega^*) \to 0$, as $\delta \to 0$. Indeed, if it is not true, then there exists v_{δ_i} so that $\rho(v_{\delta_i}, \Omega^*) \geq 0$. But that conflicts to the fact that every limit point of this sequence is the solution of the problem.
Let $\delta/\alpha(\delta) \to 0$, $\varepsilon(\delta)/\alpha(\delta) \to 0$ $\beta(\delta)/\alpha(\delta) \to 0$ be as $\delta \to 0$, then from (69) as $\delta \to 0$ we get $|v^\infty|^2 = \inf_{v^* \in \Omega^*} |v^*|^2$, i.e. v^∞ is the solution of (49) with the minimal norm.
If in addition to the conditions listed in the theorem the function $P(w)$ is a potential for the operator $\nabla_w S(v,v)$ from expansion (28) and the function $\Phi(v,w)$ is convex in w for any $v \in \Omega_0$, then by virtue of condition (24) the obtained point v^* will be the solution of problem (1). □

We shall in summary remark that the regularization method described above determines the operator R_δ, which to each set $(R_\delta(v,w), \delta)$ in problem (34) or set $(R_\delta(v,w), g_\delta(w), \delta)$ in problem (49) puts in conformity a point $v_\delta = v_{\alpha(\delta),\delta,\epsilon(\delta)}$ from (44) or $v_\delta = v_{\alpha(\delta),\beta(\delta),\delta,\epsilon(\delta)}$ from (67). From the Theorems 1 and 2 it follows, by virtue of [20], that such an operator R_δ can be called a regularized one for the problems (34) and (49).

References

[1] A.S. Antipin, "Equilibrium Programming:Proximal Methods" *Zhurnal vychisl. matem. i matem. fiz.* 37 (1997) 11 1327-1339; English Transl. *Comp. Maths.Math.Phys.* 37 (1997) 11 1285-1296.

[2] A.S. Antipin, "Equilibrium Programming: Gradint-Type Methods" *Avtomatika i telemehanika* (1997) 8 125-137; English Transl. *Automation and Control* (1997) 8.

[3] A.S. Antipin, "Controlled Proximal Differential Systems for Solving Saddle Problems" *Differentsial,nye Uravneniya* 28 (1992) 11 1846-1861; English Transl. *Differential Equations* 28 (1992) 11 1498-1510.

[4] A.S. Antipin, "Inverse problems of nonlinear programming" *Nonlinear Dynamic Systems: the Qualitative Analysis and Control.*(Collection of Transactionses. Institute of Systems Analysis. Moscow) 1993 2 5-32. English transl.: *Computational Mathematics and Modelling* (Plenume Publish. Corp. N.Y. 7 (1996) 3 263-287.

[5] A.S. Antipin, "The Convergence of Proximal Methods to Fixed Points of Extremal Mappings and Estimates of Their Rate of Convergence" *Zhurnal vychisl. mat. i mat. fiz* 35 (1995) 5 688-704; English transl. *Comp.Maths.Math.Phys* 35 (1995) 5 539-551.

[6] A.S. Antipin, "Differential gradient prediction-type methods for computing fixed points of extremal maps" *Differentsial'nye Uravneniya* 31 (1995) 11 1786-1795; English Transl. *Differential Equations* 31 (1995) 11 1498-1510.

[7] A.S. Antipin, "Splitting of a Gradient Approach for Solving Extreme Inclisions" *Zhurnal vychisl. mat. i mat. fiz* 38 (1998) 7 1118-1132; English transl. *Comp.Maths.Math.Phys.* 38 (1998) 7 1069-1082.

[8] A.S. Antipin, "Regularization Method for Convex Programming Problems" *Economika i matem. metody* 11 (1975) 2 336-342 (in Russian).

[9] J.-P. Aubin, H. Frankowska, *Set Valued Analysis* (Birkhauser. 1990).

[10] V.Z. Belen'ky, V.A. Volkonsky, *Iterative Methods at Game Theory and Programming* (Nauka, 1974) (in Rissian).

[11] E. Blum, W. Oettli, " From optimization and variational inequalities to equilibrium problems" *The Mathematics Student* 63 (1993) 1-4, 1-23.

[12] Fan Ky, " A minimax inequality and applications" *Inequalities III,* (Acad. Press. 1972) 103-113.

[13] C.B. Garcia, W.I. Zangwill, *Pathways to Solutions, Fixed Points, and Equilibria* (Prentice-Hall, Englewood Cliffs. NJ 1981).

[14] A. Kaplan, R. Tichatschke, *Stable Methods for Ill-Posed Variational Problems* (Akademie Verlag, Berlin 1994).

[15] D. Monderer, L.S. Shapley, " Potential games" *Games and economic behavior* (1996) 14 124-143.

[16] H. Nikaido, K.Isoda, "Note on noncooperative convex games" *Pacific J. Math.* 5 (1955) 1 807-815.

[17] J. Ortega, W. Rheinbold, *Iterative Solution of Nonlinear Equations in Several Variables* (Academ. Press 1970).

[18] R. Rockafellar, *Convex Analysis* (Princeton Univ. Press 1970).

[19] J.B. Rosen, "Existence and uniqueness of equilibrium points for concave n-person games" *Econometrica* 33 (1965) 3 520-534.

[20] A.N. Tikhonov, V.Ya Arsenin, *Methods of Solving for Ill-Posed Problems* (Nauka. 1986).

[21] F.P. Vasil'ev, *Numerical Methods for Extremum Value Problems* (Nauka, 1988).

[22] F.P. Vasil'ev, A.S. Antipin, "Regularization Methods of calculating of Fixed Point for Extremal Maps" *Computational Mathematics and Cybernetics. Seriya 15* (1998) 1 11-14 (in Russian).

[23] F.P. Vasil'ev, *Methods of Solving for Extreme Problems* (Nauka, 1981).

[24] F.P. Vasil'ev, "On Regularization for Ill-Posed Optimization Problems over Sets Given with Inaccuracy" *Zhurnal vychisl. mat. i mat. fiz* 20 (1980) 1 38-50 (in Russian).

[25] F.P. Vasil'ev, "On Regularization for Ill-Posed Extreme Problems" *DAN USSR* 241 (1978) 5 1001-1004 (in Russian).

[26] F.P. Vasil'ev, O. Obradovich, "A Regularized Proximal Method for Minimizing Problems with Inaccurate Initial Data" *Zhurnal vychisl. mat. i mat. fiz* 33 (1993) 2 179-188; English transl. *Comp.Maths.Math.Phys* 33 (1993) 2 157-164.

L^p Regularization of the Non-Parametric Minimal Surface Problem

H. Attouch * T. Champion *

Abstract

The non-parametric minimal surface problem is an ill-posed variational problem, even in its relaxed form. Indeed, the relaxed problem is an L^1 type problem, and it is not strictly convex so that it may have more than one solution. Following [2], we use the L^p regularization technique with $p \to 1$. Under fairly general assumptions, we show that the approximate solutions $(u_p)_p$ converge strongly in $W^{1,1}$ to a particular solution of the relaxed problem. Indeed, the so-selected solution is characterized as the unique solution of an auxiliary variational problem involving the integrand $t \mapsto |t|ln|t|$.

Keywords: L^p regularization, minimal surface problem, selection principle.

AMS subject classification: 35J70, 41A50, 41A65, 46E30, 49J30, 58E30

1 Introduction

In this paper, we are interested in a particular case of the minimal surface problem, namely the non-parametric minimal surface problem. This problem originated from the Plateau problem, which consists in finding the minimal surface (i.e. the surface of least area) bounded by a given closed curve in $\mathbb{R}^3$. In the so-called non-parametric minimal surface problem one has to determine a function u defined on a domain Ω of $\mathbb{R}^N$ whose graph is of least area among all the graphs of functions taking the same boundary values. This problem together with the minimal surface problem has been studied by many authors among which Radò, Douglas, De Giorgi, Federer, Fleming... For more precise and complete historical comments, we refer to [5].
The non-parametric minimal surface problem (P) is an ill-posed problem in its original statement, because it may have no solution taking the prescribed boundary values. Anyway, the boundary constraint can be relaxed so as to

*Laboratoire d'Analyse Convexe, Département de Mathématiques, case courrier 051, Université Montpellier II, Place Eugène Bataillon, 34095 Montpellier cedex, France. e-mails: attouch@math.univ-montp2.fr, champion@math.univ-montp2.fr obtain a problem (P_1) which admits at least one solution, which can be viewed as a generalized solution of (P). But then (P_1) is still ill-posed because in general it has more than one solution. The aim of this paper is to provide with a way to regularize (P_1) in order to select a particular solution. This is done by defining a family of approximating well-posed problems $(P_p)_{p>1}$, which in our case will be the L^p-regularized problems of (P_1), with the property that if we denote u_p the solution of (P_p), then $(u_p)_p$ is a minimizing sequence of (P_1) and converges to a particular solution of (P_1). In fact, the limit of $(u_p)_p$ is shown to solve an auxiliary well-posed minimization problem related to the regularization technique employed to define $(P_p)_p$.

2 Presentation of the Problem and the L^p Approximation

Throughout this paper, Ω is an open bounded connected subset of $I\!R^N$, with a Lipschitz continuous boundary that we shall denote $\partial\Omega$.
Let $g \in \mathrm{W}^{1,1}(\Omega) \cap \mathrm{L}^\infty(\Omega)$. Then the so-called non parametric minimal surface problem is:

$$(P) \qquad \mathrm{Inf}\left\{\int_\Omega \sqrt{1+|\nabla u(x)|^2}dx \;:\; u \in g + \mathrm{W}_0^{1,1}(\Omega)\right\}$$

This problem is in general ill-posed in the sense that it doesn't always have a solution (see [3], chapter V example 2.1). In [3], Ekeland and Temam proposed a direct study of this problem via duality arguments, and defined a notion of generalized solution. In this paper, we follow De Giorgi, Giusti and Miranda (in [4]) and consider the following relaxed form of the non parametric minimal surface problem:

$$(P_1) \quad \mathrm{Inf}\left\{J(u) = \int_\Omega \sqrt{1+|\nabla u(x)|^2}dx + \int_{\partial\Omega} |u-g|\, d\mathrm{H}_{N-1} \;:\; u \in \mathrm{W}^{1,1}(\Omega)\right\}$$

where H_{N-1} is the $N-1$-dimensional Hausdorff measure. We recall that each function u in $\mathrm{W}^{1,1}(\Omega)$ has a trace on $\partial\Omega$ (that we still denote by u) belonging to $\mathrm{L}^1(\partial\Omega)$.
Now the set $S(P_1)$ of optimal solutions of (P_1) is nonempty. This is not obvious because one would rather expect a minimizer of J to belong to $\mathrm{BV}(\Omega)$: when applying the direct method of the calculus of variations, one is led to imbed $\mathrm{W}^{1,1}(\Omega)$ into $\mathrm{BV}(\Omega)$ and to take the lsc-closure of J on $\mathrm{BV}(\Omega)$ (here we set $J(u) = +\infty$ for $u \in \mathrm{BV}(\Omega) \setminus \mathrm{W}^{1,1}(\Omega)$) so as to be able to extract a weakly convergent subsequence from a minimizing sequence of (P_1). The trick is that a solution obtained by this method (which is also a generalized solution in the sense of [3]) is analytic in Ω and belongs to $\mathrm{W}^{1,1}(\Omega)$, as shown in [5] (or [3]). The

set $S(P_1)$ is also obviously convex, but may not be reduced to a singleton. For a counterexample, we refer to [5], example 15.12. Anyway, if g is continuous on $\partial\Omega$, then (P_1) has a unique optimal solution (see [5] and the reference therein for more details). In fact, as discussed in [3], if u_1 and u_2 are two optimal solutions of (P_1) then $\nabla u_1 = \nabla u_2$ because the member of J depending on the gradient is strictly convex. So $S(P_1)$ is a segment, and two optimal solutions of (P_1) differ only by a real constant. This also implies that if u_1 and u_2 are in $S(P_1)$, then $\int_{\partial\Omega} |u_1 - g|\, d\mathrm{H}_{N-1} = \int_{\partial\Omega} |u_2 - g|\, d\mathrm{H}_{N-1}$.

As explained in the introduction, the aim of this paper is to show that under suitable hypotheses, we can construct a family $(P_p)_{p>1}$ of well-posed variational problems such that if u_p is the solution of (P_p), then $(u_p)_p$ is a minimizing sequence of (P_1) which converges to a particular solution of (P_1).

One may first think of an elliptic type regularization (see [6]), and consider the family of problems:

$$(P_\varepsilon) \qquad \mathrm{Inf}\left\{ J_\varepsilon(u) = J(u) + \varepsilon \, ||u||^2_{\mathrm{H}^1(\Omega)} \ : \ u \in \mathrm{H}^1(\Omega) \right\}$$

The trick is that for $\varepsilon > 0$, (P_ε) is elliptic on $\mathrm{H}^1(\Omega)$. Then if we denote by u_ε the unique solution of (P_ε), and suppose that $S(P_1) \cap \mathrm{H}^1(\Omega) \neq \emptyset$, it is easy to show that $(u_\varepsilon)_{\varepsilon>0}$ converges as $\varepsilon \to 0$ in $\mathrm{H}^1(\Omega)$ to the unique element $\tilde{u} \in S(P_1)$ which minimizes $||.||_{\mathrm{L}^2(\Omega)}$ over $S(P_1)$. Anyway, the condition $S(P_1) \cap \mathrm{H}^1(\Omega) \neq \emptyset$ does not seem natural since problem (P_1) is defined on $\mathrm{W}^{1,1}(\Omega)$. Moreover, when $S(P_1) \cap \mathrm{H}^1(\Omega) = \emptyset$ the convergence of $(u_\varepsilon)_{\varepsilon>0}$ to a particular solution of (P_1) is still an open problem. Noticing that, when $N = 2$, $\mathrm{W}^{1,1}(\Omega)$ is continuously imbedded in $\mathrm{L}^2(\Omega)$, an alternative approach to the elliptic regularization is to consider the following variational problem:

$$\mathrm{Inf}\left\{ J(u) + \varepsilon \, ||u||^2_{\mathrm{L}^2(\Omega)} \ : \ u \in \mathrm{W}^{1,1}(\Omega) \right\}$$

Here, we will be interested in an other type of regularization, namely the L^p regularization, for which we can weaken the sufficient condition for the convergence of the net of minimizers. Following [2], we consider the family of well-posed approximating problems (P_p) related to (P_1):

$$(P_p) \quad \mathrm{Inf}\left\{ J_p(u) = \int_\Omega \left(\sqrt{1 + |\nabla u|^2} \right)^p dx + \int_{\partial\Omega} |u - g|^p \, d\mathrm{H}_{N-1} \ : u \in \mathrm{W}^{1,p}(\Omega) \right\}$$

Now the functional J_p that appears in problem (P_p) is strictly convex, continuous and coercive on $\mathrm{W}^{1,p}(\Omega)$ for $p > 1$, so (P_p) has a unique solution, that we will denote by u_p.We aim to show that $(u_p)_p$ is a minimizing sequence for (P_1), so that every cluster point of $(u_p)_p$ in $BV(\Omega)$ is an optimal solution of (P_1). Then, we will provide with a sufficient condition under which the whole net $(u_p)_p$ converges to a particular optimal solution. To this end, we need to define $\Phi : I\!R \to I\!R \cup \{+\infty\}$ by $\Phi(r) = \begin{cases} +\infty & \text{if } r < 0 \\ 0 & \text{if } r = 0 \\ r\ln(r) & \text{if } r > 0 \end{cases}$. Then Φ is convex on $I\!R^+$ and is everywhere not less than $-1/e$. For simplicity, we will use -1 as a lower bound for Φ in the calculations. Notice that $r \to \Phi(|r|)$ is *not* convex on $I\!R$.

The following theorem establishes that the family $(P_p)_p$ is a variational approximation of (P_1):

Theorem 2.1 *The sequence (u_p) is a minimizing sequence of (P_1) as $p \to 1$. Moreover,*

$$\lim_{p \to 1} J_p(u_p) = \lim_{p \to 1} J(u_p) = \mathit{Inf}(P_1)$$

Proof For every positive r, the function $p \mapsto r^p$ is convex on $I\!R^+$, with derivative $p \mapsto r^p \ln(r)$. Thus,

$$\forall r > 0 \quad \forall p > 1 \qquad r^p \geq r + (p-1) r \ln(r)$$

Now, let $u \in \mathrm{H}^1(\Omega)$ and $1 < p \leq 2$, then by the definition of u_p and the previous inequality

$$\begin{aligned} J_p(u) &\geq J_p(u_p) \\ &\geq J(u_p) + (p-1) \left[\int_\Omega \Phi\left(\sqrt{1 + |\nabla u_p|^2}\right) dx + \int_{\partial\Omega} \Phi\left(|u_p - g|\right) d\mathrm{H}_{N-1} \right] \\ &\geq J(u_p) - (p-1)\left(\mathrm{L_N}(\Omega) + \mathrm{H_{N-1}}(\partial\Omega)\right) \\ &\geq \mathrm{Inf}(P_1) - (p-1)\left(\mathrm{L_N}(\Omega) + \mathrm{H_{N-1}}(\partial\Omega)\right) \end{aligned}$$

where $\mathrm{L_N}(\Omega)$ is the $N-1$-dimensional Lebesgue measure of Ω. Notice that $\mathrm{L_N}(\Omega)$ and $\mathrm{H}_{N-1}(\partial\Omega)$ are finite by hypothesis.

As $u \in \mathrm{H}^1(\Omega)$, we infer $\lim_{p \to 1} J_p(u) = J(u)$. By taking the limsup and the liminf as $p \to 1$ in the preceding inequalities and then taking the infimum over u in $\mathrm{H}^1(\Omega)$, which is dense in $\mathrm{W}^{1,1}(\Omega)$, we easily get our claim. □

Remark In fact, we can even show that J_p epiconverges to J on $\mathrm{W}^{1,1}(\Omega)$.

To be able to go further, we need to suppose some more regularity for the solutions of (P_1).

3 A Sufficient Condition for the Convergence of $(u_p)_p$

As in [2], we introduce the following subspace of $\mathrm{W}^{1,1}(\Omega)$:

Definition 3.1 *We will denote* $\mathrm{W}^{1,1+}(\Omega) = \bigcup_{p>1} \mathrm{W}^{1,p}(\Omega)$

Remark The function $t \to 1/\ln(t)$ belongs to $\mathrm{W}^{1,1}(0,1/2) \setminus \mathrm{W}^{1,1+}(0,1/2)$, so in general $\mathrm{W}^{1,1+}(\Omega) \not\subset \mathrm{W}^{1,1}(\Omega)$.

With this in hand, we can state the main result of this paper, which is that if the elements of $S(P_1)$ are regular enough, then the whole net $(u_p)_p$ converges to a particular solution of (P_1). This result may be compared to the one we stated for the elliptic regularization in the preceding section, for which the natural hypothesis for the convergence of $(u_\varepsilon)_{\varepsilon>0}$ is $S(P_1) \cap \mathrm{H}^1(\Omega) \neq \emptyset$.

Theorem 3.2 *Assume that* $S(P_1) \cap \mathrm{W}^{1,1+}(\Omega) \neq \emptyset$.
Then the net $(u_p)_p$ *strongly converges in* $\mathrm{W}^{1,1}(\Omega)$ *to the unique solution* $\overline{u} \in S(P_1)$ *of the following auxilliary minimization problem:*

$$(P_1^+) \qquad Inf\left\{ \int_{\partial\Omega} \Phi\left(|u-g|\right) dH_{N-1} \ : \ u \in S(P_1) \right\}$$

Remark Problem (P_1^+) can be considered as the selection principle linked to the L^p regularization of (P_1). Indeed, the solution of (P_1) selected as the limit of the approximate solutions $(u_p)_p$ is characterized as being the unique solution of (P_1^+).

Proof of theorem 3.2 As two solutions of (P_1) differ only by a real constant, $S(P_1) \subset \mathrm{W}^{1,1+}(\Omega)$. Let $q > 1$ be such that $S(P_1) \subset \mathrm{W}^{1,q}(\Omega)$. Throughout the proof, we will assume $1 < p < q$.
Let $u \in S(P_1)$, then recalling the inequalities used in the proof of theorem 2.1, we get for all $1 < p < q$:

$$\begin{aligned} J_p(u) &\geq J_p(u_p) \\ &\geq J(u_p) + (p \quad 1)\left[\int_\Omega \Phi\left(\sqrt{1+|\nabla u_p|^2}\right) dx + \int_{\partial\Omega} \Phi\left(|u_p - g|\right) d\mathrm{H}_{N-1}\right] \\ &\geq J(u) + (p-1)\left[\int_\Omega \Phi\left(\sqrt{1+|\nabla u_p|^2}\right) dx + \int_{\partial\Omega} \Phi\left(|u_p - g|\right) d\mathrm{H}_{N-1}\right] \end{aligned}$$

where, in the last inequality, we use that $u \in S(P_1)$. Let us rewrite the above inequality as

$$\int_\Omega \Phi\left(\sqrt{1+|\nabla u_p|^2}\right) dx + \int_{\partial\Omega} \Phi\left(|u_p - g|\right) d\mathrm{H}_{N-1} \leq \frac{J_p(u) - J(u)}{p-1} \qquad (1)$$

$$\leq \frac{J_q(u) - J(u)}{q-1} \qquad (2)$$

the last inequality being a consequence of the convexity of $p \mapsto r^p$ for positive r.
Since the left hand side of (1) is uniformly bounded by the right hand side of (2), which is finite because $u \in S(P_1) \cap \mathrm{W}^{1,q}(\Omega)$, we may apply the Dunford-Pettis theorem to $(\nabla u_p)_p$. But as the family $(u_p)_p$ is also bounded in $\mathrm{W}^{1,1}(\Omega)$, we infer that $(u_p)_p$ is weakly $\mathrm{W}^{1,1}(\Omega)$-relatively compact. Let $(u_{p(k)})_{k\in\mathbb{N}}$,where $p(k) \to 1$, be a subsequence weakly converging in $\mathrm{W}^{1,1}(\Omega)$ to a function $u_1 \in \mathrm{W}^{1,1}(\Omega)$. As $(u_{p(k)})_{k\in\mathbb{N}}$ is a minimizing sequence of (P_1) and J is convex continuous on $\mathrm{W}^{1,1}(\Omega)$, we obtain $u_1 \in S(P_1)$.
Let us now show that u_1 is a solution of (P_1^+). Thanks to lemma 4.1, we know that $\left(\sqrt{1+|\nabla u_{p(k)}|^2}\right)_k$ weakly converges in $\mathrm{L}^1(\Omega)$ to $\sqrt{1+|\nabla u_1|^2}$ and $(|u_{p(k)} - g|)_k$ weakly converges in $\mathrm{L}^1(\partial\Omega)$ to $|u_1 - g|$.
As Φ is convex continuous on $\mathbb{R}^+$, we may pass to the liminf on the left hand side of (1) and obtain that for all $u \in S(P_1)$ and $p \in]1,q[$

$$\int_\Omega \Phi\left(\sqrt{1+|\nabla u_1|^2}\right) dx + \int_{\partial\Omega} \Phi\left(|u_1 - g|\right) d\mathrm{H}_{N-1} \leq \frac{J_p(u) - J(u)}{p-1}$$

Applying Lebesgue's monotone convergence theorem to the right hand side of this inequality, we get, for all u in $S(P_1)$

$$\int_\Omega \Phi\left(\sqrt{1+|\nabla u_1|^2}\right) dx + \int_{\partial\Omega} \Phi\left(|u_1 - g|\right) d\mathrm{H}_{N-1}$$
$$\leq \int_\Omega \Phi\left(\sqrt{1+|\nabla u|^2}\right) dx + \int_{\partial\Omega} \Phi\left(|u - g|\right) d\mathrm{H}_{N-1}$$

As two solutions of $S(P_1)$ only differ by a constant, this results in

$$\forall u \in S(P_1) \qquad \int_{\partial\Omega} \Phi\left(|u_1 - g|\right) d\mathrm{H}_{N-1} \leq \int_{\partial\Omega} \Phi\left(|u - g|\right) d\mathrm{H}_{N-1}$$

So u_1 is a solution of (P_1^+). We claim that such a solution is unique. Indeed, let u and v be two optimal solutions of (P_1^+), then as $J(u) = J(v)$ and $\nabla u = \nabla v$, we easily get

$$||u - g||_{\mathrm{L}^1(\partial\Omega)} = ||v - g||_{\mathrm{L}^1(\partial\Omega)} = \left\|\frac{u+v}{2} - g\right\|_{\mathrm{L}^1(\partial\Omega)} \qquad (3)$$

As $S(P_1)$ and Φ are convex, we also obtain

$$\int_{\partial\Omega} \Phi\left(|u-g|\right) dx = \int_{\partial\Omega} \Phi\left(|v-g|\right) dx = \int_{\partial\Omega} \Phi\left(\left|\frac{u+v}{2}-g\right|\right) dx < \infty \tag{4}$$

Now, lemma 4.2 implies $u = v$ in $\mathrm{L}^1(\partial\Omega)$, so $u = v$ in $\mathrm{W}^{1,1}(\Omega)$. This proves our claim, and we obtain $u_1 = \overline{u}$, where $\overline{u}$ is the unique solution of (P_1^+).
This implies that the whole net $(u_p)_p$ weakly converges to $\overline{u}$ in $\mathrm{W}^{1,1}(\Omega)$. It remains to show that it strongly converges to $\overline{u}$. To this end, we use inequality (1): we apply it with $u = \overline{u}$ and pass to the limsup as $p \to 1$ to get

$$\begin{aligned}\limsup_{p\to 1} &\left[\int_\Omega \Phi\left(\sqrt{1+|\nabla u_p|^2}\right) dx + \int_{\partial\Omega} \Phi\left(|u_p-g|\right) d\mathrm{H}_{N-1}\right] \\ &\leq \int_\Omega \Phi\left(\sqrt{1+|\nabla \overline{u}|^2}\right) dx + \int_{\partial\Omega} \Phi\left(|\overline{u}-g|\right) d\mathrm{H}_{N-1}\end{aligned}$$

Now, as $(u_p)_p$ weakly converges to $\overline{u}$ in $\mathrm{W}^{1,1}(\Omega)$, we apply lemmas 4.1 and 4.4 to obtain

$$\int_\Omega \Phi\left(\sqrt{1+|\nabla \overline{u}|^2}\right) dx \leq \liminf_{p\to 1} \int_\Omega \Phi\left(\sqrt{1+|\nabla u_p|^2}\right) dx$$

and

$$\int_{\partial\Omega} \Phi\left(|\overline{u}-g|\right) d\mathrm{H}_{N-1} \leq \liminf_{p\to 1} \int_{\partial\Omega} \Phi\left(|u_p-g|\right) d\mathrm{H}_{N-1}$$

so that

$$\lim_{p\to 1} \int_\Omega \Phi\left(\sqrt{1+|\nabla u_p|^2}\right) dx = \int_\Omega \Phi\left(\sqrt{1+|\nabla \overline{u}|^2}\right) dx$$

Then lemma 4.3 allows us to conclude that $(u_p)_p$ strongly converges to $\overline{u}$ in $\mathrm{W}^{1,1}(\Omega)$, thus finishing the proof. □

Remark Notice that problem (P_1^+) always makes sense since $S(P_1) \subset \mathrm{L}^\infty(\partial\Omega)$ (it is a consequence of $g \in \mathrm{L}^\infty(\Omega)$), and (P_1^+) is also always well-posed, even if $S(P_1) \cap \mathrm{W}^{1,1+}(\Omega) = \emptyset$. As shown in the previous proof, the unicity of the solution is a consequence of lemma 4.2. The existence of an optimal solution $\overline{u}$ of (P_1^+) can be easily shown by applying the direct method of the calculus of variations. Indeed, let $(u_n)_n$ be a minimizing sequence of (P_1^+), then by the Dunford-Pettis theorem we may extract a subsequence that weakly converges in $\mathrm{L}^1(\partial\Omega)$ to some $\overline{u}$, which then belongs to $S(P_1)$. Now the same trick as the one used in the proof of lemma 4.1 shows that $(|u_n - g|)_n$ also weakly converges in $L^1(\partial\Omega)$ to $|\overline{u}-g|$, so that $\overline{u}$ is an optimal solution of (P_1^+) thanks to lemma 4.4. This naturally suggests the following question: is theorem 3.2 still valid without assuming $S(P_1)\cap W^{1,1+}(\Omega)\neq\emptyset$?

4 Technical Lemmas

We gather here some lemmas needed in the proof of theorem 3.2.

Lemma 4.1 *Suppose that* $(u_p)_{p>1}$ *weakly converges in* $W^{1,1}(\Omega)$ *to* $u\in S(P_1)$. Then $\left(\sqrt{1+|\nabla u_p|^2}\right)_{p>1}$ *weakly converges in* $L^1(\Omega)$ *to* $\sqrt{1+|\nabla u|^2}$ *and* $(|u_p-g|)_{p>1}$ *weakly converges in* $L^1(\partial\Omega)$ *to* $|u-g|$.

Proof We first notice that $\lim_{p\to1}\int_\Omega\sqrt{1+|\nabla u_p|^2}dx=\int_\Omega\sqrt{1+|\nabla u|^2}dx$ and $\lim_{p\to1}\int_{\partial\Omega}|u_p-g|d\mathrm{H}_{N-1}=\int_{\partial\Omega}|u-g|d\mathrm{H}_{N-1}$. Indeed, this is an easy consequence of the weak lower semicontinuity of the functionals $\int_\Omega\sqrt{1+|\nabla .|^2}$ and $\int_{\partial\Omega}|.-g|\,d\mathrm{H}_{N-1}$ on the space $W^{1,1}(\Omega)$ and of theorem 2.1 which asserts that $\lim_{p\to1}J(u_p)=\mathrm{Inf}(P_1)=\int_\Omega\sqrt{1+|\nabla u|^2}dx+\int_{\partial\Omega}|u-g|d\mathrm{H}_{N-1}$.

Now, let A be a Borel subset of $\partial\Omega$, then

$$\int_A|u-g|d\mathrm{H}_{N-1}\leq\liminf_{p\to1}\int_A|u_p-g|d\mathrm{H}_{N-1}$$

and

$$\limsup_{p\to1}\int_A|u_p-g|d\mathrm{H}_{N-1}=\limsup_{p\to1}\left[\int_{\partial\Omega}|u_p-g|d\mathrm{H}_{N-1}-\int_{\partial\Omega\setminus A}|u_p-g|d\mathrm{H}_{N-1}\right]$$

$$=\int_{\partial\Omega}|u-g|d\mathrm{H}_{N-1}-\liminf_{p\to1}\int_{\partial\Omega\setminus A}|u_p-g|d\mathrm{H}_{N-1}$$

so by the previous inequality, we get

$$\begin{aligned}\limsup_{p\to1}\int_A|u_p-g|d\mathrm{H}_{N-1}&\leq\int_{\partial\Omega}|u-g|d\mathrm{H}_{N-1}-\int_{\partial\Omega\setminus A}|u-g|d\mathrm{H}_{N-1}\\&=\int_A|u-g|d\mathrm{H}_{N-1}\end{aligned}$$

thus proving that $\lim_{p\to 1}\int_A |u_p - g| d\mathrm{H}_{N-1} = \int_A |u-g| d\mathrm{H}_{N-1}$. As this is true for every Borel subset A of $\partial\Omega$, the sequence $\left(|u_p - g|\right)_{p>1}$ weakly converges in $\mathrm{L}^1(\partial\Omega)$ to $|u-g|$.

We apply the same argument for $\left(\sqrt{1+|\nabla u_p|^2}\right)_{p>1}$. □

The three last lemmas can be stated in a more general setting, lemmas 4.3 and 4.4 being easy adaptations to the vectorial case of lemmas 6 and 7 in [7]. The following lemma implies the uniqueness of the solution (P_1^+).

Lemma 4.2 *Let (Ω, A, μ) be a measure space with a positive measure μ, and $\Psi : I\!R^+ \to I\!R$ a strictly convex function. Let u, v belong to $\mathrm{L}^1(\Omega)$ and satisfy (3') and (4').*

$$\|u\|_{\mathrm{L}^1(\Omega)} = \|v\|_{\mathrm{L}^1(\Omega)} = \left\|\frac{u+v}{2}\right\|_{\mathrm{L}^1(\Omega)} \qquad (3')$$

$$\int_\Omega \Psi\left(|u|\right) d\mu = \int_\Omega \Psi\left(|v|\right) d\mu = \int_\Omega \Psi\left(\left|\frac{u+v}{2}\right|\right) d\mu < \infty \qquad (4')$$

Then $u = v$ in $L^1(\Omega)$.

Proof It is easy to show that (3′) implies $\frac{|u(x)+v(x)|}{2} = \frac{|u(x)|+|v(x)|}{2}$ for μ-almost every $x \in \Omega$ (this is equivalent to say that $u(x)$ and $v(x)$ are of the same sign μ-almost everywhere on Ω). But then, as Ψ is strictly convex on $I\!R^+$, this and (4′) imply that $|u(x)| = |v(x)|$ for μ-almost every $x \in \Omega$. As a consequence, $u(x) = v(x)$ for μ-almost every $x \in \Omega$, which is our claim. □

Lemma 4.3 *Let (Ω, A, μ) be a measure space with a positive finite measure μ, and $C \subset I\!R^N$ be a closed convex set. Assume that $(h_n)_{n\in I\!N}$ is a sequence in $\mathrm{L}^1(\Omega)$ satisfying $h_n(\Omega) \subset C$ for all n. Let us assume that $(h_n)_n$ weakly converges to h in $\mathrm{L}^1(\Omega)$ and that there exists a continuous and strictly convex function $\Psi : C \to I\!R$ such that*

$$\limsup_{n\to+\infty} \int_\Omega \Psi \circ h_n d\mu \leq \int_\Omega \Psi \circ h d\mu < +\infty$$

Then (h_n) strongly converges to h in $\mathrm{L}^1(\Omega)$.

Lemma 4.4 *Let (Ω, A, μ) be a measure space with a positive finite measure μ, and $C \subset I\!R^N$ be a closed convex set. Let also $\Psi : C \to I\!R$ be a continuous convex function.*

Assume that $(h_n)_{n\in I\!N}$ is a sequence in $\mathrm{L}^1(\Omega)$ satisfying $h_n(\Omega) \subset C$ for all n, and weakly converging to h in $\mathrm{L}^1(\Omega)$. Then

$$\int_{\Omega} \Psi \circ h d\mu \leq \liminf_{n \to +\infty} \int_{\Omega} \Psi \circ h_n d\mu$$

Remark In the proof of theorem 3.2, we apply these lemmas with $C = \mathbb{R}^+$ and $C = \mathbb{R}^N$.

References

[1] H. Attouch, *Variational convergence for functions and operators* (Pitman, London, 1984)

[2] H. Attouch and R. Cominetti, L^p Approximation of variational problems in L^1 and L^∞, *Nonlinear Analysis* 36 (1999) 373-399.

[3] I. Ekeland and R. Temam, *Analyse convexe et problèmes variationnels* (Dunod Gauthiers-Villars, Paris, 1974)

[4] E. DeGiorgi, Nouveaux résultats dans la théorie des hypersurfaces minima, *Conférences faites au Collège de France* juin 1970.

[5] E. Giusti, *Minimal Surfaces and Functions of Bounded Variation* (Birkhauser, Boston, 1984)

[6] J.-L. Lions, *Quelques Méthodes de résolution des problèmes aux limites non linéaires* (Dunod et Gauthier-Villars, Paris, 1969)

[7] D. Landers and L. Rogge, Natural choice of L^1-Approximants, *Journal of Approximation Theory* 33 (1981) 268-280.

Coupling the Logarithmic-Quadratic Proximal Method and the Block Nonlinear Gauss-Seidel Algorithm for Linearly Constrained Convex Minimization

Alfred Auslender * Marc Teboulle ** Sami Ben-Tiba ‡

Abstract

We present a new variant of the block nonlinear Gauss-Seidel algorithm for solving convex minimization problems over polyhedral constraints. The method is based on coupling the recent logarithmic-quadratic proximal method introduced by the authors, with the coordinate descent strategie for problems with block structure. Our approach allows for eliminating the constraints, and removing the usual strict convexity assumption used in a Gauss-Seidel type algorithm. Thus, for each block the algorithm consists of solving an unconstrained convex problem, which can be solved efficiently via Newton type method, and leads to an attractive decomposition scheme. We prove under mild assumptions on the problems data, that all limit points of the sequence generated by the proposed algorithm are optimal solutions.

Keywords: Block Gauss-Seidel, Interior Proximal Regularization, Minimization over Cartesian products, Decomposition algorithms.

AMS subject classification: 90C25

*Laboratoire d'Econometrie de L'Ecole Polytechnique, 1 Rue Descartes, Paris 75005, France

**School of Mathematical Sciences, Tel-Aviv University, Ramat-Aviv 69978, Israel. Research partially supported by the Israeli Ministry of Science under Grant No. 9636-1-96

‡Laboratoire d'Econometrie de L'Ecole Polytechnique, 1 Rue Descartes, Paris 75005, France

1 Introduction

Recent advances in computer technology such as parallel and distributed computing systems have considerably increased computing power and motivated the search for decomposition techniques and related algorithms that can exploit it. This paper will concentrate on optimization problems with block structures and analyze a new variant of the block nonlinear Gauss-Seidel method under linear constraints.

Let $X \subset \mathbb{R}^n$, be the Cartesian product of m closed nonempty and convex subsets $X_i \subset \mathbb{R}^{n_i}$, i.e, $X = \prod_{i=1}^{m} X_i$, with $n_1 + n_2 + \ldots + n_m = n$. Any vector $x \in X$ is accordingly decomposed as $x = (x_1, x_2, \ldots, x_m)$ with, $x_i \in X_i$. Let $f : \mathbb{R}^n \to \mathbb{R}$ be a given function and consider the problem:

$$\min\{f(x) : x \in X\}. \tag{1.1}$$

Given the current iterate $x^k = (x_1^k, x_2^k, \ldots, x_m^k)$, the block-nonlinear Gauss-Seidel (**GS**) method generates sequentially the next iterate $x^{k+1} = (x_1^{k+1}, x_2^{k+1}, \ldots, x_m^{k+1})$ via the iteration

$$x_i^{k+1} = \operatorname*{argmin}_{x \in X_i}\{f(x_1^{k+1}, \ldots, x_{i-1}^{k+1}, x, x_{i+1}^k, \ldots, x_m^k)\},\ i = 1, \ldots, m. \tag{1.2}$$

At each iteration, the function f is minimized with respect to each of the block-coordinate vectors x_i^k, taken in cyclic order. Thus, the **GS** method naturally exploits the Cartesian product structure and is attractive when the minimization in (1.2) can be solved efficiently.

Convergence results for the **GS** method can be found in several works, see e.g, [1], [5], [8], [13]. One of the key assumption necessary to prove convergence is that the minimum in (1.2) is uniquely attained, see e.g., [15]. Otherwise as shown by an example given in Powell [9], the **GS** method may cycle indefinitely without converging. A basic convergence result for the **GS** method applied to convex minimization problems can be obtained under the following assumptions:

(i) f is continuously differentiable and convex over $X = \prod_{i=1}^{m} X_i$.

(ii) For each i, f is a strictly convex function of x_i, when the values of the other components of X are held constant.

Then, under assumptions (i) and (ii), every limit point of the sequence $\{x^k\}$ generated by the **GS** method (1.2) minimizes f over X, see e.g., [5].

In this work we concentrate on the important class of convex minimization problems over polyhedral constraints with block structure, namely when the sets X_i are given in the form

$$X_i = \{x \in \mathbb{R}^{n_i} : A^i x \leq b^i\}, \tag{1.3}$$

where A^i is a $p_i \times n_i$ matrix and $b^i \in \mathbb{R}^{p_i}$.

We introduce a new Gauss-Seidel type algorithm to solve (1.1) under the constraints (1.3). Our main objective is two fold. Firstly, to remove the strict convexity assumption (ii). Secondly, and more importantly, to eliminate the constraints in each minimization step of the form (1.2), replacing it by a suitable *unconstrained* minimization step. Indeed, as mentioned earlier while the **GS** method is useful for decomposition, yet one still have to solve in (1.2) for each block, a constrained minimization problem which can be time consuming and quite inefficient, as opposed to an unconstrained problem, which could be solved via an efficient Newton type algorithm (given a smooth enough objective function).

To achieve the first task, namely removing the strict convexity assumption, one can consider the coupling of the proximal algorithm [6], [12] with the **GS** method and generate a sequence $\{x^k\}$ via the iteration:

$$x_i^{k+1} = \operatorname*{argmin}_{x_i \in X_i}\{f(x_1^{k+1}, \ldots, x_{i-1}^{k+1}, x_i, x_{i+1}^k, \ldots, x_m^k) + \frac{\lambda_i^k}{2}\|x_i - x_i^k\|^2\}, \tag{1.4}$$

where $\lambda_i^k > 0$. In fact such a variant of the **GS** method was already suggested by Auslender in [2].

However, in (1.4) one still has to solve a constrained minimization problem for each $i = 1, \ldots, m$. To achieve our second task, namely to eliminate the constraints, we propose in the same spirit to couple the Gauss-Seidel algorithm with a new *proximal interior* algorithm, called the " proximal logarithmic-quadratic method", which was recently introduced by the authors in [3]. This allows for achieving both tasks at once (removing the strict convexity assumption and the constraints) by generating a sequence $\{x^k\}$ via the scheme:

$$x_i^{k+1} = \operatorname*{argmin}_{x_i \in \mathbb{R}^{n_i}}\{f(x_1^{k+1}, \ldots, x_{i-1}^{k+1}, x_i, x_{i+1}^k, \ldots, x_m^k) + \lambda_i^k D(x_i, x_i^k)\}, \tag{1.5}$$

where $D(.,.)$ is a proximal logarithmic-quadratic term which forces the iterates $\{x_i^k\}$ to stay in the interior of the feasible set $\text{int}X_i$ (see Section 3 for details). The new algorithm (1.5) not only overcomes the difficulties present in the **GS** method but also seems promising from the numerical point of view since the logarithmic-quadratic proximal term satisfies the so-called self-concordance properties [7], which appear fundamental to solve problems of the form (1.5) via efficient Newton methods.

In the next section, we state precisely the problem to be analyzed and assumptions. In fact, we allow for considering also nonsmooth functions by considering an objective of the form

$$f(x) := h_0(x_1, \ldots, x_m) + \sum_{i=1}^{m} h_i(x_i), \tag{1.6}$$

where h_0 is convex and differentiable on X and $h_i : \mathbb{R}^{n_i} \to \mathbb{R} \cup \{+\infty\}$ are closed convex functions. Optimization problems modeled with an objective of the form (1.6) represent a broad class of problems that arise in various contexts, see e.g., [2] and reference therein, and as we shall see can be handled within our algorithm at little expense. In section 3, we recall the basic properties of the logarithmic-quadratic proximal functional and we state our algorithm. In section 4, we prove convergence of the algorithm and we conclude the paper in section 5 with some remarks, extensions and applications. Notation and/or definitions not explicitly given in the paper can be found in Rockafellar's book [11].

2 Problem Formulation and Assumptions

Let X_i be a polyhedral set on $\mathbb{R}^{n_i}$ defined by

$$X_i = \{x \in \mathbb{R}^{n_i} : A^i x \leq b^i\},$$

where A^i is a $p_i \times n_i$ matrix, $b^i \in \mathbb{R}^{p_i}$, $p_i \geq n_i$.

We suppose that the matrix A^i is of maximal rank, i.e., $\operatorname{rank} A^i = n_i$. Let $X = \prod_{i=1}^{m} X_i$ and set $n_1 + \ldots + n_m = n$, so that X is also a polyhedral set in $\mathbb{R}^n$. The vector x is partitioned as $x = (x_1, \ldots, x_m)$, with $x_i \in \mathbb{R}^{n_i}$ so that $x \in X$ if and only if $x_i \in X_i, i = 1, \ldots, m$.

Consider the following convex minimization problem:

$$(P) \quad \min\{h(x) : x \in X\},$$

where $h(x) := h_0(x_1, \ldots, x_m) + \sum_{i=1}^{m} h_i(x_i)$.

throughout this paper we make the following assumptions for problem (P):
(A1) $h_0 : \mathbb{R}^n \to \mathbb{R}$ is continuously differentiable and convex on the set $X = X_1 \times \ldots \times X_m$.
(A2) For each $i = 1, \ldots, m$, $h_i : \mathbb{R}^{n_i} \to \mathbb{R} \cup \{+\infty\}$ is a closed proper convex function with $\mathrm{dom} h_i \supset X_i$.
(A3) The interior of X_i, $\mathrm{int} X_i := \{x \in \mathbb{R}^{n_i} : A^i x < b^i\}$ is nonempty for each $i = 1, \ldots, m$.
(A4) $\mathrm{dom}\partial h_i \cap \mathrm{int} X_i \neq \emptyset$, $i = 1, \ldots, m$, where ∂h_i denotes the subdifferential of h_i.
An important special case of problem (P) is when $h \equiv h_0$, i.e., with differentiable data only reducing to the problem

$$(P_0) \quad \min\{h_0(x_1, \ldots, x_m) : x \in X\},$$

and in that case, only assumptions (A1), (A3) are necessary.

3 The Gauss-Seidel-Log-Quadratic Proximal Method

Our algorithm to solve (P) is based on coupling the Gauss-Seidel scheme with the logarithmic-quadratic proximal method. We thus start this section by recalling the definition and properties of the later, as recently introduced in [3]. Let φ be the convex function defined on $\mathbb{R}$ by

$$\varphi(t) = \begin{cases} t^2 - t - \log t & \text{if } t > 0 \\ +\infty & \text{otherwise.} \end{cases} \tag{3.1}$$

Associated with φ we define for any $v \in \mathbb{R}^p_{++}$

$$d(u, v) = \begin{cases} \sum_{i=1}^{p} u_i^2 - u_i v_i - v_i^2 \log(\frac{u_i}{v_i}) & \text{if } u \in \mathbb{R}^p_{++} \\ +\infty & \text{otherwise.} \end{cases} \tag{3.2}$$

The functional d defined in (3.2) can be rewritten as

$$d(u, v) = \sum_{i=1}^{p} v_i^2 \varphi(u_i v_i^{-1}) \quad \forall u, v > 0. \tag{3.3}$$

We note that $\varphi = \varphi_1 + \varphi_2$ with $\varphi_1(t) = (t-1)^2$, $\varphi_2(t) = -\log t + t - 1$ if $t > 0$, $+\infty$ otherwise.

It can be easily seen that $\varphi_2(t) \geq 0$ so that the following useful lower bound holds:

$$d(u,v) \geq \|u-v\|^2. \tag{3.4}$$

Second-order homogeneous functionals of the form (3.3) with other choices of the kernel φ, and their role in the development of interior multiplier methods have been recently studied in [4]. Here we consider only the important special case (3.3) with the logarithmic-quadratic kernel (3.1). The next proposition collects some basic properties of φ and d which will be useful in our analysis, for a proof see [3].

Proposition 3.1 *Let φ be given in (3.1) and d as defined in (3.2).*
(i) φ is a differentiable strongly convex function on $\mathbb{R}_{++}$ with modulus 2 and $\lim_{t\to 0+} \varphi'(t) = -\infty$.
(ii) $d(.,v)$ is a closed proper convex function $\forall v \in \mathbb{R}^p_{++}$. Moreover, $d(u,v) \geq \|u-v\|^2$ and $d(u,v) = 0$ if and only if $u = v$, $\forall (u,v) \in \mathbb{R}^p_{++} \times \mathbb{R}^p_{++}$.
(iii) For any $s > 0$, $t > 0$ and $u \geq 0$ we have

$$(t-u)(2t-s-t^{-1}s^2) \geq (s-t)(3u-2t-s)$$

Proposition 3.1 (ii) shows that d can be viewed as a distance-like functional for vectors in the nonnegative orthant, while property (i) allows for generating a well-defined proximal method, forcing each iterate to stay positive, and leading to proximal interior methods. For recent developments, results and applications of proximal interior algorithms, see [14] and references therein.
The basic Logarithmic-Proximal method to solve a linearly constrained optimization problem of the form

$$(CL)\ \ \min\{f(x) : x \in C\},\ C := \{x \in \mathbb{R}^N : Ax \leq b\},$$

with $f : \mathbb{R}^N \to \mathbb{R} \cup \{+\infty\}$, A is an $p \times N$ matrix, $b \in \mathbb{R}^p$, $\text{int}C \neq \emptyset$ is as follows

The LQP Method: Start with $x^0 \in \text{int}C$ and generate the sequence $\{x^k\} \subset \text{int}C$ satisfying

$$(\mathbf{LQP})\ x^{k+1} \in \underset{x\in\mathbb{R}^N}{\text{argmin}}\{f(x) + \lambda_k D(L(x), L(x^k))\},$$

with $\lambda_k \geq \lambda > 0$, and $D(x,y) := d(L(x), L(y))$ where

$$\begin{aligned} l_j(x) &= b_j - < a_j, x >, \qquad j = 1, \dots, p, \\ L(x) &= (l_1(x), \dots, l_p(x))^T, \end{aligned}$$

and a_j denotes the rows of the matrix A.
In [3, Proposition 2.1 and Theorem 2.1] the following result was proved.

Theorem 3.1 *Assume that the following hold for problem* (CL)*:*
(i) A *is of maximal rank and* f *is closed proper convex.*
(ii) The optimal set of problem (CL) *is nonempty.*
(iii) $dom\partial f \cap intC \neq \emptyset$*, where* ∂f *denotes the subdifferential of* f*.*
Then, for each $\lambda_k > 0$, $x^k \in intC$*, there exists a unique* $x^{k+1} \in intC$ *solving* (**LQP**)*, and the sequence* $\{x^k\}$ *converges to an optimal solution* x^*.□

To define our algorithm for solving problem (P), we need to define the following quantities:

Let a^i_j denotes the rows of the matrix A^i and define for $i = 1, \dots, m$; $j = 1, \dots, p_i$,

$$\begin{aligned} l^i_j(x_i) &= b^i_j - < a^i_j, x_i >, \\ L^i(x_i) &= (l^i_1(x_i), \dots, l^i_{p_i}(x_i))^T, \\ D(x_i, y_i) &= d(L^i(x_i), L^i(y_i)). \end{aligned}$$

Here $< ., . >$ denotes the inner product between vectors in $\mathbb{R}^{n_i}$.
For each $x_i \in \text{int}X_i$, $y_i \in \text{int}X_i$ we have

$$\nabla_1 D(x_i, y_i) = -\sum_{j=1}^{p_i} a^i_j (2l^i_j(x_i) - l^i_j(y_i) - \frac{l^i_j(y_i)^2}{l^i_j(x_i)}), \tag{3.5}$$

where $\nabla_1 D$ denotes the gradient of D in $\mathbb{R}^{n_i}$ with respect the first argument. We propose now to solve (P) by coupling the usual block Gauss-Seidel algorithm with the logarithmic proximal method as follows

GS-LQP Method:
Step 0. Set $k = 0$, $x^0 \in \text{int}X$, $0 < \bar{\lambda} \leq \lambda^k_i \leq \lambda \; \forall i = 1, \dots, m$.

Step k. For $i = 1, \dots, m$, generate $x^{k+1}_i \in \text{int}X_i$ satisfying

$$x^{k+1}_i = \operatorname*{argmin}_{x_i \in \mathbb{R}^{n_i}} \{h(x^{k+1}_1, \dots, x^{k+1}_{i-1}, x_i, x^k_{i+1}, \dots, x^k_m) + \lambda^k_i D(x_i, x^k_i)\} \tag{3.6}$$

and set $x^{k+1} = (x_1^{k+1}, \ldots, x_m^{k+1})$.
For convenience, we will also use the following notations:

$$h_i^k(z) := h(x_1^{k+1}, \ldots, x_{i-1}^{k+1}, z, x_{i+1}^k, \ldots, x_m^k), \tag{3.7}$$

$$x_i(k) := (x_1^{k+1}, \ldots, x_{i-1}^{k+1}, x_i^{k+1}, x_{i+1}^k, \ldots, x_m^k), \tag{3.8}$$

with the obvious conventions: $h_1^k(z) = h(z, x_2^k, \ldots, x_m^k)$, $x_m(k) = (x_1^{k+1}, \ldots, x_m^{k+1}) \equiv x^{k+1}$ and $x_0(k) = x^k$.
Note that from Proposition 3.1 (i) the functional

$$x_i \to h_i^k(x_i) + \lambda_i^k D(x_i, x_i^k)$$

is strongly convex, thus if a minimizer x_i^{k+1} in step (3.6) exists, it must be unique, see Proposition 4.1, below.

4 Convergence Analysis

Recall that our blanket assumption (A1)-(A4) for problem (P) is in force.
We denote by X_* the set of optimal solutions of problem (P). We first show that the algorithm **GS-LQP** is well-defined and then prove its convergence.

Proposition 4.1 *Let $\{x^k\}$ be the sequence generated by* **GS-LQP** *and suppose that X_* is a nonempty compact set. Then, for each k and for each $i = 1, \ldots, m$, there exists a unique point x_i^{k+1} solving (3.6) and such that $x_i^{k+1} \in \text{int}X_i$.*

Proof: Starting with $x^0 \in \text{int}X$, suppose by induction that $x^k \in \text{int}X$. Since X_* is nonempty and compact it follows from the Cartesian product structure of the set X that the optimal set of solutions of each problem

$$(P_i^k) \qquad \alpha_i^k := \inf\{h_i^k(x_i) : x_i \in X_i\},$$

is also nonempty and compact. Therefore, since by (A4) we have $\text{dom}\partial h_i \cap \text{int}X_i \neq \emptyset$, invoking Theorem 3.1 to problem (P_i^k) it follows that for each k and $i = 1, \ldots, m$, there exists a unique $x_i^{k+1} \in \text{int}X_i$ solving (3.6).□

Theorem 4.1 *Let $\{x^k\}$ be the sequence generated by* **GS-LQP** *and suppose that X_*, the set of optimal solutions of (P) is nonempty and compact. Then, the sequence $\{x^k\}$ is bounded and all of its limit points are in X_*.*

Proof: We first show that the sequence $\{x^k\}$ is bounded.
Since $D(x_i^k, x_i^k) = 0$, it follows from (3.6) and using the notation (3.7) that

$$h_i^k(x_i^{k+1}) + \lambda_i^k D(x_i^{k+1}, x_i^k) \leq h_i^k(x_i^k), \tag{4.1}$$

and hence, since $D(.,.) \geq 0$ the above inequality implies $\forall k$

$$\begin{aligned} h(x^{k+1}) &\leq h(x_{m-1}(k)) \leq \ldots \leq h(x_2(k)) \\ &\leq h(x_1(k)) \leq h(x^k). \end{aligned} \tag{4.2}$$

Now, since $x^k \in \text{int}X$, and by assumption X_* is a nonempty compact set, and since by (4.2) we have

$$x^k \in \mathcal{L}(x^0) := \{x \in X : h(x) \leq h(x^0)\},$$

it follows that the sequence $\{x^k\}$ is bounded.
Now define the following quantities:

$$\begin{aligned} \beta_i^k &:= h(x_1^{k+1}, \ldots, x_{i-1}^{k+1}, x_i^k, x_{i+1}^k, \ldots, x_m^k) \\ &\equiv h_i^k(x_i^k), \\ \gamma_i^k &:= h(x_1^{k+1}, \ldots, x_{i-1}^{k+1}, x_i^{k+1}, x_{i+1}^k, \ldots, x_m^k) \\ &\equiv h_i^k(x_i^{k+1}). \end{aligned}$$

Then by (4.1) and (4.2) it follows that

$$\begin{aligned} 0 \leq \lambda_i^k D(x_i^{k+1}, x_i^k) &\leq \beta_i^k - \gamma_i^k \\ &\leq h(x^k) - h(x^{k+1}). \end{aligned} \tag{4.3}$$

Since by (4.2) the sequence $\{h(x^k)\}$ is decreasing, and hence convergent, together with (4.3) we thus obtain (recalling that $0 < \bar{\lambda} \leq \lambda_i^k$)

$$\lim_{k \to \infty} D(x_i^{k+1}, x_i^k) = 0. \tag{4.4}$$

Therefore, using (3.4) we obtain in particular that

$$\lim_{k \to \infty} \sum_{j=1}^{p_i} (l_j^i(x_i^k) - l_j^i(x_i^{k+1}))^2 = 0 \quad \forall i = 1, \ldots, m,$$

so that

$$\lim_{k \to \infty} ||A^i(x_i^k - x_i^{k+1})|| = 0, \quad \forall i = 1, \ldots, m.$$

Since A^i is of maximal rank it follows that

$$\lim_{k \to \infty} ||x_i^k - x_i^{k+1}|| = 0, \quad \forall i = 1, \ldots, m. \tag{4.5}$$

Now, writing down the optimality condition for (3.6) we obtain together with Proposition 4.1 that the sequence $\{x^k\}$ generated by **GS-LQP** satisfies for each $i = 1, \dots, m$

$$x_i^{k+1} \in \operatorname{int} X_i, \tag{4.6}$$

$$\delta_i^k + g_i^k + \lambda_i^k \nabla_1 D(x_i^{k+1}, x_i^k) = 0, \tag{4.7}$$

$$g_i^k \in \partial h_i(x_i^{k+1}) + N_{X_i}(x_i^{k+1}), \tag{4.8}$$

where

$$\delta_i^k := \nabla_i h_0(x_1^{k+1}, \dots, x_{i-1}^{k+1}, x_i^{k+1}, x_{i+1}^k, \dots, x_m^k)$$

and $\nabla_i h_0$ denotes the partial derivative of h_0 with respect to x_i.
Let $\partial H_i := \partial h_i + N_{X_i}$, where N_{X_i} denotes the normal cone of X_i. Since we assumed that $\partial h_i \cap \operatorname{int} X_i \neq \emptyset$ (cf. assumption A4), and since N_{X_i} is maximal monotone, by [12] it follows that ∂H_i is also maximal monotone, for each $i = 1, \dots, m$. Furthermore, since by (4.6) $x_i^{k+1} \in \operatorname{int} X_i$, then $N_{X_i}(x_i^{k+1}) = \{0\}$ and we have by (4.8)

$$g_i^k \in \partial H_i(x_i^{k+1}), \quad i = 1, \dots, m.$$

Using the definition of l_j^i and the monotonicity of ∂H_i, it follows from (3.5) and (4.7) that

$$< x_i - x_i^{k+1}, \delta_i^k + g_i > \geq \lambda_i^k \sum_{j=1}^{p_i} R_{i,j}^k, \tag{4.9}$$

$$\forall (x_i, g_i) \in \text{ Graph}(H_i),$$

where, for $i = 1, \dots, m$, $j = 1, \dots, p_i$

$$R_{i,j}^k := (l_j^i(x_i^{k+1}) - l_j^i(x_i))(2l_j^i(x_i^{k+1}) - l_j^i(x_i^k) - \frac{l_j^i(x_i^k)^2}{l_j^i(x_i^{k+1})}),$$

Invoking Proposition 3.1 with

$$s := l_j^i(x_i^k), \quad t := l_j^i(x_i^{k+1}) \quad \text{and} \quad u := l_j^i(x_i)$$

and summing over $j = 1, \dots, p_i$ we obtain

$$\sum_{j=1}^{p_i} R_{i,j}^k \geq \sum_{j=1}^{p_i} (l_j^i(x_i^k) - l_j^i(x_i^{k+1}))(3l_j^i(x_i) - 2l_j^i(x_i^{k+1}) - l_j^i(x_i^k)). \tag{4.10}$$

Since $\{x^k\}$ is bounded, $\{l_j^i(x_i^k) - l_j^i(x_i^{k+1})\}$ converges to zero and $0 < \lambda_i^k \leq \lambda$, passing to the limit in (4.10) we obtain that

$$\liminf_{k \to \infty} \lambda_i^k \sum_{j=1}^{p_i} R_{i,j}^k \geq 0. \tag{4.11}$$

Let x^∞ be a limit point of the sequence $\{x^k\}$. Using (4.5) and the fact that the gradient $\nabla_i h_0$ is continuous on X_i, passing to the limit in (4.9), it follows that $\forall i = 1, \ldots, m$

$$< x_i - x_i^\infty, \nabla_i h_0(x^\infty) + g_i > \geq 0 \quad \forall (x_i, g_i) \in \text{Graph}(H_i). \tag{4.12}$$

To complete the proof, once again we use the Cartesian product structure. Let $\partial H := \partial H_1 \times \ldots \times \partial H_m$ and set $T := \partial H + \nabla h_0(x^\infty)$. Then, T is maximal monotone. Furthermore, from (4.12) we obtain that

$$< x - x^\infty, \theta > \geq 0 \quad \forall (x, \theta) \in \text{Graph}(T).$$

But T is maximal monotone, so it follows from the above inequality that $0 \in T(x^\infty)$, which in turns is equivalent to $0 \in \partial H(x^\infty) + \nabla h_0(x^\infty)$. But since $\partial H_i = \partial h_i + N_{X_i}$, the latter inclusion is equivalent to

$$0 \in \partial h(x^\infty) + N_X(x^\infty),$$

i.e., x^∞ is an optimal solution of problem (P).□

5 Concluding Remarks

We have proposed a new Gauss-Seidel type algorithm leading to decomposition for block structured optimization problems. The key advantage of the proposed method is that each iteration of **GS-LQP** given in (3.6) is an *unconstrained* problem, which in turn can be solved efficiently via Newton-type algorithms, given sufficiently smooth data. This is in sharp contrast with classical Gauss-Seidel methods, which not only required the strict convexity of the objective function in problem (P) (an hypothesis not needed here), but also necessitate to solve a constrained problems at each iteration. The logarithmic-quadratic proximal term used here allows for eliminating both of these drawbacks.

Gauss-Seidel type methods are providing a natural framework for serial decomposition schemes. However, following a similar analysis one could also obtain the same kind of results by extending the proposed approach to derive Jacobi-like methods leading to parallel decomposition schemes. More precisely, to solve problem (P) one can devise the following Jacobi type algorithm coupled with the logarithmic proximal method:

J-LQP Method:
Step 0. Set $k = 0$, $x^0 \in \text{int}X$, $0 < \bar{\lambda} \le \lambda_i^k \le \lambda \ \forall i = 1, \ldots, m$.

Step k. For $i = 1, \ldots, m$, generate $x_i^{k+1} \in \text{int}X_i$ satisfying

$$x_i^{k+1} = \operatorname*{argmin}_{x_i \in \mathbb{R}^{n_i}} \{h(x_1^k, \ldots, x_{i-1}^k, x_i, x_{i+1}^k, \ldots, x_m^k) + \lambda_i^k D(x_i, x_i^k)\} \tag{5.1}$$

and set $x^{k+1} = \sum_{i=1}^m \mu_i^k x_i^{k+1}$, with $\sum_{i=1}^m \mu_i^k = 1$, $\mu_i^k \ge \mu > 0$.

Each minimization step (5.1) in the above algorithm can be performed separately for each x_i, resulting in a parallel decomposition scheme for solving problem (P).

References

[1] A. Auslender, Méthodes Numériques pour la decomposition et la minimisation de fonctions non differentiables. *Numerish Mathematik*, **18**, (1971), 213-223.

[2] A. Auslender, Asymptotic properties of the Fenchel dual functional and applications to decomposition problems. *J. of Optimization Theory and Applications*, **73**, 1992, 427-449.

[3] A. Auslender , M. Teboulle and S. Ben-Tiba, A Logarithmic-Quadratic proximal method for variational inequalities. *Computational Optimization and Applications*, (1998), In press.

[4] A. Auslender , M. Teboulle and S. Ben-Tiba, Interior proximal and multiplier methods based on second order homogeneous functionals. February 1998. *Mathematics of Operations Research*, Submitted for publication.

[5] D. P. Bertsekas and J. N. Tsitsiklis, *Parallel and Distributed Computation: Numerical Methods* (Prentice-Hall, New Jersey, 1989).

[6] B. Martinet, Régularisation d'inéquations variationnelles par approximations successives, *Revue Francaise d'Automatique et Informatique Recherche Opérationnelle*, **4**, 1970, 154-159.

[7] Y. Nesterov, A. Nemirovski, *Interior point polynomial algorithms in convex programming*, SIAM Publications, Philadelphia, PA, 1994.

[8] J. M. Ortega and W.C. Rheinboldt, *Iterative Solution of Nonlinear Equations in Several Variables.* Academic Press, New-York, 1970.

[9] M. J. D. Powell, On search directions for minimization algorithms. *Mathematical Programming*, **4**, (1973), 193-201.

[10] R.T. Rockafellar, On the maximality of sums of nonlinear monotone operators, it Transactions of the American Mathematical Society, **149**, (1970), 75-88.

[11] R. T. Rockafellar, *Convex Analysis*, Princeton University Press, Princeton , NJ, 1970.

[12] R.T. Rockafellar, Monotone operators and the proximal point algorithm, *SIAM J. of Control and Optimization*, **14**, (1976), 877- 898.

[13] P. Tseng. Decomposition algorithm for convex differentiable minimization. *J. of Optimization Theory and Applications*, **70**, (1991), 109-135.

[14] M. Teboulle, Convergence of Proximal-like Algorithms, *SIAM J. of Optimization*, **7**, 1997, 1069-1083.

[15] W. I. Zangwill, *Nonlinear programming: A unified Approach.* Prentice Hall, Englewood Cliffs, 1969.

Bundle Methods for Maximal Monotone Operators

Regina S. Burachik * Claudia Sagastizábal ** B. F. Svaiter ‡

Abstract

To find a zero of a maximal monotone operator T we use an enlargement T^ε playing the role of the ε-subdifferential in nonsmooth optimization. We define a convergent and implementable algorithm which combines projection ideas with bundle-like techniques and a transportation formula. More precisely, first we separate the current iterate x^k from the zeros of T by computing the direction of minimum norm in a polyhedral approximation of $T^{\varepsilon_k}(x^k)$. Then suitable elements defining such polyhedral approximations are selected following a bundle strategy. Finally, the next iterate is computed by projecting x^k onto the corresponding separating hyperplane.

Keywords: bundle methods, maximal monotone operators, enlargements of maximal monotone operators, transportation formulæ.

AMS subject classification: Primary: 65K05; Secondary: 90C30, 52A41, 90C25.

1 Introduction and Motivation

Consider the problem

$$0 \in T(x)\,, \tag{1}$$

where T is a maximal monotone operator on $\mathbb{R}^N$ with nonempty solution set $\mathcal{S}$. When T is single-valued, well-known algorithms for solving (1) are Korpelevich's method [10], Khobotov's [7] and others.

In this paper we present an implementable algorithm for solving (1) when T is a multi-valued mapping. Our departing point is the following, simple, remark:

*Engenharia de Sistemas de Computação, COPPE–UFRJ, CP 68511, Rio de Janeiro, RJ, CEP 21945–970, Brazil.

**INRIA, BP 105, 78153 Le Chesnay, France and PUC, Depto. Engenharia Elétrica. Rio de Janeiro, RJ, CEP 22453-900, Brazil. Research supported by FAPERJ (Brazil) under Grant No.E26/150.205/98.

‡IMPA, Estrada Dona Castorina, 110. Rio de Janeiro, RJ, CEP 22460-320, Brazil. Research partially supported by CNPq Grant 301200/93-9(RN) and by PRONEX–Optimization.

given an arbitrary y and $v \in T(y)$, the monotonicity of T implies that $\mathcal{S}$ is contained in the halfspace

$$H_{y,v} := \{z \in \mathbb{R}^N : \langle z - y, v \rangle \leq 0\}. \tag{2}$$

We have already presented in [3] a conceptual method for finding zeros of T developing these ideas. More precisely, given a current iterate $x^k \notin \mathcal{S}$,
- first, find y^k and $v^k \in T(y^k)$ such that $x^k \notin H_{y^k,v^k}$,
- then, project x^k onto $H_{y^k,v^k} \supseteq \mathcal{S}$ to obtain a new iterate

$$x^{k+1} := Proj_{H_{y^k,v^k}}(x^k) = x^k - \frac{\langle x^k - y^k, v^k \rangle}{\|v^k\|^2} v^k . \tag{3}$$

Being a projection, x^{k+1} is closer to $\mathcal{S}$ than x^k. However, to have a significant progress from x^k to x^{k+1}, adequate choices of (y^k, v^k) are required. Since $v^k \in T(y^k)$ is given by an oracle, the control can only be done when generating y^k. In our proposal we analyzed points of the form

$$y^k = x^k - t^k s^k , \quad \text{with} \quad s^k = Proj_{T^{\varepsilon_k}(x^k)}(0) , \tag{4}$$

for positive t^k and ε_k. The sets $T^\varepsilon(x)$ are the ε-enlargements of T from [2], further studied in [3]. These enlargements behave in many ways as the ε-subdifferential of a closed convex function, see § 2 below. In particular, they are also Lipschitz-continuous multi-functions of x and ε, provided ε is positive. The scheme (3)-(4) is useful for theoretical purposes. Actually, it allows to clearly identify the crucial elements to obtain convergence: (3) and (4) have to be combined in order to drive ε_k to 0, to generate a convergent sequence. At the same time, ε^k should not be driven to 0 too fast, otherwise the resulting multifunction $x \hookrightarrow T^{\varepsilon^k}$ would not be smooth enough and the improvement when passing from x^k to x^{k+1} would become negligible. Such a "loss of smoothness" is reflected by the sensitivity of s^k to variations in ε_k and x^k. More precisely, when $\varepsilon_k = 0$, s^k ceases to depend continuously on x^k. In this respect, we say that the problem of finding s^k in (4) becomes *ill-posed*.
When coming to implementation concerns, it appears that s^k in (4) cannot be computed without having a *full* knowledge of $T^{\varepsilon_k}(x^k)$, a fairly bold (if not impossible) assumption. Instead, we assume that an oracle, giving one element in $T(z)$ for any z, is available. Then s^k can be approached by projecting 0 onto a polyhedral approximation of $T^{\varepsilon_k}(x^k)$. A suitable polyhedral approximation is obtained by using the transportation formula for T^ε, proved in [3], together with bundle techniques like in [8], [18], [13]: having at the current iteration a raw bundle with all the oracle information collected so far, $\{(z^i, w^i \in T(z^i))\}_{i \leq p}$, the convex hull of some *selected* w^i's is a good approximation of $T^{\varepsilon_k}(x^k)$. Again, special attention has to be put when selecting the sub-bundle, in order to control ε_k and preserve convergence.

The paper is organized as follows. In Section 2 we start with some notations and assumptions. Then we recall the definition of T^ε, together with some continuity properties and the transportation formula. Section 3 describes an implementable algorithm combining the projections ideas and bundle techniques outlined above. This algorithm is proved to be convergent in § 4. Finally, in § 5 we give some concluding remarks and we compare our algorithm with a closely related work by Konnov, [9].

2 Preliminary Results

We gather in this section notation and general assumptions, as well as some results, straightforward or already proved in [2] and [3].

2.1 Notation and Assumptions

Let $\mathcal{P}(\mathbb{R}^N)$ denote the set of all subsets of $\mathbb{R}^N$. Then, given a multifunction $F: \mathbb{R}^N \to \mathcal{P}(\mathbb{R}^N)$ and a set $A \subseteq \mathbb{R}^N$:

- the closure of A is denoted by $\overline{A}$ and
- the domain of F is $D(F) := \{x \in \mathbb{R}^N : F(x) \neq \emptyset\}$.
- We define the set $F(A) := \bigcup_{a \in A} F(a)$.
- $B(x, \rho)$ denotes the ball centered in x with radius ρ.
- F is *locally bounded* at x if there exists a neighbourhood U of x such that the set $F(U)$ is bounded.
- F is *monotone* if $\langle u - v, x - y \rangle \geq 0$ for all $x, y \in \mathbb{R}^N$, and all $u \in F(x)$, $v \in F(y)$.
- T is *maximal monotone* if it is monotone and, additionally, whenever there is some monotone F such that $T(x) \subset F(x)$ for all $x \in \mathbb{R}^N$, this implies $T = F$. □

Recall that any maximal monotone operator is locally bounded in the interior of its domain ([17, Theorem 1]).
All along this paper, we suppose that T in (1) is defined in the whole of $\mathbb{R}^N$, so that it maps bounded sets in bounded sets. Finally, recall that the solution set $\mathcal{S}$ is assumed to be not empty.

2.2 The Enlargement T^ε. Some Useful Properties

Given $T: \mathbb{R}^N \to \mathcal{P}(\mathbb{R}^N)$ maximal monotone and $\varepsilon \geq 0$, the ε-enlargement of T at x, introduced in [2] and thoroughly studied in [3], is defined by

$$\begin{array}{rccl} T^\varepsilon: & \mathbb{R}^N & \to & \mathcal{P}(\mathbb{R}^N) \\ & x & \hookrightarrow & \{u \in \mathbb{R}^N : \langle v - u, y - x \rangle \geq -\varepsilon \text{ for all } y \in \mathbb{R}^N,\ v \in T(y)\}. \end{array} \tag{5}$$

The idea of using a smearing parameter ε to obtain better regularity properties can be traced back to [1] and [4] for the subdifferential of convex and non convex functions, respectively. For monotone operators, not necessarily maximal, relaxed monotonicity was explored in [20]. Other related works are [11], [19], [15] and [16].

2.2.1 Continuity Properties

The ε-enlargement is proved to be locally bounded in [3]. Also, in [2] it was proved that the application $(x, \varepsilon) \hookrightarrow T^\varepsilon(x)$ is continuous (upper and lower semicontinuous). This result was improved in [3, Lemma 2.5], where T^ε is shown to be Lipschitz-continuous whenever $\varepsilon > 0$ (in [3] we also show that the Lipschitz constant diverges when $\varepsilon \to 0$).
In the sequel, we will just make use of the following result, establishing the closedness of the graph of T^ε:

Proposition 2.1 *[2, Propositions 1(iv) and 2(i)]* $T^0 = T$, *and for any sequence* $\{(\varepsilon_i, x^i, w^i \in T^{\varepsilon_i}(x^i))\}_i$ *such that* $\varepsilon_i \geq 0$ *for all* i,

$$\lim_{i\to\infty} x^i = x, \quad \lim_{i\to\infty} \varepsilon_i = \varepsilon, \quad \lim_{i\to\infty} w^i = v \quad \Longrightarrow \quad v \in T^\varepsilon(x)\,.$$

□

2.2.2 Transportation Formula

In our algorithm, we build polyhedral approximations of $T^\varepsilon(x)$ using elements in a certain bundle of information $\{(z^i, w^i \in T(z^i))\}_{i=1}^m$. To prove convergence, it is crucial to relate elements in the bundle with some $v \in T^\varepsilon(x)$. This is the transportation formula for T^ε proved in [3, Theorem 2.3].

Theorem 2.2 *Consider a set of* m *triplets* $\{(\varepsilon_i, z^i, w^i \in T^{\varepsilon_i}(z^i))\}_{i=1,\dots,m}$, *and take the convex sum:*

$$(\hat{x}, \hat{s}) := (\sum_{i=1}^m \alpha_i z^i, \sum_{i=1}^m \alpha_i w^i)\,,$$

where $\alpha \in \mathbb{R}^m$ *is such that* $\alpha_i \geq 0$ *and* $\sum_{i=1}^m \alpha_i = 1$. *Then* $\hat{s} \in T^{\hat{\varepsilon}}(\hat{x})$, *with*

$$\hat{\varepsilon} := \sum_{i=1}^m [\alpha_i \varepsilon_i + \alpha_i \langle z^i - \hat{x}, w^i - \hat{s}\rangle]\,.$$

□

Note that when compared to the result obtained for $\partial_\varepsilon f$ (see, for example, [5, Proposition XI.4.2.2.]), our formula is weaker, because it only transports convex sums.
In this paper, we make use of the transportation formula above only when $\varepsilon_i = 0$, that is, when $w^i \in T(z^i)$ is given by an oracle for all z^i.
As a consequence of Theorem 2.2, $\hat{\varepsilon}$ can be uniformly bounded on α. This result will be very useful in the sequel.

Corollary 2.3 *Consider the notations and assumptions of Theorem 2.2. Suppose that $\varepsilon_i = 0$ for all $i \leq m$. In addition, let $\tilde{x} \in \mathbb{R}^N$ and $\rho > 0$ be such that $\|z^i - \tilde{x}\| \leq \rho$ for all $i \leq m$. Then, the convex sum $(\hat{x}, \hat{s}) := (\sum_1^m \alpha_i z^i, \sum_1^m \alpha_i w^i)$ satisfies the following:*

$$\|\hat{x} - \tilde{x}\| \leq \rho\,,$$
$$\hat{s} \in T^{\hat{\varepsilon}}(\hat{x}) \quad \textit{with} \quad \hat{\varepsilon} := \textstyle\sum_1^m \alpha_i \langle z^i - \hat{x}, w^i - \hat{s}\rangle \leq 2\rho M\,,$$

where $M := \max\{\|w^i\| \mid i = 1, \ldots, m\}$.

Proof. The convexity of the norm implies that $\|\hat{x} - \tilde{x}\| \leq \rho$, and also $\|\hat{s}\| \leq M$. Theorem 2.2 establishes that $\hat{s} \in T^{\hat{\varepsilon}}(\hat{x})$, for $\hat{\varepsilon} = \sum_1^m \alpha_i \langle z^i - \hat{x}, w^i - \hat{s}\rangle$. The last expression can be rewritten as follows:

$$\begin{aligned}
\textstyle\sum_1^m \alpha_i \langle z^i - \hat{x}, w^i - \hat{s}\rangle &= \textstyle\sum_1^m \alpha_i \langle z^i - \tilde{x}, w^i - \hat{s}\rangle + \sum_1^m \alpha_i \langle \tilde{x} - \hat{x}, w^i - \hat{s}\rangle \\
&= \textstyle\sum_1^m \alpha_i \langle z^i - \tilde{x}, w^i - \hat{s}\rangle + 0 \\
&= \textstyle\sum_1^m \alpha_i \langle z^i - \tilde{x}, w^i - \hat{s}\rangle\,.
\end{aligned}$$

Because $\|\hat{s}\| \leq M$, together with the Cauchy-Schwarz inequality and our assumptions, the proof is done. □

3 Building Bundle-Like Methods

We present first the conceptual algorithm scheme from [3], working directly with the convex sets $T^\varepsilon(x)$. Further on, we use a bundle technique to build adequate polyhedral approximations of the sets $T^\varepsilon(x)$.

3.1 The Conceptual Algorithm Scheme (CAS)

The scheme from [3] has the following form.

INITIALIZATION:Choose parameters $\tau, R, \varepsilon > 0$ and $\sigma \in (0,1)$. Set $k := 0$ and take $x^0 \in \mathbb{R}^N$.

k-STEP:
Step 0: If $0 \in T(x^k)$, then STOP. **(stopping test)**
Step 1: **(computing search direction)**
Compute $s^k := \operatorname{argmin}\{\|v\|^2 \mid v \in T^{\varepsilon 2^{-j}}(x^k)\}$, where j is such that $\|s^k\| > \tau 2^{-j}$.
Step 2: **(line search)**
Define $y^k := x^k - R 2^{-l} s^k$ and take $v^k \in T(y^k)$, for l such that $\langle v^k, s^k\rangle > \sigma \|s^k\|^2)$.
Step 3: **(projection step)**
Define $x^{k+1} := x^k - \langle v^k, x^k - y^k\rangle v^k / \|v^k\|^2$.
Set $k := k + 1$ and LOOP to **k**-STEP. □

Indices j and l are nonnegative integers; in [3] we prove that (CAS) is well defined, with no infinite loops in Steps 1 or 2. (CAS) is also convergent in the following sense: it either stops with a last $x^k \in \mathcal{S}$, or it generates an infinite sequence $\{x^k\}$ converging to a solution of (1). We will see that the same convergence result holds for an implementable version of (CAS) (cf. Theorem 4.8).

3.2 The Implementable Bundle Strategy (BS)

We already mentioned that a major drawback of (CAS) is the expensive, when not impossible, computation of an element s^k of minimum norm in $T^\varepsilon(x^k)$. A proposal to get round this difficulty, inspired from our guiding model $T = \partial f$, is to apply a bundle-like strategy. However, since no functional values are available when $T \neq \partial f$, we need to extend and adapt dual-bundle methods (see [14], [6], also Chapters XI, XIII in [5]) to this more general context.
Let us describe how a bundle strategy can be used in our setting. As usual in bundle methods, we suppose that an *oracle* computes $v = v(y) \in T(y)$ for any given y. In addition, we denote by

$$\Delta_I := \{\lambda \in \mathbb{R}^I_+ : \sum_{i \in I} \lambda_i = 1\}$$

the unit-simplex associated to a set of indices I.
The following mechanism makes the "k-step" in (CAS) implementable. The convergence is then preserved by an appropriate choice of the current bundle in step (1.b) below.

Bundling Strategy (BS)

INITIALIZATION: Choose parameters $\tau, R > 0$ and $\sigma \in (0,1)$. Set $k := 0$, $p := 0$ and take $x^0 \in \mathbb{R}^N$.

IMPLEMENTABLE **k**-STEP:

Step 0:
- (0.a) Compute $u^k \in T(x^k)$, if $u^k = 0$, then STOP. **(stopping test)**
- (0.b) Else, set $p := p+1$, $(z^p, w^p) := (x^k, u^k)$. Set $n := 0$.

Step 1: **(computing search direction)**
- (1.a) Set $j := 0$.
- (1.b) Define $I_{k,n,j} := \{1 \leq i \leq p \mid \|z^i - x^k\| \leq R2^{-j}\}$.
- (1.c) Compute $\alpha^{k,n,j} := \operatorname{argmin}\{\|\sum_{i \in I_{k,n,j}} \alpha_i w^i\|^2 \mid \alpha \in \Delta_{I_{k,n,j}}\}$.
- (1.d) Take $s^{k,n,j} := \sum_{i \in I_{k,n,j}} \alpha_i^{k,n,j} w^i$.
- (1.e) If $\|s^{k,n,j}\| \leq \tau 2^{-j}$ then set $j := j+1$ and LOOP to (1.b).
- (1.f) Else, define $j_{k,n} := j$ and $s^{k,n} := s^{k,n,j_{k,n}}$.

Step 2: **(line search)**
- (2.a) Set $l := 0$.
- (2.b) Define $y^{k,n,l} := x^k - R2^{-l} s^{k,n}/\|s^{k,n}\|$ and take $v^{k,n,l} \in T(y^{k,n,l})$.
- (2.c) If $\left(\langle v^{k,n,l}, s^{k,n}\rangle \leq \sigma \|s^{k,n}\|^2 \text{ and } l < j_{k,n}+1\right)$, then

Set $l := l + 1$ and LOOP to (2.b).

(2.d) Else, define $l_{k,n} := l$ and $y^{k,n} := y^{k,n,l_{k,n}}$, $v^{k,n} := v^{k,n,l_{k,n}}$.

Step 3: **(evaluating the pair (y, v))**

(3.a) If $\langle v^{k,n}, s^{k,n} \rangle \leq \sigma \|s^{k,n}\|^2$, then **(null step)**

Set $p := p + 1$, $(z^p, w^p) := (y^{k,n}, v^{k,n})$.

Set $n := n + 1$ and LOOP to (1.b).

(3.b) Else **(serious step)**

Define $n_k := n$, $j_k := j_{k,n_k}$ and $l_k := l_{k,n_k}$,
$s^k := s^{k,n_k}$, $y^k := y^{k,n_k}$, $v^k := v^{k,n_k}$.

Define $x^{k+1} := x^k - \langle v^k, x^k - y^k \rangle v^k / \|v^k\|^2$.

Set $k := k + 1$ and LOOP to IMPLEMENTABLE **k**-STEP.

□

We comment here some features of this implementable version of (CAS).

Remark 3.1

- Observe that index p is a counter of elements in the raw bundle $\{(z^i, w^i)\}_{i \leq p}$. Namely, serious points are added at (0.b), while the information obtained in null steps is incorporated to the bundle in (3.a).
- In *Step 0* a stopping test of the kind $\|u^k\| \leq$ TOL could have been used in (0.a). In such a case, the convergence analysis would essentially remain the same by setting the tolerance TOL to 0 in the results (to make the algorithm loop forever and produce an infinite sequence $\{x^k\}$).
- In *Step 1* the search direction is computed. The set $T^{\varepsilon 2^{-j}}(x^k)$ used in (CAS) is now replaced by the convex hull generated by some *selected* elements in the range of T, belonging to a reduced bundle. Observe that, by definition of the index set $I_{k,n,j}$, the vector $u^k \in T(x^k)$ is always in the associated (reduced) bundle. Hence this sub-bundle is not empty and $\alpha^{k,n,j}$ is well defined. We show in Lemma 4.2(i) that if x^k is not a solution, then the loop (1.e)↔(1.b) is always finite.
- In *Step 2* we evaluate the pair $(y^{k,n,l}, v^{k,n,l})$ provided by the direction $s^{k,n}$. Step (2.c) makes a loop (to (2.b)) which eventually ends. The two possible endings are:

  1.*null step:* For $l = j + 1$ it holds that $\langle v^{k,n}, s^{k,n} \rangle \leq \sigma \|s^{k,n}\|^2$.

  2.*serious step:* When for some $l \leq j+1$ the direction $v^{k,n}$ satisfies $\langle v^{k,n}, s^{k,n} \rangle > \sigma \|s^{k,n}\|^2$.

  In the first case, which leads to step (3.a), the line search performed in step (2.c) gives a pair $(y^{k,n}, v^{k,n})$ such that $\|y^{k,n} - x^k\| = R2^{-j-1}$. Then $v^{k,n}$ is added to the current sub-bundle (recall that v^i is incorporated only when $\|y^i - x^k\| \leq R2^{-j}$), and no progress is made in k. Lemma 4.2(ii) shows that, when x^k is not a solution, there is no infinite loop in (3.a)↔(1.b).
- If a serious step is done, (3.b) is visited. In this case, the direction defined by $v^{k,n}$ defines a separating hyperplane far enough from the current iterate x^k.

By "far enough" we mean that the stepsize $||x^k - x^{k+1}||$ does not go to zero too fast. A new iterate x^{k+1} is generated by projecting x^k onto the halfspace H_{y^k,v^k} from (2). In addition, note that at step (3.b) the following holds:

$$\begin{array}{lcl} l_k \leq j_k + 1\,, & & \langle v^k, s^k\rangle > \sigma||s^k||^2\,, \\ ||s^k|| > \tau 2^{-j_k}\,, & \text{and} & y^k = x^k - R2^{-l_k} s^k/||s^k||\,. \end{array} \tag{6}$$

□

4 Convergence Analysis

We show in this section that either (BS) generates a finite sequence, whose last iterate solves (1), or it generates an infinite sequence, converging to a solution of (1).

The bundle strategy (BS) provides us with a constructive device for (CAS). For this reason the convergence analysis is close to the one in [3]. The main difference appears when analyzing null steps, which lead to an enrichment of the bundle. For proving that the convergence is preserved when replacing $T^\varepsilon(x^k)$ by the current polyhedral approximation, we use the following technical result:

Lemma 4.1 *[5, Lemma IX.2.1.1] Let $\gamma > 0$ fixed. Consider two infinite sequences sequences $\{v^m\}$ and $\{\hat{v}^m\}$ satisfying for $m = 1, 2, \ldots$:*

$$\langle v^i - v^{m+1}, \hat{v}^m\rangle \geq \gamma||\hat{v}^m||^2 \quad \textit{for all } i = 1, \ldots, m\,. \tag{7}$$

If $\{v^i\}$ is bounded then $\hat{v}^m \to 0$ when $m \to \infty$.

□

Each k-step of (BS) has two ending points: (0.a) and (3.b). If some implementable k-step ends at (0.a), then the algorithm stops at a solution of (1). If the implementable k-step ends at (3.b), then x^{k+1} is generated and a new k-step is started with k replaced by $k+1$.

There are three inner loops: (1.e)↔(1.b) on j indices; (3.a)↔(1.b), iterating along n indices and finally (2.c)↔(2.b), incrementing l.

We prove first that infinite loops do not occur on these indices when $x^k \notin \mathcal{S}$.

Lemma 4.2 *Let x^k be the current iterate in (BS) and suppose $x^k \notin \mathcal{S}$. Then the following holds:*

(i) Relative to the loop (1.e)↔(1.b), there exists a finite $j = j_{k,n}$ such that (1.f) is reached:

$$||s^{k,n}|| > \tau 2^{-j_{k,n}}\,.$$

Furthermore, the loop (2.c)↔(2.b) is finite: (2.d) is reached with $l_{k,n} \leq j_{k,n} + 1$.

(ii) Relative to the loop (3.a)↔(1.b), there exists a finite $n = n_k$ such that (3.b) is reached.

Proof. By assumption, $0 \notin T(x^k)$. To prove (*i*), suppose, for contradiction, that (BS) loops forever in (1.e)$\leftrightarrow$(1.b). Then $j \to \infty$ and an infinite sequence $\{s^{k,n,j}\}_{j \in \mathbb{N}}$ is generated, satisfying $\|s^{k,n,j}\| \leq \tau 2^{-j}$. Therefore, there exist two subsequences $\{n_q\}$, $\{j_q\}$ such that

$$\|s^{k,n_q,j_q}\| \leq \tau 2^{-j_q}, \tag{8}$$

with $\lim_{q \to \infty} j_q = \infty$. For such indices, define $I_q := I_{k,n_q,j_q}$. Because of step (1.b), for all $i \in I_q$, $\|z^i - x^k\| \leq R2^{-j_q}$. Consider the convex sum given by $\alpha^q := \alpha^{k,n_q,j_q}$ from step (1.c):

$$(\hat{x}^q, \hat{s}^q) := \Big(\sum_{i \in I_q} \alpha_i^q z^i, s^{k,n_q,j_q}\Big).$$

Corollary 2.3 applies, with $\rho = R2^{-j_q}$ and $\tilde{x} = x^k$, and we have

$$\hat{s}^q \in T^{\hat{\varepsilon}_q}(\hat{x}^q), \quad \text{with} \quad \hat{\varepsilon}_q \leq 2R2^{-j_q}M, \tag{9}$$

where $M := \sup\{\|u\| \mid u \in T(\overline{B(x^k, R)})\}$. In addition,

$$\|\hat{x}^q - x^k\| \leq R2^{-j_q}. \tag{10}$$

Altogether, letting $q \to \infty$ in (8), (9) and (10), Proposition 2.1 yields:

$$(\hat{\varepsilon}_q, \hat{x}^q, \hat{s}^q \in T^{\hat{\varepsilon}_q}(\hat{x}^q)) \longrightarrow (0, x^k, 0) \quad \Longrightarrow \quad 0 \in T^0(x^k) = T(x^k),$$

a contradiction. Hence, there exists a finite j such that the loop (1.e)$\leftrightarrow$(1.b) ends. For this index j, step (1.f) is reached, $j_{k,n} = j$ is defined with $\|s^{k,j_{k,n}}\| > \tau 2^{-j_{k,n}}$, and the first part in (*i*) holds.
Furthermore, the test in (2.c) will eventually be false and the loop (2.c)$\leftrightarrow$(2.b) ends, with a value of $l = l_{k,n} \leq j_{k,n} + 1$.
Now we prove (*ii*). If an infinite loop occurs at (3.a)$\leftrightarrow$(1.b), then $n \to \infty$. Thus, at step (2.d) an infinite sequence $\{(y^{k,n}, v^{k,n} \in T(y^{k,n})\}_{n \in \mathbb{N}}$ is generated. We have that for each n, the loop (1.e)$\rightarrow$(1.b) ends with an index j such that

$$s^{k,n,j} = s^{k,n,j_{k,n}} = s^{k,n} \quad \text{and} \quad \|s^{k,n}\| > \tau 2^{-j_{k,n}}. \tag{11}$$

We proved in (*i*) that j eventually reaches its final value, say J. Therefore, there exists $\bar{n}$ such that $j_{k,n} = j_{k,\bar{n}} = J$ for any $n \geq \bar{n}$. Consider now the sequence $\{(y^{k,n}, v^{k,n})\}_{n \geq \bar{n}}$. At Step (3.a) infinite null steps are made:

$$\langle v^{k,n}, s^{k,n} \rangle \leq \sigma \|s^{k,n}\|^2 \tag{12}$$

with

$$\|y^{k,n} - x^k\| = R2^{-l_{k,n}} = R2^{-j_{k,n}-1} = R2^{-J-1}, \tag{13}$$

for all $n \geq \bar{n}$ (if (2.a) leads to (3.a) then $l_{k,n} = j_{k,n} + 1$). This means that $(y^{k,n}, v^{k,n})$ is incorporated to the sub-bundle associated to $I_{k,n,j_{k,n}}$ for any $n \geq$

$\bar{n}$. In particular, choose an index $\tilde{n}$ such that $\bar{n} < \tilde{n} < n$: since $j_{k,n} = J$ in (13), $(y^{k,\tilde{n}}, v^{k,\tilde{n}})$ is incorporated to the sub-bundles defining $s^{k,n}$, the projection of 0 onto the convex hull of $\{w^i\}_{i \in I_{k,n,j}}$ (cf. steps (1.c)-(1.f)). Therefore, by the classical projection property (see for instance [21])

$$\langle v^{k,\tilde{n}}, s^{k,n} \rangle \geq \|s^{k,n}\|^2 , \qquad \text{for any } n > \tilde{n} > \bar{n} \ .$$

Together with (12) written for $n > \tilde{n}$, we obtain

$$\langle v^{k,\tilde{n}} - v^{k,n}, s^{k,n} \rangle \geq (1 - \sigma)\|s^{k,n}\|^2 \text{ for any } n > \tilde{n} > \bar{n} \ . \tag{14}$$

We claim that the assumptions of Lemma 4.1 hold, after a suitable renaming is done. Actually, define, for all $i \geq 1$,

$$t^i := v^{k,\tilde{n}+i}, \quad \tilde{t}^i := s^{k,\tilde{n}+i+1} \ .$$

Using now (14), we obtain

$$\langle t^i - t^{m+1}, \tilde{t}^m \rangle \geq (1 - \sigma)\|\tilde{t}^m\|^2 \text{ for any } i = 1, \ldots, m \, .$$

Therefore, the sequences $\{t^m\}$ and $\{\tilde{t}^m\}$ satisfy condition (7) in Lemma 4.1 with $\gamma = 1 - \sigma$. Moreover, because $\{t^i\} \subseteq T(B(x^k, R))$ and T is locally bounded, the last part in Lemma 4.1 also applies:

$$\tilde{t}^m \to 0 \quad \text{when} \quad m \to \infty \, .$$

However, (11) and the choice of $\tilde{n}$ yields

$$\|\tilde{t}^m\| = \|s^{k,\tilde{n}+m+1}\| > \tau\, 2^{-J} > 0 \, ,$$

a contradiction. Altogether, the loop (3.a)$\leftrightarrow$(1.b) must eventually finish with a finite value of n and (ii) is proved. □

In the next result we analyze all the possibilities for an iteration of (BS).

Proposition 4.3 *Let x^k be the current iterate in (BS). Then*

(i) if x^k is a solution, either the oracle answers $u^k = 0$ and (BS) stops in (0.a), or (BS) loops forever after this last serious step, without updating k.

(ii) Else, x^k is not a solution, and (BS) reaches step (3.b) after finitely many inner iterations. Furthermore,

$$\|s^{k,n_k^*,j_k-1}\| \leq \tau 2^{-j_k+1} , \tag{15}$$

where n_k^* is the smallest value of n equating $j_{k,n} = j_k$, whenever $j_k > 0$.

Proof. Suppose first that x^k is a solution. If the oracle gives $u^k = 0$, the stopping test holds in (0.a) and (BS) stops. Otherwise, $u^k \neq 0$. Then, suppose for contradiction that (3.b) is reached. Recall that by (6), we have

$$\begin{array}{rcl}\langle x^k - y^k, v^k\rangle & = & R2^{-l_k}\langle s^k, v^k\rangle/\|s^k\| \\ & > & 2^{-l_k-j_k}\sigma\tau R > 0\,, \quad \text{with} \quad v^k \in T(y^k)\,.\end{array}$$

Because $0 \in T(x^k)$, the inequality above contradicts the monotonicity of T and (i) is proved.
Let us prove (ii). Suppose $j_k > 0$. If x^k is not a solution, Lemma 4.2 shows that no infinite loop occurs inside iteration k and (3.b) is eventually reached. To prove (15), define

$$n_k^* := \min\{n \leq n_k | \, j_{k,n} = j_k\}\,.$$

Then in (1.e)↔(1.b), the indices

$$j := j_{k,n_k^*} - 1 < j_k \quad \text{and} \quad j + 1 = j_{k,n_k^*} = j_k\,,$$

are such that (1.e) holds for index j and (1.f) holds for index $j+1$:

$$\|s^{k,n_k^*,j_{k,n_k^*}-1}\| \leq \tau 2^{-(j_{k,n_k^*}-1)}\,, \quad \text{and} \quad \|s^{k,n_k^*,j_{k,n_k^*}}\| > \tau 2^{-(j_{k,n_k^*})}\,,$$

and the conclusion follows. □
As a consequence of the last result, the sequence of serious points $\{x^k\}$ generated by (BS) is either finite, ending at a solution; or infinite, with no iterate being a solution. Before proving the convergence for the infinite sequence, we need some preliminary technical results.

Proposition 4.4 *Let x^k be the current iterate in (BS) and assume $x^k \notin \mathcal{S}$. Then, after x^{k+1} is generated in (3.b), the following holds:*

(i) Let H_{y^k,v^k} be the halfspace defined in (2)*, written with $(y, v) := (y^k, v^k)$. Then $x^k \notin H_{y^k,v^k}$ and $x^{k+1} = P_{H_{y^k,v^k}}(x^k)$.*

(ii) For all $x^* \in \mathcal{S}$, $\|x^{k+1} - x^*\|^2 \leq \|x^k - x^*\|^2 - \|x^{k+1} - x^k\|^2$.

(iii) Finally, $\|x^{k+1} - x^k\| > R\sigma\tau 2^{-2j_k-1}/\|v^k\|$.

Proof. To prove (i), recall from (6) that the pair $(y^k = x^k - R2^{-l_k}s^k, v^k = v^{k,n_k})$ is such that

$$\langle x^k - y^k, v^k\rangle = R2^{-l_k}\langle s^k, v^k\rangle/\|s^k\| > 2^{-l_k-j_k}\sigma\tau R > 0\,,$$

so $x^k \notin H_{y^k,v^k}$. To see that x^{k+1} is its projection, just recall the definition of x^{k+1} in step (3.a) of (BS). Because x^{k+1} is an orthogonal projection onto H_{y^k,v^k} and $\mathcal{S} \subset H_{y^k,v^k}$, (ii) follows from the properties of orthogonal projections (see e.g. [22].) As for (iii), it is also straightforward from (6) and the definition of x^{k+1}. □
The boundedness of the variables generated by (BS) now follows from the Fejér convergence of $\{x^k\}$ to $\mathcal{S}$, via the following obvious statement.

Proposition 4.5 *Consider a sequence $\{x^k\}$ be such that*

$$\|x^{k+1} - x\| \leq \|x^k - x\|, \quad \textit{for any } x \in \mathcal{S}.$$

with $\mathcal{S} \neq \emptyset$. Then

(i) $\{x^k\}$ is bounded.

(ii) If $\{x^k\}$ has an accumulation point which is in $\mathcal{S}$, then the full sequence converges to a limit in $\mathcal{S}$.

□

All the variables generated by (BS), namely x^k, s^k, y^k, v^k, $\{(y^{k,n}, v^{k,n})\}$ and $\{(z^p, w^p)\}$ are bounded. Our final theorem only needs the boundedness of the last two sequences.

Lemma 4.6 *The sequences $\{(y^{k,n}, v^{k,n})\}$ and $\{(z^p, w^p)\}$ generated by (BS) in Steps (2.d) and (0.b), respectively, are bounded.*

Proof. First we show that the sequence $\{x^k\}$ is bounded. If the sequence is finite, the boundedness is trivial. If $k \to \infty$, by Proposition 4.4(*ii*) the assumptions of Proposition 4.5 hold. Hence, using Proposition 4.5(*i*), we obtain that $\{x^k\}$ is bounded.
Therefore, there exists some compact set K_0 such that the (bounded) sequence $\{x^k\} \subset K_0$. Define $K_1 := K_0 + \overline{B(0,R)}$. Then, from steps (2.b), (2.d) and (3.b) in (BS), it follows that the variables $y^{k,n}$ and y^k are contained in K_1. Since z^p is extracted from either $\{x^k\}$ or $\{y^k\}$, the sequence is also contained in K_1. Finally, because T is locally bounded, $\{v^k\}$ and $\{w^p\}$ are bounded too and the proof is finished. □
In our last Lemma we show that index j_k in (BS) goes to infinity together with k. Along the lines of [3, Section 3.3], convergence of (BS) will be proved using Proposition 4.5 (ii), by exhibiting a subsequence of triplets $(\hat{\varepsilon}_q, \hat{x}^q, \hat{s}^q \in T^{\varepsilon_q}(\hat{x}^q))$ tending to $(0, \bar{x}, 0)$ as q is driven by j_k to infinity.

Lemma 4.7 *Suppose (BS) loops forever on k (i.e., $k \to \infty$). Then $\lim_{k\to\infty} j_k = +\infty$.*

Proof. Combine Proposition 4.4(*iii*) and (*ii*) to obtain

$$R\sigma\tau 2^{-2j_k-1}/\|v^k\| < \|x^{k+1} - x^k\| \to 0, \quad \text{when } k \to \infty.$$

Since $\|v^k\|$ is bounded (Lemma 4.6), the result follows. □
Finally, we state our main convergence result.

Theorem 4.8 *Consider the sequence $\{x^k\}$ generated by (BS). Then the sequence is either finite with last element in $\mathcal{S}$, or it converges to a solution of (1).*

Proof. We already dealt with the finite case in Proposition 4.3.
If there are infinitely many x^k, keeping Proposition 4.5(*ii*) in mind, we only need to show that some accumulation point of the bounded sequence $\{x^k\}$ is a solution of (1). Let $\{x^{k_q}\}$ be a convergent subsequence, with limit point $\bar{x}$. Because of Lemma 4.7, we can suppose $j_{k_q} > 0$, for q large enough. Then Proposition 4.3(*ii*) applies: for n_k^* defined therein, we have

$$||s^{k_q,n^*_{k_q},j_{k_q}-1}|| \leq \tau 2^{-j_{k_q}+1} \tag{16}$$

Consider the associated index set $I_q := I_{k_q,n^*_{k_q},j_{k_q}-1}$. By developing an argument similar to the one used in the proof of Lemma 4.2, mutatis mutandis, define

$$\begin{aligned} \alpha^q &:= \alpha^{k_q,n^*_{k_q},j_{k_q}-1}, \\ \hat{x}^q &:= \textstyle\sum_{i\in I_q} \alpha^q{}_i z^i, \\ \hat{s}^q &:= s^{k_q,n^*_{k_q},j_{k_q}-1} = \textstyle\sum_{i\in I_q} \alpha^q_i w^i. \end{aligned}$$

We have that

$$||\hat{x}^q - x^{k_q}|| \leq R 2^{-j_{k_q}+1}. \tag{17}$$

Let M be an upper bound for $||w^p||$ (these variables are bounded by Lemma 4.6). Then Corollary 2.3 yields

$$\hat{s}^q \in T^{\hat{\varepsilon}_q}(\hat{x}^q) \text{ with } \hat{\varepsilon}_q \leq 2R2^{-j_{k_q}+1}M. \tag{18}$$

Using Lemma 4.7 we have $\lim_{q\to\infty} j_{k_q} = \infty$. Hence, by (16), (17), (18) we conclude that

$$(\hat{\varepsilon}_q, \hat{x}^q, \hat{s}^q \in T^{\hat{\varepsilon}_q}(\hat{x}^q)) \longrightarrow (0, \bar{x}, 0),$$

when $q \to \infty$. Now Proposition 2.1 applies, implying that $0 \in T(\bar{x})$. □

5 Conclusions and Perspectives

We presented a convergent bundle-like method that can be used for finding a zero of maximal monotone operators. Although it already has an implementable form, further modifications are possible:

1. Parameters R and τ in (BS) can be replaced by $||u^k||$, in order to deal with "relative" values.

2. To avoid memory overload, a variant of (BS) working with sub-bundles of limited size could be defined. Using techniques of compression and selection which are now standard for bundle methods in nonsmooth optimization (NSO) (see, for instance, Chapter XIV in [5]), Theorem 2.2 ensures that after solving (1.c) in (BS), the *aggregated* triplet

$$(\hat{\varepsilon}, \hat{x} := \sum_{i\in I_{k,n,j}} \alpha_i^{k,n,j} z^i, \hat{s} := s^{k,n,j})$$

synthesizes the most essential information contained in the sub-bundle (we have shortened $\hat{\varepsilon} = \sum_i \alpha_i \langle z^i - \hat{x}, w^i - \hat{s} \rangle$).

3. We insisted on the importance of driving ε to 0 at an appropriate speed to avoid in Step (1.c) what we have called an *ill-posed* problem. Because of Corollary 2.3, we achieve this goal indirectly through j-indices. In NSO, it is well known that the practical efficiency of bundle methods depends strongly on an appropriate management of the parameters. For example, (BS) could be modified in order to define ε as a *parameter* whose value is reduced along iterations. To achieve this direct control on ε, a possibility (not explored yet) could be to solve the following subproblem in *Step 1*:

$$\begin{cases} \min_{\alpha \in \Delta_I} \frac{1}{2} \| \sum_I \alpha_i w^i \|^2 \\ \sum_I \langle z^i - \hat{x}, w^i - \hat{s} \rangle \leq \varepsilon \,, \end{cases}$$

where $\hat{x}, \hat{s}$ have been computed in the previous iteration, and I defines a sub-bundle.

Finally, let us mention the recent paper [9], describing an algorithm for solving variational inequality problems (VIP) with convex feasible sets. The methodology presented there has been developed independently and can be related to our (BS). Roughly speaking, **Procedure P** therein corresponds to our computation of $s^{k,n,j}$ using an "economic" bundle formed by only two elements, namely $v^{k,n}$ and $s^{k,n,j-1}$ (or $s^{k,n-1,j}$). In his paper, Konnov himself refers to the conjugate subgradients method for NSO in [21] (see also [12]). Since these very first "economic" (conjugate subgradient) formulations a huge progress has been done on bundle methods for NSO. In particular, it is known that the use of "rich" bundles (properly managed to avoid overload) improves dramatically the speed of convergence. In this sense, our contribution, based on "rich" sub-bundles, may open the way to defining new variants, hopefully more efficient and robust. Moreover, since our development is supported by the theoretical background of ε-"subgradients" in $T^\varepsilon(x)$, it gives a different insight of the mechanism that makes algorithms of this type work.

References

[1] A. Brøndsted and R.T. Rockafellar. On the subdifferentiability of convex functions. *Proc. of the Amer. Math. Soc.*, 16:605–611, 1965.

[2] R.S. Burachik, A.N. Iusem, and B.F. Svaiter. Enlargements of maximal monotone operators with application to variational inequalities. *Set Valued Analysis*, 5:159–180, 1997. Also (extended version) Tech. Rep. B-110/97, IMPA, Rio de Janeiro, Brazil.

[3] R.S. Burachik, C.A. Sagastizábal, and B. F. Svaiter. ε-Enlargements of maximal monotone operators: Theory and Application. In *Reformulation –*

Nonsmooth, Piecewise Smooth, Semismooth and Smoothing Methods, pages 25–43. Kluwer, 1998.

[4] I. Ekeland and G. Lebourg. Sous-gradients approchés et applications. *C.R. Acad. Sc. Paris, série A*, 281:219–222, 1975.

[5] J.-B. Hiriart-Urruty and C. Lemaréchal. *Convex Analysis and Minimization Algorithms.* Number 305-306 in Grund. der math. Wiss. Springer-Verlag, 1993. (two volumes).

[6] J.-J.Strodiot and V.H. Nguyen. On the numerical treatment of the inclusion $0 \in \partial f(x)$. In J. J. Moreau, P. D. Panagiotopolus, and G. Strang, editors, *Topics in Nonsmooth Mechanics*, pages 267–294. Birkhäuser Verlag, 1988.

[7] E. N. Khobotov. Modifications of the extragradient method for solving variational inequalities and certain optimization problems. *USSR Computational Mathematics and Mathematical Physics*, 27(5):120–127, 1987.

[8] K.C. Kiwiel. Proximity control in bundle methods for convex nondifferentiable minimization. *Mathematical Programming*, 46:105–122, 1990.

[9] I.V. Konnov. A combined relaxation method for variational inequalities with nonlinear constraints. *Mathematical Programming*, 80:239–252, 1997.

[10] G.M. Korpelevich. The extragradient method for finding saddle points and other problems. *Matecon*, 12:747–756, 1976.

[11] J. E. Martínez Legaz and M. Théra. ϵ-subdifferentials in terms of subdifferentials. *Set-Valued Analysis*, 4:327–332, 1996.

[12] C. Lemaréchal. An extension of Davidon methods to nondifferentiable problems. *Mathematical Programming Study*, 3:95–109, 1975.

[13] C. Lemaréchal and C. Sagastizábal. Variable metric bundle methods: from conceptual to implementable forms. *Mathematical Programming*, 76:393–410, 1997.

[14] C. Lemaréchal, J.-J. Strodiot, and A. Bihain. On a bundle method for nonsmooth optimization. In O.L. Mangasarian, R.R. Meyer, and S.M. Robinson, editors, *Nonlinear Programming 4*, pages 245–282. Academic Press, 1981.

[15] M. Nisipeanu. *Somme variationnelle d'opérateurs et applications.* PhD thesis, Université de Limoges - France, 1997.

[16] J. P. Revalski and M. Théra. Enlargements and sums of monotone operators. Working paper, 1998.

[17] R.T. Rockafellar. Local boundedness of nonlinear monotone operators. *Michigan Mathematical Journal*, 16:397–407, 1969.

[18] H. Schramm and J. Zowe. A version of the bundle idea for minimizing a nonsmooth function: conceptual idea, convergence analysis, numerical results. *SIAM Journal on Optimization*, 2(1):121–152, 1992.

[19] D. Torralba. *Convergence épigraphique et changements d'échelle en Analyse Variationnelle et Optimisation*. PhD thesis, Université de Montpellier II - France, 1996.

[20] L. Veselý. Local uniform boundedness principle for families of ε-monotone operators. *Nonlinear Analysis, Theory, Methods & Applications*, 24(9):1299–1304, 1995.

[21] P. Wolfe. A method of conjugate subgradients for minimizing nondifferentiable functions. *Mathematical Programming Study*, 3:145–173, 1975.

[22] E. H. Zarantonello. Projections on convex sets in Hilbert spaces and spectral theory. In E. H. Zarantonello, editor, *Contributions to Nonlinear Functional Analysis*, pages 237–424. Academic Press, 1971.

Nonlinear Averages and Convergence of Penalty Trajectories in Convex Programming *

Roberto Cominetti **

Abstract

We investigate the convergence of optimal trajectories associated with a class of penalty methods for solving convex programming problems. We establish conditions ensuring existence and uniqueness of such trajectories, as well as their convergence towards a particular solution of the convex program. This particular limit point is characterized as the unique solution of a nested hierarchy of optimization problems. The construction leading to this hierarchy is based on asymptotic nonlinear averages, a notion which is described and studied in some detail.

Keywords: Penalty Methods, Convex Programming, Asymptotic Analysis

AMS subject classification: 90C25, 90C31

1 Introduction

Consider the convex mathematical program

$$(P) \qquad v := \min\{f_0(x) : x \in F\}$$

with feasible set $F \subset \mathbb{R}^n$ defined by a finite number of inequalities

$$F := \{x \in \mathbb{R}^n : f_i(x) \leq 0, i = 1, \ldots, m\}.$$

In this paper we investigate the convergence of optimal trajectories associated with a class of penalty methods for solving (P). The methods are built with the help of a non-decreasing convex function $\theta : \mathbb{R} \to \mathbb{R} \cup \{\infty\}$ as well as a function $\alpha : (0, \infty) \to (0, \infty)$, to which we associate the one-parameter family of optimization problems

$$(P_r) \qquad v(r) := \min_{x \in \mathbb{R}^n} p_r(x)$$

*Partially supported by Fondecyt 1981052 and FONDAP-Matemáticas Aplicadas
**Departamento de Ingeniería Matemática, Universidad de Chile, Casilla 170/3, Correo 3, Santiago, Chile. e-mail: rcominet@dim.uchile.cl

where $r > 0$ is a small *penalty parameter* and

$$p_r(x) := f_0(x) + \alpha(r) \sum_{i=1}^{m} \theta(f_i(x)/r).$$

Problem (P_r) is intended to be an approximation scheme for (P) in the sense that when $r \to 0^+$, the optimal value and optimal solutions of (P_r) are expected to converge to a solution of (P).

Our main result establishes a set of assumptions on (P), $\theta(\cdot)$ and $\alpha(\cdot)$, ensuring that (P_r) has a unique optimal solution $x(r)$ which converges when $r \to 0$ towards an optimal solution x^θ of (P). The point x^θ is characterized as the unique solution of a nested hierarchy of auxiliary optimization problems. We stress the fact that the whole trajectory $x(r)$ is shown to converge to a particular solution of $S(P)$ (not only convergence of subsequences) even if (P) has multiple solutions.

This result extends and unifies several convergence results for penalty methods. In particular it covers the convergence of the central path associated with the logarithmic barrier in linear programming [7, 11, 8] as well as in quasi-analytic convex programming [9]. It also covers the exponential penalty trajectory in linear and quasi-analytic convex programming [4, 1, 2], as well as the convergence results presented in [3] for wide classes of penalty methods in linear programming. Not only we recover and extend these previous results, but we provide a simple and unified framework that encompasses all of them at once. In particular this is an advantage with respect to the analysis presented in [3] which uses different particular techniques to study different classes of penalty functions.

Our analysis is based on ideas that generalize [4, 3], and exploits the notion of nonlinear average. As we shall see, this notion is tightly linked with penalty methods and enjoys many remarkable properties which shall be described in some detail.

The paper is organized as follows. In §2 we state the general assumptions that will hold throughout the paper, while in §3 we describe the basic results on existence and uniqueness of an optimal trajectory $x(r)$. Section §4 is completely devoted to the study of nonlinear averages, which are then used in §5 to establish our main result on convergence of optimal trajectories.

We follow the standard notations in convex analysis (see *e.g.* [10]), except for the recession function of a closed proper convex function $g : \mathbb{R} \to \mathbb{R} \cup \{\infty\}$, which is denoted $g^\infty(\cdot)$ and is defined as usual by

$$g^\infty(d) := \lim_{t \to \infty} \frac{g(x + td) - g(x)}{t}$$

where x is an arbitrary point in the domain of g.

2 Basic Assumptions

Throughout the paper we suppose the following:

- $f_0 : I\!R^n \to I\!R \cup \{\infty\}$ is closed, proper and convex.
- $f_i : I\!R^n \to I\!R$ is finite and convex for $i = 1, \ldots, m$.
- $S(P)$, the optimal solution set of (P), is nonempty and compact.
- $\theta : I\!R \to I\!R \cup \{\infty\}$ is convex non-decreasing with $\theta^\infty(-1) = 0 < \theta^\infty(1)$.
- If $\theta(0) = \infty$ we assume the existence of a Slater point $\tilde{x}$, that is to say

$$f_0(\tilde{x}) < \infty \quad \text{and} \quad f_i(\tilde{x}) < 0 \quad i = 1, \ldots, m.$$

- $\alpha : (0, \infty) \to (0, \infty)$ is such that $\alpha(r) \to 0$ when $r \to 0^+$, with

$$\beta := \liminf_{r \to 0^+} \alpha(r)/r > 0.$$

- $\beta\theta^\infty(1) = \infty$ (*e.g.* if $\theta^\infty(1) = \infty$ we may take $\alpha(r) = r$).

3 Existence and Uniqueness of Optimal Paths

In this section we prove the existence and uniqueness of optimal solutions for (P_r), and a first convergence result showing that (P_r) is a reasonable approximation scheme for (P). The existence and convergence results are slight extensions of [3, Theorems 2.2 and 2.3] (which assumed $\theta \geq 0$ at some places). For the sake of completeness we include a short proof. The uniqueness result is new.

We will use the following well known fact [10]: if $g : I\!R^n \to I\!R \cup \{\infty\}$ is closed proper and convex, then $\arg\min(g)$ is non-empty and bounded if and only if $g^\infty(d) > 0$ for all $d \in I\!R^n, d \neq 0$. Using this, the assumption "$S(P)$ nonempty and compact" can be restated as:

$$(\forall\, d \neq 0) \quad f_i^\infty(d) \leq 0 \quad \forall\, i = 1, \ldots, m \quad \Longrightarrow \quad f_0^\infty(d) > 0. \tag{1}$$

Theorem 3.1 (Existence [3]) *$S(P_r)$ is nonempty and bounded for all $r > 0$ small enough. If either $f_0^\infty(\cdot) \geq 0$ or $\theta^\infty(1) = \infty$, this holds for every $r > 0$.*

Proof. Using the calculus rule $(\theta \circ g)^\infty(d) = \theta^\infty(g^\infty(d))$ (see for instance [3, Proposition 2.1]), we may compute p_r^∞ as

$$p_r^\infty(d) = f_0^\infty(d) + \frac{\alpha(r)}{r} \sum_{i=1}^{m} \theta^\infty(f_i^\infty(d)). \tag{2}$$

If $f_0^\infty(\cdot) \geq 0$ or $\theta^\infty(1) = \infty$, (1) and (2) imply $p_r^\infty(d) > 0$ for all $d \neq 0$, hence $S(P_r)$ is nonempty and bounded. Otherwise, this holds for r small enough. Indeed, if this was not the case we could find sequences $r_k \to 0^+$ and $d_k \to d \neq 0$

with $p_{r_k}^\infty(d_k) \leq 0$. Using (2), the lower semicontinuity of f_i^∞ and $\alpha(r)/r \to \beta = \infty$, we get $\sum_{i=1}^m \theta^\infty(f_i^\infty(d)) \leq 0$ and $f_0^\infty(d) \leq 0$, contradicting (1). □

In general (P_r) need not have a unique solution, even if θ is strictly convex. The following regularity property will help to ensure uniqueness.

Definition 3.1 [1] *We let Q be the class of convex functions $g : \mathbb{R}^n \to \mathbb{R}\cup\{\infty\}$ with the property that, for every $x \neq y$, if g is constant over the segment $[x, y]$ then g is constant over the whole line passing through x and y.*

Clearly Q contains every strictly convex function, as well as linear, quadratic and analytic convex functions. The class Q is stable by composition with a linear mapping $g \circ A$ and also by (pre) composition $\sigma \circ g$ with a strictly increasing convex function $\sigma : \mathbb{R} \to \mathbb{R}$. However, Q is not stable under addition nor pointwise maxima. The main fact we need about Q is the following:

Lemma 3.2 *Suppose $f_0, \ldots, f_m \in Q$ and let $C \subset \mathbb{R}^n$ be nonempty and convex with all functions f_i finite and constant over C. Then C is a singleton.*
Proof. Let x, y be distinct points in C and set $d := x - y$. Since the f_i's are constant over $[x, y] \subset C$ with $f_i \in Q$, it follows that $f_i^\infty(d) = 0$ for all i's, contradicting (1). □

Theorem 3.3 (Uniqueness) *Suppose $f_0, \ldots, f_m \in Q$ and θ strictly convex. Then (P_r) admits at most one solution.*
Proof. By Lemma 3.2 it suffices to prove that all the f_i's are constant over $C = S(P_r)$. Since $f_0(\cdot) = p_r(\cdot) - \alpha(r)\sum_{i=1}^m \theta(f_i(\cdot)/r)$ and $p_r(\cdot)$ is obviously constant over $S(P_r)$, it is enough to prove this for $i = 1, \ldots, m$.

If this was not the case, we could find $x, y \in S(P_r)$ such that $f_i(x) \neq f_i(y)$ for some constraint i. Letting $\bar{x} := (x + y)/2$, the convexity of the f_i's and the strict convexity of $\theta(\cdot)$ yield the contradiction

$$p_r(\bar{x}) < \frac{1}{2}[p_r(x) + p_r(y)] = v(r).$$

□

We conclude this section with an easy convergence result.

Theorem 3.4 (Convergence [3]) *Every optimal trajectory $x(r) \in S(P_r)$ remains bounded as $r \to 0$, and its accumulation points belong to $S(P)$. Moreover, we have convergence of optimal values*

$$\lim_{r \to 0+} v(r) = v.$$

Proof. Let $\hat{x} \in S(P)$. If $\theta(0) = \infty$ we let $x_\alpha = \alpha\hat{x} + (1 - \alpha)\tilde{x}$ where $\tilde{x}$ is a Slater point and $\alpha \in (0, 1)$. Otherwise we simply take $x_\alpha = \hat{x}$. The inequality $p_r(x(r)) \leq p_r(x_\alpha)$ leads to

$$\limsup_{r \to 0+} p_r(x(r)) \leq \limsup_{r \to 0+} p_r(x_\alpha) \leq f_0(x_\alpha)$$

so that letting $\alpha \to 1$ we get

$$\limsup_{r\to 0^+} p_r(x(r)) \leq v. \tag{3}$$

To prove boundedness of $x(r)$ we proceed by contradiction. Suppose $r_k \to 0^+$ is such that $\|x(r_k)\| \to \infty$, and let $x_k := x(r_k)$ and $d_k := x_k/\|x_k\|$. With no loss of generality we may assume $d_k \to d \neq 0$, so that

$$\lim_{k\to\infty} \frac{f_0(x_k)}{\|x_k\|} = f_0^\infty(d) > -\infty,$$

while for all $i = 1, \ldots, m$

$$\lim_{k\to\infty} \frac{\theta(f_i(x_k)/r_k)}{\|x_k\|/r_k} = \theta^\infty(f_i^\infty(d)).$$

¿From this and (3) it follows

$$0 \geq \limsup_{k\to\infty} \frac{p_{r_k}(x_k)}{\|x_k\|} \geq f_0^\infty(d) + \beta \sum_{i=1}^m \theta^\infty(f_i^\infty(d)) \geq f_0^\infty(d),$$

and since $\beta\theta^\infty(1) = \infty$ we get $f_i^\infty(d) \leq 0$ for all $i = 0, \ldots, m$, contradicting (1).

We prove next that the accumulation points of $x(r)$ belong to $S(P)$. Let $r_k \to 0^+$ be such that $x(r_k) \to x^*$. Since

$$\lim_{k\to\infty} r_k \theta\left(\frac{f_i(x(r_k))}{r_k}\right) = \theta^\infty(f_i(x^*)),$$

we deduce

$$v \geq \limsup_{k\to\infty} p_{r_k}(x(r_k)) \geq f_0(x^*) + \beta \sum_{i=1}^m \theta^\infty(f_i(x^*)) \geq f_0(x^*) \tag{4}$$

which combined with $\beta\theta^\infty(1) = \infty$ implies $x^* \in S(P)$.

It remains to establish convergence of optimal values. Take $r_k \to 0^+$ such that $v(r_k) \to \liminf_{r\to 0^+} v(r)$. Passing to a subsequence we may assume that $x(r_k) \to x^*$ for some $x^* \in S(P)$, and then (4) implies

$$v = \lim_{k\to\infty} v(r_k) = \liminf_{r\to 0^+} v(r)$$

which combined with (3) gives the desired conclusion. □

When (P) has a unique optimal solution x^*, this result implies that every optimal trajectory $x(r) \in S(P_r)$ converges towards x^* as $r \to 0^+$. We shall consider the case where (P) has a multiple optimal solutions, establishing a set of conditions which ensure the convergence of the optimal trajectories towards a particular point in $S(P)$. In the next section we develop the basic tool for establishing such result.

REMARK. As in [3], the results in this section still hold when the constraints $f_i : \mathbb{R}^n \to \mathbb{R} \cup \{\infty\}$ are extended real valued, closed, proper and convex. However, our main result in §5 requires these functions to be finite, so we have chosen this more limited setting for the paper. This is not very restrictive though, since infinite values may still be incorporated in the objective function f_0 to handle explicit or hard constraints.

4 Nonlinear Averages

Suppose $\theta : \mathbb{R} \to \mathbb{R} \cup \{\infty\}$ is strictly increasing and convex. We may use the function $\theta(\cdot)$ to compute a (nonlinear) average of a set of real numbers $y_1, \ldots, y_m$ through the formula

$$M_\theta(y) := \theta^{-1}\left(\frac{1}{m}\sum_{i=1}^{m}\theta(y_i)\right)$$

where we adopt the convention $\theta^{-1}(\infty) = m_\theta$ with $m_\theta := \sup\{u : \theta(u) < \infty\}$.

These *generalized means* were introduced and studied in [6, Chapter 3], where the reader may find a wealth of results and details. In our context however, because of the penalty parameter r which tends to 0, we shall rather be concerned with a related concept which is an asymptotic version of the generalized means, namely

Definition 4.1 *Let $\theta : \mathbb{R} \to \mathbb{R} \cup \{\infty\}$ be strictly increasing and convex. We define the θ-average as the mapping $A_\theta : \mathbb{R}^m \to \mathbb{R}$ given by*

$$A_\theta(y_1, \ldots, y_m) := \limsup_{r \to 0+} r\theta^{-1}\left(\frac{1}{m}\sum_{i=1}^{m}\theta(y_i/r)\right).$$

The notation is a little ambiguous since in fact for each $m \in \mathbb{N}$ we have a different θ-average A_θ^m. Since the m will always be clear from the context, we shall keep the simpler notation A_θ.

The function A_θ looks formally as the recession function of M_θ. However, we remark that M_θ is not convex in general so that one has to be careful with this interpretation. In particular we are not allowed to replace "lim sup" by "lim" and in fact, in our main result in §5, we will have to assume that the upper limit is attained as a limit.

Let us point out that when $m_\theta < \infty$ we have $A_\theta(y) = 0$ as soon as $y_i > 0$ for some i, so that A_θ will only be nontrivial on $\mathbb{R}^m_-$. In other words, our attention is primarily oriented to nonlinear averages of *non-positive* quantities. One may develop a theory of nonlinear averages over $\mathbb{R}^m_+$ by considering convex *decreasing* functions $\theta(\cdot)$. However, in the present setting our choice is motivated by the fact that we consider constraints in the form $f_i(x) \leq 0$ for $i = 1, \ldots, m$.

Example 4.1 *The harmonic mean is obtained by taking* $\theta_1(u) = -1/u$ *if* $u < 0$ *and* $\theta_1(u) = \infty$ *otherwise. Indeed, in such case we have*

$$A_{\theta_1}(y) = \begin{cases} \left[\frac{1}{m}\sum_{i=1}^m 1/y_i\right]^{-1} & \text{if } y_i < 0 \text{ for all } i = 1, \dots, n \\ 0 & \text{otherwise.} \end{cases}$$

The geometric mean is obtained with $\theta_2(u) = -\ln(-u)$ *if* $u < 0$ *and* $\theta_2(u) = \infty$ *otherwise, namely*

$$A_{\theta_2}(y) = \begin{cases} -\left[\Pi_{i=1}^m(-y_i)\right]^{1/m} & \text{if } y_i < 0 \text{ for all } i = 1, \dots, m \\ 0 & \text{otherwise.} \end{cases}$$

For $\theta_3(u) = \exp(u)$ *we have*

$$A_{\theta_3}(y) = \max\{y_i : i = 1, \dots, m\},$$

while taking $\theta_4(u) = -\sqrt{-u}$ *if* $u \leq 0$ *and* $\theta_4(u) = \infty$ *otherwise, we obtain*

$$A_{\theta_4}(y) = \begin{cases} -\left[\frac{1}{m}\sum_{i=1}^m \sqrt{-y_i}\right]^2 & \text{if } y_i \leq 0 \text{ for all } i = 1, \dots, m \\ 0 & \text{otherwise.} \end{cases}$$

We remark that θ_1, θ_2 *and* θ_4 *are such that the asymptotic mean* A_θ *coincides with the generalized mean* M_θ. *A detailed study of these* M_θ*'s can be found in* [6, Chapters 2 and 3]. *In particular, using* [6, Theorem 84] *and the positive homogeneity of* A_θ, *we deduce that* $A_\theta = M_\theta$ *only for* θ_2 *and all power functions such as* θ_1, θ_4.

The previous examples illustrate already some properties of θ-averages. In all cases we have componentwise monotonicity, convexity and continuity on the negative orthant $I\!R^m_-$. Let us prove these in general (see also [6] for the algebraic properties of generalized means).

Theorem 4.1 *Let* $\theta : I\!R \to I\!R \cup \{\infty\}$ *be a strictly increasing convex function. Then* A_θ *is positively homogeneous, quasi-convex, symmetric, componentwise non-decreasing and satisfies*

$$\frac{1}{m}\sum_{i=1}^m y_i \leq A_\theta(y) \leq \max_{i=1\dots m} y_i.$$

Moreover, A_θ *is convex and continuous on int(*$I\!R^m_-$*).*

Proof. The algebraic properties are straightforward consequences of the definition of A_θ and the convexity and monotonicity of θ, and are left to the reader. The convexity of A_θ over int($I\!R^m_-$) follows from the following lemma, while the continuity is a consequence of the convexity and finiteness of A_θ. □

Lemma 4.2 [5] *Let P be a convex cone and $\delta : P \to \mathbb{R}$ positively homogeneous and quasi-convex, with $\delta(y) < 0$ (or $\delta(y) > 0$) for all $y \in P$. Then δ is convex.*

Proof. Take $y_1, y_2 \in P$ and $\alpha \in [0,1]$. Letting $\bar{\delta} = \alpha\delta(y_1) + (1-\alpha)\delta(y_2)$, we have

$$\alpha y_1 + (1-\alpha)y_2 = \beta z_1 + (1-\beta)z_2$$

with $\beta = \alpha\delta(y_1)/\bar{\delta}$, $z_1 = [\bar{\delta}/\delta(y_1)]y_1$ and $z_2 = [\bar{\delta}/\delta(y_2)]y_2$. By positive homogeneity we get $\delta(z_1) = \delta(z_2) = \bar{\delta}$, so that quasi-convexity implies

$$\delta(\alpha y_1 + (1-\alpha)y_2) \leq \max\{\delta(z_1), \delta(z_2)\} = \bar{\delta} = \alpha\delta(y_1) + (1-\alpha)\delta(y_2).$$

□

The function A_θ may be discontinuous on the boundary of $\mathbb{R}^m_-$. Such an unstable behavior is not essential however since the mapping A_θ admits a continuous extension from $\mathrm{int}(\mathbb{R}^m_-)$ to $\mathbb{R}^m_-$.

Proposition 4.3 *Let $\bar{A}_\theta : \mathbb{R}^m_- \to \mathbb{R}$ be defined as the monotone limit*

$$\bar{A}_\theta(y) = \lim_{\epsilon \to 0^+} A_\theta(y - \epsilon\mathbb{1})$$

where $\mathbb{1} = (1, \ldots, 1)^t \in \mathbb{R}^m$. Then $\bar{A}_\theta$ is the (unique) continuous extension of A_θ from $\mathrm{int}(\mathbb{R}^m_-)$ to $\mathbb{R}^m_-$. As such, $\bar{A}_\theta$ is positively homogeneous, convex, continuous, symmetric, componentwise non-decreasing, and satisfies

$$\frac{1}{m}\sum_{i=1}^{m} y_i \leq \bar{A}_\theta(y) \leq \max_{i=1\ldots m} y_i.$$

Proof. The algebraic properties of $\bar{A}_\theta$ are easily derived from the corresponding properties of A_θ. Moreover, since A_θ is continuous over $\mathrm{int}(\mathbb{R}^m_-)$, it is clear that $\bar{A}_\theta = A_\theta$ on that set. Hence all we must check is the continuity of $\bar{A}_\theta$ on the boundary of $\mathbb{R}^m_-$. Let $y_k \in \mathbb{R}^m_-$ converging towards some y as $k \to \infty$. For each $\epsilon \in (0,1)$ there exists $k(\epsilon) \in \mathbb{N}$ such that $y - \epsilon\mathbb{1} \leq y_k \leq (1-\epsilon)y$ for all $k \geq k(\epsilon)$. The monotonicity and positive homogeneity of $\bar{A}_\theta$ imply that

$$\bar{A}_\theta(y - \epsilon\mathbb{1}) \leq \bar{A}_\theta(y_k) \leq (1-\epsilon)\bar{A}_\theta(y)$$

and since $\bar{A}_\theta(y - \epsilon\mathbb{1}) = A_\theta(y - \epsilon\mathbb{1})$ we deduce

$$A_\theta(y - \epsilon\mathbb{1}) \leq \liminf_{k\to\infty} \bar{A}_\theta(y_k) \leq \limsup_{k\to\infty} \bar{A}_\theta(y_k) \leq (1-\epsilon)\bar{A}_\theta(y).$$

Letting $\epsilon \to 0^+$ we conclude $\lim_{k\to\infty} \bar{A}_\theta(y_k) = \bar{A}_\theta(y)$ as required. □

In the rest of this section we develop some techniques which facilitate the task of computing θ-averages.

Definition 4.2 *Let $\theta, \tilde{\theta} : I\!R \to I\!R \cup \{\infty\}$ be strictly increasing and convex. We say that θ and $\tilde{\theta}$ are asymptotically equivalent, if both functions have the same infimum (eventually $-\infty$) and*

$$\lim_{u \to -\infty} \frac{\theta^{-1}(\tilde{\theta}(u))}{u} > 0. \tag{5}$$

Denoting $\gamma = \inf \theta = \inf \tilde{\theta}$, condition (5) may be equivalently stated as

$$\lim_{w \to \gamma^+} \frac{\theta^{-1}(w)}{\tilde{\theta}^{-1}(w)} > 0,$$

from which it is easy to see that this is an equivalence relation.

Proposition 4.4 *If $\theta, \tilde{\theta}$ are asymptotycally equivalent, then for all $y \in \mathrm{int}(I\!R^m_-)$ we have $A_\theta(y) = A_{\tilde{\theta}}(y)$, and $A_\theta(y)$ exists as a limit if and only if $A_{\tilde{\theta}}(y)$ does.*

Proof. Let $\alpha := \lim_{u \to -\infty} \theta^{-1}(\tilde{\theta}(u))/u$, so that for each $\rho > 1$ there exists $r(\rho) > 0$ such that

$$\alpha/\rho \le \frac{\theta^{-1}(\tilde{\theta}(y_i/r))}{y_i/r} \le \alpha\rho$$

for all $r < r(\rho)$ and $i = 1, \dots, m$, and therefore

$$r\theta^{-1}\left(\frac{1}{m}\sum_{i=1}^{m}\theta((\alpha\rho)\frac{y_i}{r})\right) \le r\theta^{-1}\left(\frac{1}{m}\sum_{i=1}^{m}\tilde{\theta}(\frac{y_i}{r})\right) \le r\theta^{-1}\left(\frac{1}{m}\sum_{i=1}^{m}\theta((\alpha/\rho)\frac{y_i}{r})\right).$$

The upper limits of the expressions on the left and on the right of this inequality are $(\alpha\rho)A_\theta(y)$ and $(\alpha/\rho)A_\theta(y)$ respectively. For the expression in the middle, we observe that

$$\lim_{r \to 0^+} \frac{\theta^{-1}(\frac{1}{m}\sum_{i=1}^{m}\tilde{\theta}(y_i/r))}{\tilde{\theta}^{-1}(\frac{1}{m}\sum_{i=1}^{m}\tilde{\theta}(y_i/r))} = \alpha$$

and therefore we get $(\alpha\rho)A_\theta(y) \le \alpha A_{\tilde{\theta}}(y) \le (\alpha/\rho)A_\theta(y)$. Letting $\rho \to 1^+$ we deduce $A_\theta(y) = A_{\tilde{\theta}}(y)$ as claimed.

A similar argument shows that when $A_\theta(y)$ exists as a limit, the same holds for $A_{\tilde{\theta}}(y)$. Since the situation is completely symmetric, the converse also holds and the proof is complete. □

Corollary 4.5 *Let $\theta : I\!R \to I\!R \cup \{\infty\}$ be strictly increasing and convex, and let $y \in \mathrm{int}(I\!R^m_-)$.*
(a) If $\lim_{u \to -\infty} u\theta(u) < 0$ then $A_\theta(y) = [\frac{1}{m}\sum_{i=1}^{m} 1/y_i]^{-1}$.
(b) If $\lim_{u \to -\infty} u\exp(\theta(u)) < 0$ then $A_\theta(y) = -\left[\Pi_{i=1}^{m}(-y_i)\right]^{1/m}$.
(c) If $\lim_{u \to -\infty} \ln[\theta(u)]/u > 0$ then $A_\theta(y) = \max\{y_i : i = 1, \dots, m\}$.
In all three cases $A_\theta(y)$ exists as a limit.

Proof. It suffices to observe that θ is asymptotically equivalent to θ_1, θ_2 and θ_3 respectively. □

Example 4.2 *The function $\theta_5(u) = 1/(1-u)$ if $u < 1$ and $\theta_5(u) = \infty$ otherwise, satisfies (a) in the above corollary and we have*

$$A_{\theta_5}(y) = \begin{cases} \left[\frac{1}{m}\sum_{i=1}^m 1/y_i\right]^{-1} & \text{if } y_i < 0 \text{ for all } i = 1, \ldots, m \\ 0 & \text{otherwise.} \end{cases}$$

Similarly, for $\theta_6(u) = -\ln(1-u)$ if $u < 1$ and $\theta_6(u) = \infty$ otherwise, property (b) holds and we get

$$A_{\theta_6}(y) = \begin{cases} -\left[\Pi_{i=1}^m(-y_i)\right]^{1/m} & \text{if } y_i < 0 \text{ for all } i = 1, \ldots, m \\ 0 & \text{otherwise.} \end{cases}$$

Finally, the function $\theta_7(u) = \ln(1+\exp(u))$ satisfies the criteria (c). Moreover, since $\theta_7^\infty(1) = 1$, we find that

$$A_{\theta_7}(y) = \begin{cases} \max\{y_i : i = 1, \ldots, m\} & \text{if } y_i < 0 \text{ for all } i = 1, \ldots, m \\ \frac{1}{m}\sum_{i=1}^m [y_i]_+ & \text{otherwise.} \end{cases}$$

REMARKS. (1) We observe that $\theta_5(u) = \theta_1(u-1)$ and $\theta_6(u) = \theta_2(u-1)$. In general, the function $\tilde{\theta}(u) := \theta(u+a)$ with constant a, is asymptotycally equivalent to θ and therefore will generate the same θ-average.
(2) If $\theta^\infty(1) < \infty$ it can be checked that $A_\theta(y) = \frac{1}{m}\sum_{i=1}^m [y_i]_+$ for $y \notin I\!R_-^m$.

Corollary 4.6 *Let $\theta : I\!R \to I\!R \cup \{\infty\}$ be a strictly increasing convex function and let $\{a_k : k \in I\!N\}$ be a sequence decreasing to $-\infty$. We denote θ_d the polyhedral function obtained by linear interpolation of θ at the points a_k, that is to say, on the interval $[a_{k+1}, a_k]$ the function θ_d is given by*

$$\theta_d(u) = \theta(a_k) + \frac{\theta(a_{k+1}) - \theta(a_k)}{a_{k+1} - a_k}(u - a_k).$$

If $\lim_{k\to\infty} a_{k+1}/a_k = 1$ then for all $y \in \operatorname{int}(I\!R_-^m)$ we have $A_\theta(y) = A_{\theta_d}(y)$, and $A_{\theta_d}(y)$ exists as a limit if and only if $A_\theta(y)$ does.

Proof. For each $u < 0$ let $k \in I\!N$ be such that $a_{k+1} \le u \le a_k$. It follows easily that $u \le \theta^{-1}(\theta_d(u)) \le a_k$, hence

$$1 \ge \theta^{-1}(\theta_d(u))/u \ge a_k/u \ge a_k/a_{k+1}.$$

Since $a_{k+1}/a_k \to 1$ as $k \to \infty$, we get $\theta^{-1}(\theta_d(u))/u \to 1$ as $u \to -\infty$ and we may conclude by the previous proposition. □

The preceeding results allow to compute the asymptotic average associated with a given θ by reduction to a simpler equivalent function $\tilde{\theta}$. In contrast, the following proposition gives a simple condition on the function θ which ensures that A_θ is the "max" function.

Proposition 4.7 *Let $\theta : I\!R \to I\!R \cup \{\infty\}$ be a twice differentiable strictly increasing convex function, bounded below and such that $L := \lim_{u\to-\infty} \theta''(u)/\theta'(u)$ exists and is strictly positive. Then, for all $y \in I\!R^m_-$ we have*

$$\bar{A}_\theta(y) = A_\theta(y) = \lim_{r\to 0^+} r\theta^{-1}\left(\frac{1}{m}\sum_{i=1}^{m}\theta(y_i/r)\right) = \max_{i=1\ldots m} y_i.$$

Proof. Assume with no loss of generality that $\inf\{\theta(u) : u \in I\!R\} = 0$. Then

$$r\theta^{-1}\left(\frac{1}{m}\theta(\max_{i=1\ldots m} y_i/r)\right) \le r\theta^{-1}\left(\frac{1}{m}\sum_{i=1}^{m}\theta(y_i/r)\right) \le \max_{i=1\ldots m} y_i.$$

Therefore, it suffices to show that $\lim_{u\to-\infty} \theta^{-1}(\alpha\theta(u))/u = 1$ for all $\alpha \in (0,1)$, which follows easily by using l'Hôpital rule twice. □

5 Convergence of Optimal Trajectories

Let us go back to the analysis of the penalty method (P_r). In Theorem 3.4 we established that every optimal trajectory $x(r) \in S(P_r)$ remains bounded and its accumulation points belong to $S(P)$. In this section, we prove the convergence of the optimal paths $x(r)$ towards a particular optimal solution in $S(P)$. In order to describe and establish the existence of this special solution, we shall need the following additional properties to hold for every $u, v \in I\!R^m_-$:

(H_0) *θ is strictly increasing and $A_\theta(u)$ exists as a limit for all $u \in \operatorname{int}(I\!R^m_-)$*

(H_1) $$\max_i v_i < \max_i u_i = 0 \Longrightarrow \inf_{w\in[u,v]} \bar{A}_\theta(w) < \bar{A}_\theta(u)$$

(H_2) $$\max_i v_i \ne \max_i u_i \Longrightarrow \inf_{w\in[u,v]} \bar{A}_\theta(w) < \max\{\bar{A}_\theta(u), \bar{A}_\theta(v)\}.$$

These properties hold in all the examples of the previous section. In particular (H_1) holds trivially if $\bar{A}_\theta$ vanishes on the boundary of $I\!R^m_-$, while (H_2) holds trivially if A_θ is strictly convex or when $A_\theta(y) = \max_i y_i$.

The following Lemma will be our main tool.

Lemma 5.1 *Suppose θ satisfies (H_0), (H_1) and (H_2). Let $g_1, \ldots, g_k \in \mathcal{Q}$ be non-positive and not identically 0 over a nonempty compact convex set C. If S denotes the optimal solution set of*

$$\min_{x \in C} \bar{A}_\theta(g_1(x), \ldots, g_k(x))$$

then there is a constant $M < 0$ and at least one index $j \in \{1, \ldots, k\}$ such that

$$g_j(x) = \max_{i=1 \ldots k} g_i(x) = M \quad \forall\, x \in S.$$

Proof. Let $g(x) := (g_1(x), \ldots, g_k(x))$. For each $i \in \{1, \ldots, k\}$ we may find $x^i \in C$ such that $g_i(x^i) < 0$, so that $\bar{x} = \frac{1}{k}(x^1 + \ldots + x^k) \in C$ satisfies $\max\{g_i(\bar{x}) : i = 1, \ldots, k\} < 0$. It follows that every optimal solution $x \in S$ satisfies $\max\{g_i(x) : i = 1, \ldots, k\} < 0$. Indeed, if this was not the case, using the convexity of the g_i's, the componentwise monotonicity of $\bar{A}_\theta$, and assumption (H_1) we would have

$$\min_{z \in [x, \bar{x}]} \bar{A}_\theta(g(z)) \leq \min_{w \in [g(x), g(\bar{x})]} \bar{A}_\theta(w) < \bar{A}_\theta(g(x))$$

contradicting the optimality of x.

Using (H_2) we may similarly prove that $\max\{g_i(x) : i = 1, \ldots, k\}$ is constant over S. Indeed, if this was not the case we could find $x, y \in S$ such that $\max\{g_i(x) : i = 1, \ldots, k\} \neq \max\{g_i(y) : i = 1, \ldots, k\}$, and then

$$\min_{z \in [x, y]} \bar{A}_\theta(g(z)) \leq \min_{w \in [g(x), g(y)]} \bar{A}_\theta(w) < \max\{\bar{A}_\theta(g(x)), \bar{A}_\theta(g(y))\}$$

contradicting the optimality of x and y.

Let $M < 0$ be the constant value of $\max\{g_i(x) : i = 1, \ldots, k\}$ over S.

We now prove the existence of $j \in \{1, \ldots, k\}$ such that $g_j(x) = M$ for all $x \in S$. Otherwise, for each i we could find $x^i \in S$ such that $g_i(x^i) < M$, and the point $\bar{x} = \frac{1}{k}(x^1 + \ldots + x^k) \in S$ would satisfy $\max\{g_i(\bar{x}) : i = 1, \ldots, k\} < M$ which is impossible. □

We may now describe the point $x^\theta \in S(P)$ that will be the limit of $x(r)$.

Definition 5.1 *In addition to the assumptions in §2, suppose $f_0, \ldots, f_m \in \mathcal{Q}$ and (H_0), (H_1), (H_2). Starting with $(P^0) = (P)$, we consider the hierarchy of minimization problems constructed recursively as*

$$(P^{k+1}) \qquad \min_{x \in S(P^k)} \bar{A}_\theta(f_i(x) : i \notin I_k).$$

with $I_k = \{i : f_i \text{ is constant over } S(P^k)\}$.

By Lemma 5.1 *the sequence I_k is strictly increasing, so that after finitely many steps s we have $I_s = \{0, 1, \ldots, m\}$. It follows that all the f_i's are constant over $S(P^s)$ and* Lemma 3.2 *implies that $S(P^s)$ is reduced to a single point which we denote x^θ and call the θ-center of problem (P).*

Let us remark that according to Lemma 5.1 and since $x^\theta \in S(P^1)$, we have $f_i(x^\theta) < 0$ for all $i \notin I_0$. We are now ready to prove our main result.

Theorem 5.2 *In addition to the assumptions in §2, suppose $f_0, \ldots, f_m \in \mathcal{Q}$ and (H_0), (H_1), (H_2). Then every optimal path $x(r) \in S(P_r)$ converges towards the θ-center x^θ when r goes to 0. If θ is strictly convex, the optimal trajectory $x(r)$ is unique.*

Proof. Suppose that $x(r)$ has an accumulation point x^* different from x^θ. We will prove inductively that x^* solves the problems $(P^0), (P^1), \ldots, (P^s)$, which is absurd since $S(P^s) = \{x^\theta\}$.

Let $r_j \to 0^+$ with $x_j := x(r_j) \to x^*$. Set $x_j^\theta := x_j + x^\theta - x^*$ so that $x_j^\theta \to x^\theta$.

¿From Theorem 3.4 we already know that $x^* \in S(P) = S(P^0)$. For the induction step suppose we have already established that $x^* \in S(P^k)$. Since x^θ also belongs to $S(P^k)$ it follows that $\alpha x^* + (1-\alpha)x^\theta \in S(P^k)$ for all $\alpha \in (0,1)$. By the very definition of I_k this implies that f_i is constant on the segment $[x^*, x^\theta]$ for all $i \in I_k$, and since the f_i's are in $\mathcal{Q}$ we get in fact that $d := x^\theta - x^*$ is a direction of constancy for these functions. It follows that $f_i(x_j) = f_i(x_j^\theta)$ for all $i \in I_k$, which can be used to simplify the inequality

$$f_0(x_j) + \alpha(r_j)\sum_{i=1}^{m} \theta(f_i(x_j)/r_j) \leq f_0(x_j^\theta) + \alpha(r_j)\sum_{i=1}^{m} \theta(f_i(x_j^\theta)/r_j)$$

by cancelling out those terms corresponding to $i \in I_k$. We get

$$\alpha(r_j)\sum_{i \notin I_k} \theta(f_i(x_j)/r_j) \leq \alpha(r_j)\sum_{i \notin I_k} \theta(f_i(x_j^\theta)/r_j)$$

and, letting $m_k = |I_k^c|$, we deduce

$$r_j\theta^{-1}\left(\frac{1}{m_k}\sum_{i \notin I_k} \theta(f_i(x_j)/r_j)\right) \leq r_j\theta^{-1}\left(\frac{1}{m_k}\sum_{i \notin I_k} \theta(f_i(x_j^\theta)/r_j)\right).$$

Let $\epsilon > 0$ small enough so that $f_i(x^\theta) + \epsilon < 0$ for all $i \notin I_k$. For j sufficiently large we have $f_i(x_j) > f_i(x^*) - \epsilon$ and $f_i(x_j^\theta) < f_i(x^\theta) + \epsilon$, so we may write

$$r_j\theta^{-1}\left(\frac{1}{m_k}\sum_{i \notin I_k} \theta\left(\frac{f_i(x^*) - \epsilon}{r_j}\right)\right) \leq r_j\theta^{-1}\left(\frac{1}{m_k}\sum_{i \notin I_k} \theta\left(\frac{f_i(x^\theta) + \epsilon}{r_j}\right)\right).$$

Passing to the limit with $j \to \infty$ and since $A_\theta = \bar{A}_\theta$ over $\mathrm{int}(\mathbb{R}_-^{m_k})$, we get

$$\bar{A}_\theta(f_i(x^*) - \epsilon : i \notin I_k) \leq \bar{A}_\theta(f_i(x^\theta) + \epsilon : i \notin I_k)$$

and letting $\epsilon \to 0^+$ it follows that x^* solves (P^{k+1}). This completes the induction step and the result is established. □

References

[1] F. Alvarez, *Métodos continuos en optimización paramétrica: el método de Newton y aplicaciones a la optimización estructural*, (Engineering thesis, Universidad de Chile, Santiago, 1996).

[2] F. Alvarez, "Absolute minimizer in convex programming by exponential penalty", to appear in *Journal of Convex Analysis* (1998).

[3] A. Auslender, R. Cominetti and M. Haddou, "Asymptotic analysis for penalty methods in convex and linear programming", *Mathematics of Operations Research* 22(1) (1997) 43–62.

[4] R. Cominetti and J. San Martín, "Asymptotic analysis of the exponential penalty trajectory in linear programming", *Mathematical Programming* 67 (1994) 169–187.

[5] J.P. Crouzeix, "Continuity and differentiability properties of quasiconvex functions on $\mathbb{R}^n$", in *Generalized Concavity in Optimization and Economics*, Schaible and Ziemba Editors, Academic Press (1982) 109–130.

[6] G. Hardy, J.E. Littlewood and G. Pólya, *Inequalities* (second edition) (Cambridge Univ. Press, Cambridge, 1952).

[7] L. MacLinden, "An analogue of Moreau's proximation theorem with applications to the nonlinear complementary problem", *Pacific Journal of Mathematics* 88 (1980) 101–161.

[8] N. Megiddo, "Pathways to the optimal set in linear programming", in *Progress in Mathematical Programming: interior-point and related methods*, Springer-Verlag (1989) 131–158.

[9] R. Monteiro, "On the existence and convergence of the central path for convex programming and some duality results", Preprint (1996).

[10] R.T. Rockafellar, *Convex Analysis* (Princeton Univ. Press, Princeton, NJ, 1970).

[11] Gy. Sonnevend, "An 'analytic center' for polyhedrons and new classes of global algorithms for linear (smooth, convex) programming", in *Lecture Notes in Control and Information Sciences*, Springer-Verlag, New York (1985) 866–876.

Case Studies in Complementarity: Improving Model Formulation *

Michael C. Ferris ** Todd S. Munson ‡

Abstract

Over the past several years, many practitioners have been formulating nonlinear variational inequalities as mixed complementarity problems within modeling languages such as GAMS and AMPL. Sometimes the models generated are poorly specified, either because the function is undefined near the solution or the problem is ill-conditioned or singular. In this paper, we look at information provided by the PATH solver about the model that can be used to identify problem areas and improve formulation. Descriptions and uses of the data provided are detailed via several case studies.

Keywords: Complementarity, Modeling, Ill-Posed, Optimization

AMS subject classification: 90C33

1 Introduction

Developing a practical model of a complex situation is a difficult task in which an approximate representation is initially constructed and then iteratively refined until an accurate formulation is obtained. During the intermediate stages, the models generated have a tendency to be ill-defined, ill-conditioned, and/or singular. Information generated by a solver can help the modeler to detect these problems, quickly locate the source, and make appropriate modifications to the model. By preventing the propagation of errors to successive models, the development cycle shortens and the final product becomes easier to solve with a more meaningful solution. In this paper, we consider equilibrium problems built using the mixed complementarity framework and discuss information provided by the PATH solver that can be used to find potential difficulties. Descriptions and uses of the data provided are detailed via several case studies.

---

*This material is based on research supported in part by National Science Foundation Grant CCR-9619765 and Air Force Office of Scientific Research Grant F49620-98-1-0417.

**Computer Sciences Department, University of Wisconsin, 1210 West Dayton Street, Madison, WI 53706 (ferris@cs.wisc.edu)

‡Computer Sciences Department, University of Wisconsin, 1210 West Dayton Street, Madison, WI 53706 (tmunson@cs.wisc.edu)

We begin by introducing the mixed complementarity problem (MCP) and briefly review pertinent information about modeling languages and the PATH solver in Section 2. We also discuss merit functions used by solvers to indicate if an iterate is close to the solution set of the given MCP. We utilize the merit function information in Section 3 when we present the first of our problem areas, ill-defined models. Section 4 addresses poorly-scaled models and the difficulties encountered in such circumstances. Information provided by the solver to detect scaling problems are mentioned. Section 5 concludes our trilogy of problems areas by looking at singular models. Finally, we present a detailed case study in Section 6 on the Von Thunen Land Use model, a classic and difficult problem from the economics literature. The traditional formulation is coded and analyzed using the tools provided by PATH. We conclude that the model developed is poor and use problem specific expertise to improve it.
We remark that the output generated by PATH is not a replacement for the modeler's knowledge about the application, rather it augments the available information to help them rapidly identify potential problem areas.

2 MCP and Merit Functions

The mixed complementarity problem is defined by a function, $F : D \to \mathbf{R}^n$ where $D \subseteq \mathbf{R}^n$ is the domain of F, and possibly infinite lower and upper bounds, ℓ and u. Let $C := \{x \in \mathbf{R}^n \mid \ell \leq x \leq u\}$, a Cartesian product of closed (possibly infinite) intervals. The problem is given as

$$MCP : \text{find } x \in C \cap D \text{ s.t. } \langle F(x),\ y - x \rangle \geq 0,\ \forall y \in C.$$

This formulation is a special case of the variational inequality problem defined by F and a (nonempty, closed, convex) set C. Special choices of ℓ and u lead to the familiar cases of a system of nonlinear equations

$$F(x) = 0$$

(generated by $\ell \equiv -\infty$, $u \equiv +\infty$) and the nonlinear complementarity problem

$$0 \leq x \perp F(x) \geq 0$$

(generated using $\ell \equiv 0$, $u \equiv +\infty$). The notation $\perp$ is used throughout this paper to denote the fact that the enclosing quantities (in this case x and $F(x)$) are complementary, that is $\langle x,\ F(x) \rangle = 0$.
Complementarity problems can be easily written within modeling languages such as AMPL[13] and GAMS[1]. Modeling languages offer an environment tailored to expressing mathematical constructs. They can efficiently manage a large volume of data and allow users to concentrate on the model rather than the solution methodology. Details of the modeling language extensions and syntax for complementarity problems can be found in [6] and [14, 9]. From an algorithmic perspective, modeling languages provide a clean interface to the problem by transforming a general model into a canonical form and providing

the solver with access to first and possibly second order derivatives via automatic differentiation.

A mathematically trivial, but useful modeling feature is when $\ell_i = u_i$; that is x_i is fixed at a finite value. Due to the product nature of C it is easy to see that $F_i(x)$ is then unrestricted at all solutions and therefore the size of the MCP can be reduced by fixing the variable x_i and removing the function F_i. Both AMPL and GAMS perform this operation automatically. For the rest of this paper we will assume that all fixed variables have been removed and $\ell < u$.

The PATH solver [3, 4] is a damped Newton method based upon the "normal map" operator

$$F(\pi(y)) + y - \pi(y) = 0$$

where $\pi(y)$ represents projection of y onto the set C. Note that if y is a zero of the normal map, the $\pi(y)$ solves the MCP. The "Newton point" is a solution of the linearization:

$$LF(\pi(y)) + y - \pi(y) = 0.$$

Several algorithmic enhancements have been made to improve the solver's robustness including non-monotone searching [8], a watchdog strategy, a crash technique [5] to rapidly identify the active set, restarts and artificial variables [10], and the ability to change merit functions and take gradient steps [7]. The current version of the code is highly successful at solving the mixed complementarity problems in the GAMSLIB and MCPLIB [2] test suites. Examples of applications that use the complementarity framework are given in [11, 12].

A solver for complementarity problems typically employs a merit function to indicate the closeness of the current iterate to the solution set. The merit function is zero at a solution to the original problem and strictly positive otherwise. Numerically, an algorithm terminates when the merit function is approximately equal to zero, thus possibly introducing spurious "solutions".

The modeler needs to be able to determine with some reasonable degree of accuracy whether the algorithm terminated at solution or if it simply obtained a point satisfying the desired tolerances that is not close to the solution set. For complementarity problems, we can provide several indicators with different characteristics to help make such a determination. If one of the indicators is not close to zero, then there is some evidence that the algorithm has not found a solution. We note that if all of the indicators are close to zero, we are reasonably sure we have found a solution. However, the modeler has the final responsibility to evaluate the "solution" and check that it makes sense for their application.

For the NCP, a standard merit function is

$$\|(-x)_+, (-F(x))_+, [(x_i)_+(F_i(x))_+]_i\|$$

with the first two terms measuring the infeasibility of the current point and the last term indicating the complementarity error. In this expression, we use $(\cdot)_+$ to represent the Euclidean projection of x onto the nonnegative orthant, that is

$(x)_+ = \max(x, 0)$. For the more general MCP, we can define a similar function:

$$\left\| x - \pi(x), \left[\left(\frac{x_i - \ell_i}{\|\ell_i\| + 1} \right)_+ (F_i(x))_+ \right]_i, \left[\left(\frac{u_i - x_i}{\|u_i\| + 1} \right)_+ (-F_i(x))_+ \right]_i \right\|$$

where $\pi(x)$ represents the Euclidean projection of x onto C. We can see that if we have an NCP, the function is exactly the one previously given and for nonlinear systems of equations, this becomes $\|F(x)\|$.
There are several reformulations of the MCP as systems of nonlinear (nonsmooth) equations for which the corresponding residual is a natural merit function. Some of these are as follows:

- Generalized Minimum Map: $x - \pi(x - F(x))$
- Normal Map: $F(\pi(y)) + y - \pi(y)$
- Fischer Function: $\Phi(x)$, where $\Phi_i(x) := \phi(x_i, F_i(x))$ with

  $$\phi(a, b) := \sqrt{a + b} - a - b.$$

  Note that $\phi(a, b) = 0$ if and only if $0 \leq a \perp b \geq 0$. A straightforward extension of Φ to the MCP format is given for example in [7].

In the context of nonlinear complementarity problems the generalized minimum map corresponds to the classic minimum map $\min(x, F(x))$. Furthermore, for NCPs the minimum map and the Fischer function are both local error bounds and were shown to be equivalent in [15]. Figure 3 in the subsequent section plots all of these merit functions for the ill-defined example discussed therein and highlights the differences between them.
The squared norm of Φ, namely $\Psi(x) := \frac{1}{2} \sum \phi(x_i, F_i)^2$, is continuously differentiable on $\mathbf{R}^n$ provided F itself is. Therefore, the first order optimality conditions for the unconstrained minimization of $\Psi(x)$, namely $\nabla \Psi(x) = 0$ give another indication as to whether the point under consideration is a solution of MCP.
The merit functions and the information PATH provides at the solution can be useful for diagnostic purposes. By default PATH 4.0 returns the best point with respect to the merit function because this iterate likely provides better information to the modeler. As detailed in [7], the default merit function in PATH 4.0 is the Fischer function. To change this behaviour the `merit_function` option can be used. Other added options to control the output discussed in this paper are described in Table 1.

3 Ill-Defined Models

A problem can be ill-defined for several different reasons. We concentrate on the following particular cases. We will call F well-defined at $\bar{x} \in C$ if $\bar{x} \in D$ and ill-defined at $\bar{x}$ otherwise. Furthermore, we define F to be well-defined near $\bar{x} \in C$

Option	Default	Explanation
output_factorization_singularities	no	Report on the linearly dependent columns encountered during factorization.
output_final_degeneracy_statistics	no	Print information regarding degeneracy at the solution.
output_final_point_statistics	yes	Output information about the point, function, and Jacobian at the final point.
output_final_scaling_statistics	no	Display matrix norms on the Jacobian at the final point.
output_final_statistics	yes	Output evaluatation of available merit functions at the final point.
output_initial_point_statistics	yes	Output information about the point, function, and Jacobian at the initial point.
output_initial_scaling_statistics	yes	Display matrix norms on the Jacobian at the initial point.
output_initial_statistics	no	Output evaluatation of available merit functions at the initial point.
output_model_statistics	yes	Turns on or off printing of all the statistics generated about the model.
return_best_point	yes	Return the best point encountered or the absolute last iterate.

Table 1: Options added to PATH 4.0

if there exists an open neighbourhood of $\bar{x}$, $\mathcal{N}(\bar{x})$, such that $C \cap \mathcal{N}(\bar{x}) \subseteq D$. By saying the function is well-defined near $\bar{x}$, we are simply stating that F is defined for all $x \in C$ sufficiently close to $\bar{x}$. A function not well-defined near $\bar{x}$ is termed ill-defined near $\bar{x}$.

We will say that F has a well-defined Jacobian at $\bar{x} \in C$ if there exists an open neighbourhood of $\bar{x}$, $\mathcal{N}(\bar{x})$, such that $\mathcal{N}(\bar{x}) \subseteq D$ and F is continuously differentiable on $\mathcal{N}(\bar{x})$. Otherwise the function has an ill-defined Jacobian at $\bar{x}$. We note that a well-defined Jacobian at $\bar{x}$ implies that the MCP has a well-defined function near $\bar{x}$, but the converse is not true.

PATH uses both function and Jacobian information in its attempt to solve the MCP. Therefore, both of these definitions are relevant. We discuss cases where the function and Jacobian are ill-defined in the next two subsections. We illustrate uses for the merit function information and final point statistics within the context of these problems.

3.1 Function Undefined

We begin with a one-dimensional problem for which F is ill-defined at $x = 0$ as follows:

$$0 \leq x \quad \perp \quad \tfrac{1}{x} \geq 0.$$

Here x must be strictly positive because $\frac{1}{x}$ is undefined at $x = 0$. This condition implies that $F(x)$ must be equal to zero. Since $F(x)$ is strictly positive for all x strictly positive, this problem has no solution.

We are able to perform this analysis because the dimension of the problem is small. Preprocessing linear problems can be done by the solver in an attempt to detect obviously inconsistent problems, reduce problem size, and identify active components at the solution. Similar processing can be done for nonlinear models, but the analysis becomes more difficult to perform. Currently, PATH only checks the consistency of the bounds and removes fixed variables and the corresponding complementary equations from the model.

A modeler might not know a priori that a problem has no solution and might attempt to formulate and solve it. GAMS code for this model is provided in Figure 1. We must specify an initial value for x in the code. If we were to not provide one, GAMS would use $x = 0$ as the default value, notice that F is undefined at the initial point, and terminate before giving the problem to PATH. The error message problem indicates that the function $\frac{1}{x}$ is ill-defined at $x = 0$, but does not determine whether the corresponding MCP problem has a solution.

After setting the starting point, GAMS generates the model, and PATH proceeds to "solve" it. A portion of the output relating statistics about the solution is given in Figure 2. PATH uses the Fischer Function indicator as its termination criteria by default, but evaluates all of the merit functions given in Section 2 at the final point. The Normal Map merit function, and to a lesser extent, the complementarity error, indicate that the "solution" found does not necessarily solve the MCP.

```
positive variable x;
equations F;

F.. 1 / x =g= 0;

model simple / F.x /;

x.l = 1e-6;

solve simple using mcp;
```

Figure 1: GAMS Code for Ill-Defined Function

```
FINAL STATISTICS
Inf-Norm of Complementarity . .  1.0000e+00 eqn: (F)
Inf-Norm of Normal Map. . . . .  1.1181e+16 eqn: (F)
Inf-Norm of Minimum Map . . . .  8.9441e-17 eqn: (F)
Inf-Norm of Fischer Function. .  8.9441e-17 eqn: (F)
Inf-Norm of Grad Fischer Fcn. .  8.9441e-17 eqn: (F)

FINAL POINT STATISTICS
Maximum of X. . . . . . . . . .  8.9441e-17 var: (X)
Maximum of F. . . . . . . . . .  1.1181e+16 eqn: (F)
Maximum of Grad F . . . . . . .  1.2501e+32 eqn: (F)
                                            var: (X)
```

Figure 2: PATH Output for Ill-Defined Function

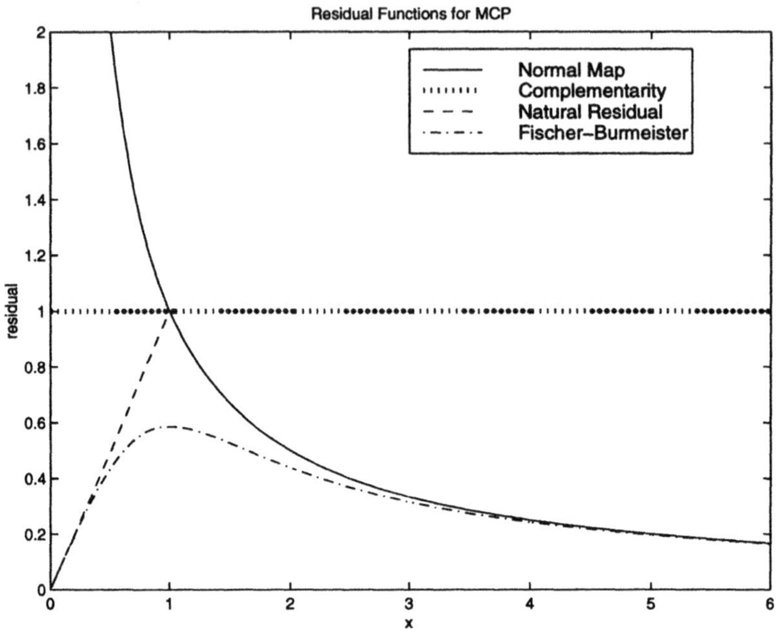

Figure 3: Merit Function Plot

To indicate the difference between the merit functions, Figure 3 plots them all for the simple example. We note that as x approaches positive infinity, numerically, we are at a solution to the problem with respect to all of the merit functions except for the complementarity error, which remains equal to one. As x approaches zero, the merit functions diverge, also indicating that $x = 0$ is not a solution.

The natural residual and Fischer function tend toward 0 as $x \downarrow 0$. ¿From these measures, we might think $x = 0$ is the solution. However, as previously remarked F is ill-defined at $x = 0$. F and ∇F become very large, indicating that the function (and Jacobian) might not be well-defined. We might be tempted to conclude that if one of the merit function indicators is not close to zero, then we have not found a solution. This conclusion is not always the case. When one of the indicators is non-zero, we have reservations about the solution, but we cannot eliminate the possibility that we are actually close to a solution. If we slightly perturb the original problem to

$$0 \leq x \quad \perp \quad \tfrac{1}{x+\epsilon} \geq 0$$

for a fixed $\epsilon > 0$, the function is well-defined over $C = \mathbf{R}^n_+$ and has a unique solution at $x = 0$. In this case, by starting at $x > 0$ and sufficiently small, all of the merit functions, with the exception of the Normal Map, indicate that we have solved the problem as is shown by the output in Figure 4 for $\epsilon = 1 * 10^{-6}$

```
FINAL STATISTICS
Inf-Norm of Complementarity . .  1.0000e-14 eqn: (G)
Inf-Norm of Normal Map. . . . .  1.0000e+06 eqn: (G)
Inf-Norm of Minimum Map . . . .  1.0000e-20 eqn: (G)
Inf-Norm of Fischer Function. .  1.0000e-20 eqn: (G)
Inf-Norm of Grad Fischer Fcn. .  1.0000e-20 eqn: (G)

FINAL POINT STATISTICS
Maximum of X. . . . . . . . . .  1.0000e-20 var: (X)
Maximum of F. . . . . . . . . .  1.0000e+06 eqn: (G)
Maximum of Grad F . . . . . . .  1.0000e+12 eqn: (G)
                                            var: (X)
```

Figure 4: PATH Output for Well-Defined Function

and $x = 1 * 10^{-20}$. In this case, the Normal Map is quite large and we might think that the function and Jacobian are undefined. When only the normal map is non-zero, we may have just mis-identified the optimal basis. By setting the `merit_function normal` option, we can resolve the problem, identify the correct basis, and solve the problem with all indicators being close to zero. This example illustrates the point that all of these tests are not infallible. The modeler still needs to do some detective work to determine if they have found a solution or if the algorithm is converging to a point where the function is ill-defined.

3.2 Jacobian Undefined

Since PATH uses a Newton-like method to solve the problem, it also needs the Jacobian of F to be well-defined. One model for which the function is well-defined over C, but for which the Jacobian is undefined at the solution is: $0 \leq x \perp -\sqrt{x} \geq 0$. This model has a unique solution at $x = 0$.

Using PATH and starting from the point $x = 1 * 10^{-14}$, PATH generates the output given in Figure 5. We can see the that gradient of the Fischer Function is nonzero and the Jacobian is beginning to become large. These conditions indicate that the Jacobian is undefined at the solution. It is therefore important for a modeler to inspect the given output to guard against such problems.

If we start from $x = 0$, PATH correctly informs us that we are at the solution. Even though the entries in the Jacobian are undefined at this point, the GAMS interpreter incorrectly returns a value of 0 to PATH. This problem with the Jacobian is therefore undetectable by PATH.

```
FINAL STATISTICS
Inf-Norm of Complementarity . .  1.0000e-07 eqn: (F)
Inf-Norm of Normal Map. . . . .  1.0000e-07 eqn: (F)
Inf-Norm of Minimum Map . . . .  1.0000e-07 eqn: (F)
Inf-Norm of Fischer Function. .  2.0000e-07 eqn: (F)
Inf-Norm of Grad FB Function. .  2.0000e+00 eqn: (F)

FINAL POINT STATISTICS
Maximum of X. . . . . . . . . .  1.0000e-14 var: (X)
Maximum of F. . . . . . . . . .  1.0000e-07 eqn: (F)
Maximum of Grad F . . . . . . .  5.0000e+06 eqn: (F)
                                            var: (X)
```

Figure 5: PATH Output for Ill-Defined Jacobian

4 Poorly Scaled Models

Problems which are well-defined can have various numerical problems that can impede the algorithm's convergence. One particular problem is a badly scaled Jacobian. In such cases, we can obtain a poor "Newton" direction because of numerical problems introduced in the linear algebra performed. This problem can also lead the code to a point from which it cannot recover.

The final model given to the solver should be scaled such that we avoid numerical difficulties in the linear algebra. The output provided by PATH can be used to iteratively refine the model so that we eventually end up with a well-scaled problem. We note that we only calculate our scaling statistics at the starting point provided. For nonlinear problems these statistics may not be indicative of the overall scaling of the model. Model specific knowledge is very important when we have a nonlinear problem because it can be used to appropriately scale the model to achieve a desired result.

We look at a contact problem, the `kyh.gms` model in MCPLIB, that has some scaling problems. The relevant output from PATH for the original code is given in Figure 6. The maximum row norm is defined as

$$\max_{1\leq i\leq n} \sum_{1\leq j\leq n} \mid (\nabla F(x))_{ij} \mid$$

and the minimum row norm is

$$\min_{1\leq i\leq n} \sum_{1\leq j\leq n} \mid (\nabla F(x))_{ij} \mid .$$

Similar definitions are used for the column norm. The norm numbers for this particular example are not extremely large, but we can nevertheless improve the scaling. We first decided to reduce the magnitude of the `DN` block of equations as indicated by PATH. Using the GAMS modeling language, we can scale particular

```
INITIAL POINT STATISTICS
Maximum of X. . . . . . . . . . 7.0711e-01 var: (X.1)
Maximum of F. . . . . . . . . . 3.6322e+01 eqn: (DN.1)
Maximum of Grad F . . . . . . . 3.1982e+04 eqn: (GAPA.1)
                                           var: (PA.1)

INITIAL JACOBIAN NORM STATISTICS
Maximum Row Norm. . . . . . . . 7.1961e+04 eqn: (DN.2)
Minimum Row Norm. . . . . . . . 1.0807e+00 eqn: (GAP7.1)
Maximum Column Norm . . . . . . 7.1962e+04 var: (Z1.1)
Minimum Column Norm . . . . . . 1.1414e+00 var: (PT.1)
```

Figure 6: PATH Output - Poorly Scaled Model

```
INITIAL POINT STATISTICS
Maximum of X. . . . . . . . . . 7.0711e-01 var: (X.1)
Maximum of F. . . . . . . . . . 1.0000e-01 eqn: (Z2.1)
Maximum of Grad F . . . . . . . 3.1982e+01 eqn: (GAPA.1)
                                           var: (PA.1)

INITIAL JACOBIAN NORM STATISTICS
Maximum Row Norm. . . . . . . . 7.1961e+01 eqn: (DN.2)
Minimum Row Norm. . . . . . . . 1.0807e+00 eqn: (GAP7.1)
Maximum Column Norm . . . . . . 7.2961e+01 var: (Z1.1)
Minimum Column Norm . . . . . . 1.4142e-03 var: (DT.1)
```

Figure 7: PATH Output - Well-Scaled Model

equations and variables using the .scale attribute. To turn the scaling on for the model we use the .scaleopt model attribute. After scaling the `DN` block, we re-ran PATH and found two additional blocks of equations that also needed scaling, `GAPA` and `GAPB`. The code added to the model follows:

```
contact.scaleopt = 1;
dn.scale(I) = 1000;
gapa.scale(I) = 1000;
gapb.scale(I) = 1000;
```

The modified model has been called `kyh-scale.gms`. After scaling these blocks of equations in the model, we have improved the scaling statistics which are given in Figure 7 for the new model. Using the scaling language features and the information provided by PATH we are able to remove some of the problem's difficulty and obtain better performance from PATH. For this particular problem PATH had difficulty solving the unscaled model, but was able to easily solve the correctly scaled problem. It is possible to get even more information on initial

```
INITIAL POINT STATISTICS
Zero column of order. . . . . .  0.0000e+00 var: (X)
Zero row of order . . . . . . .  0.0000e+00 eqn: (F)
Total zero columns. . . . . . .  1
Total zero rows . . . . . . . .  1
Maximum of F. . . . . . . . . .  1.0000e+00 eqn: (F)
Maximum of Grad F . . . . . . .  0.0000e+00 eqn: (F)
                                            var: (X)
```

Figure 8: PATH Output - Zero Rows and Columns

point scaling by inspecting the GAMS listing file. The equation row listing gives the values of all the entries of the Jacobian at the starting point. The row norms generated by PATH give good pointers into this source of information.
Not all of the numerical problems are directly attributable to poorly scaled models. Problems for which the Jacobian of the active constraints is singular or nearly singular can also cause numerical difficulty as illustrated next.

5 Singular Models

Assuming that the problem is well-defined and properly scaled, we can still have a Jacobian for which the active constraints are singular or nearly singular (i.e. it is ill-conditioned). When problems are singular or nearly singular, we are also likely to have numerical problems. As a result the "Newton" direction obtained from the linear problem solver can be very bad. In PATH, we can use proximal perturbation or add artificial variables to attempt to remove the singularity problems from the model. However, it is most often beneficial for solver robustness to remove singularities if possible.
The easiest problems to detect are those for which the Jacobian has zero rows and columns. A simple problem for which we have zero rows and columns is:

$$-2 \leq x \leq 2 \quad \perp \quad -x^2 + 1.$$

Note that the Jacobian, $-2x$, is non-singular at all three solutions, but singular at the point $x = 0$. Output from PATH on this model starting at $x = 0$ is given in Figure 8. We display in the code the variables and equations for which the row/column in the Jacobian is close to zero. These situations are problematic and for nonlinear problems likely stem from the modeler providing an inappropriate starting point or fixing some variables resulting in some equations becoming constant. We note that the solver may perform well in the presence of zero rows and/or columns, but the modeler should make sure that these are what was intended.
Singularities in the model can also be detected by the linear solver. This in itself is a hard problem and prone to error. For matrices which are poorly scaled, we

can incorrectly identify "linearly dependent" rows because of numerical problems. Setting `output_factorization_singularities yes` in an options file will inform the user which equations the linear solver thinks are linearly dependent. Typically, singularity does not cause a lot of problems and the algorithm can handle the situation appropriately. However, an excessive number of singularities are cause for concern. A further indication of possible singularities at the solution is the lack of quadratic convergence to the solution.
We now turn our attention towards using the diagnostic information provided by PATH to improve an actual model.

6 Von Thunen Land Model

In this section, we describe the Von Thunen land model, a problem renowned in the mathematical programming literature for its computational difficulty. We attempt to understand more carefully the facets of the problem that make it difficult to solve. This will enable us in the sequel to outline and identify these problems and furthermore to extend the model to a more realistic and computationally more tractable form.

6.1 Classical Model

The problem is cast in the Arrow-Debreu framework as an equilibrium problem. The basic model is a closed economy consisting of three economic agents, a landowner, a worker and a porter. There is a central market, around which concentric regions of land are located. Since the produced goods have to be delivered to the market, this is an example of a spatial price equilibrium. The key variables of the model are the prices of commodities, land, labour and transport. Given these prices, it is assumed that the agents demand certain amounts of the commodities, which are supplied so as to maximize profit in each sector. Walras' law is then a consequence of the assumed competitive paradym, namely that supply will equal demand in the equilibrium state.
We now describe the problems that the consumers and the producers face. We first look at consumption, and derive a demand function for each of the consumer agents in the economy. Each of these agents has a utility function, that they wish to maximize subject to their budgetary constraints. As is typical in such problems, the utility function is assumed to be Cobb-Douglas

$$u_a(d) = \prod_c d_c^{\alpha_{c,a}}, \quad \alpha_{c,a} \geq 0, \sum_c \alpha_{c,a} = 1,$$

where the $\alpha_{c,a}$ are given parameters dependent only on the agent. For each agent a, the variables d_c represent quantities of the desired commodities c. In the Von Thunen model, the goods are wheat, rice, corn and barley. The agents endowments determine their budgetary constraint as follows. Given current market prices, an agents wealth is the value of the initial endowment of goods at those prices. The agents problem is therefore

$$\max_d u_a(d) \text{ subject to } \langle p,\, d\rangle \leq \langle p,\, e_a\rangle, d \geq 0,$$

where e_a is the endowment bundle for agent a. A closed form solution, corresponding to demand from agent a for commodity c is thus

$$d_{c,a}(p) := \frac{\alpha_{c,a}\langle p,\, e_a\rangle}{p_c}.$$

Note that this assumes the prices of the commodities p_c are positive.
The supply side of the economy is similar. The worker earns a wage w_L for his labour input. The land is distributed around the market in rings with a rental rate w_r associated with each ring r of land. The area of land a_r in each ring is an increasing function of r. The model assumes that labour and land are substitutable via a constant elasticities of substitution (CES) function.
Consider the production $x_{c,r}$ of commodity c in region r. In order to maximize profit (or minimize costs), the labour y_L and land use y_r solve

$$\min w_L y_L + w_r y_r \text{ subject to } \phi_c y_L^{\beta_c} y_r^{1-\beta_c} \geq x_{c,r},\; y_L, y_r \geq 0, \tag{1}$$

where ϕ_c is a given cost function scale parameter, and $\beta_c \in [0,1]$ is the share parameter. The technology constraint is precisely the CES function allowing a suitable mix of labour and land use. Again, a closed form solution can be calculated. For example, the demand for labour in order to produce $x_{c,r}$ of commodity c in region r is given by

$$x_{c,r}\frac{\beta_c\left(\frac{w_L}{\beta_c}\right)^{\beta_c}\left(\frac{w_r}{1-\beta_c}\right)^{1-\beta_c}}{\phi_c w_L}.$$

Considering all such demands, this clearly assumes the prices of inputs w_L, w_r are positive. A key point to note is that input commodity (factor) demands to produce $x_{c,r}$ can be determined by first solving (1) for unit demand $x_{c,r} \equiv 1$ and then multiplying these factor demands by the actual amount desired. Let $\bar{y}_L$ and $\bar{y}_r$ denote the optimal solutions of (1) with $x_{c,r} \equiv 1$. Using this fact, the *unit* production cost $\gamma_{c,r}$ for commodity c in region r can be calculated as follows:

$$\begin{aligned}
\gamma_{c,r} &= w_L\bar{y}_L + w_r\bar{y}_r \\
&= w_L\frac{\beta_c\left(\frac{w_L}{\beta_c}\right)^{\beta_c}\left(\frac{w_r}{1-\beta_c}\right)^{1-\beta_c}}{\phi_c w_L} + w_r\frac{(1-\beta_c)\left(\frac{w_L}{\beta_c}\right)^{\beta_c}\left(\frac{w_r}{1-\beta_c}\right)^{1-\beta_c}}{\phi_c w_r} \\
&= \frac{1}{\phi_c}\left(\frac{w_L}{\beta_c}\right)^{\beta_c}\left(\frac{w_r}{1-\beta_c}\right)^{1-\beta_c}.
\end{aligned}$$

Transportation is provided by a porter, earning a wage w_p. If we denote the unit cost for transportation of commodity c by t_c, then unit transportation cost to market is

$$T_{c,r}(w_p) := t_c d_r w_p,$$

where d_r is the distance of region r to the market. Spatial price equilibrium arises from the consideration:

$$0 \le x_{c,r} \perp \gamma_{c,r}(w_L, w_r) + T_{c,r}(w_p) \ge p_c.$$

This is intuitively clear; it states that commodity c will be produced in region r only if the combined cost of production and transportation equals the market price.

The above derivations assumed that the producers and consumers acted as price takers. Walras' law is now invoked to determine the prices so that markets clear. The resulting complementarity problem is:

$$\gamma_{c,r} = \frac{1}{\phi_c}\left(\frac{w_L}{\beta_c}\right)^{\beta_c}\left(\frac{w_r}{1-\beta_c}\right)^{1-\beta_c} \tag{2}$$

$$0 \le x_{c,r} \quad \perp \quad \gamma_{c,r} + T_{c,r}(w_p) \ge p_c \tag{3}$$

$$0 \le w_L \quad \perp \quad e_L \ge \sum_{r,c} x_{c,r}\frac{\beta_c\gamma_{c,r}}{w_L} \tag{4}$$

$$0 \le w_r \quad \perp \quad a_r \ge \sum_{c} \frac{x_{c,r}(1-\beta_c)\gamma_{c,r}}{w_r} \tag{5}$$

$$0 \le w_p \quad \perp \quad e_P \ge \sum_{r,c} t_c d_r x_{c,r} \tag{6}$$

$$0 \le p_c \quad \perp \quad \sum_{r} x_{c,r} \ge \frac{\alpha_{c,P}e_P w_p + \alpha_{c,L}e_L w_L + \alpha_{c,O}\sum_r w_r a_r}{p_c} \tag{7}$$

Note that in (4), (5) and (6), the amounts of labour, land and transport are bounded from above, and hence the prices on these inputs are determined as multipliers (or shadow prices) on the corresponding constraints. The final relationship (7) in the above complementarity problem corresponds to market clearance; prices are nonnegative and can only be positive if supply equals demand. (Some modelers multiply the last inequality throughout by p_c. This removes problems where p_c becomes zero, but can also introduce spurious solutions.)

The Arrow-Debreu theory guarantees that the problem is homogeneous in prices; $(x, \lambda w, \lambda p)$ is also a solution whenever (x, w, p) solves the above. Typically this singularity in the model is removed by fixing a numeraire, that is fixing a price (for example $w_L = 1$) and dropping the corresponding complementary relationship.

Unfortunately, in this formulation even after fixing a numeraire, some of the variables p and w may go to zero, resulting in an ill-defined problem. In the case of the Von Thunen land model, the rental price of land w_r decreases as the distance to market increases, and for remote rings of land, it becomes zero. A standard modeling fix is to put artificial lower bounds on these variables. Even with this fix, the problem typically remains very hard to solve. More importantly, the homogeneity property of the prices used above to fix a numeraire no longer holds, and the corresponding complementary relationship (which was dropped from the problem) may fail to be satisfied. It therefore matters which numeriare is fixed, and many modelers run into difficulty since in many cases the solution found by a solver is invalid for the originally posed model.

In order to test our diagnostic information, we implemented a version of the above model in GAMS. The model corresponds closely to the MCPLIB model pgvon105.gms except we added more regions to make the problem even more difficult. The model file has been documented more fully, and the data rounded to improve clarity.

The first trial we attempted was to solve the model without fixing a numeraire. In this case, PATH 4.0 failed to find a solution. At the starting point, the indicators described in Section 3 are reasonable, and there are no zero rows/columns in the Jacobian. At the best point found, all indicators are still reasonable. However, the listing file indicates a lare number of division by zero problems occurring in (5). We also note that a nonzero proximal perturbation is used in the first iteration of the crash method. This is an indication of singularities. We therefore added an option to output factorization singularities, and singularities appeared in the first iteration. At this point, we decided to fix a numeraire to see if this alleviated the problem.

We chose to fix the labour wage rate to 1. After increasing the iterations allowed to 100,000, PATH 4.0 solved the problem. The statistics at the solution are cause for concern. In particular, the gradient of the Fischer function is 7 orders of magnitude larger than all the other residuals. Furthermore, the Jacobian is very large at the solution point. Looking further in the listing file, a large number of division by zero problems occur in (5).

To track down the problem further, we added an artifical lower bound on the variables w_r of 10^{-5}, that would not be active at the aforementioned solution. Resolving gave the same "solution", but resulted in the domain errors disappearing.

Although the problem is solved, there is concern on two fronts. Firstly, the gradient of the Fischer function should go to zero at the solution. Secondly, if a modeler happens to make the artificial lower bounds on the variables a bit larger, then they become active at the solution, and hence the constraint that has been dropped by fixing the price of labour at 1 is violated at this point. Of course, the algorithm is unable to detect this problem, since it is not part of the model that is passed to it, and the corresponding output looks satisfactory.

We are therefore led to the conclusion that the model as posulated is ill-defined. The remainder of this section outlines two possible modeling techniques to overcome the difficulties with ill-defined problems of this type.

6.2 Intervention Pricing

The principal difficulty is the fact that the rental prices on land go to zero as proximity to the market decreases, and become zero for sufficiently remote rings. Such a property is unlikely to hold in a practical setting. Typically, a landowner has a minimum rental price (for example, land in fallow increases in value). As outlined above, a fixed lower bound on the rental price violates

the well-establised homogeneity property. A suggestion postulated by Professor Thomas Rutherford is to allow the landowner to intervene and "purchase-back" his land whenever the rental cost gets smaller than a certain fraction of the labour wage.
The new model adds a (homogeneous in price) constraint

$$0 \leq i_r \quad \perp \quad w_r \geq 0.0001 * w_L$$

and modifies (5) and (7) as follows:

$$0 \leq w_r \quad \perp \quad a_r - i_r \geq \sum_c \frac{x_{c,r}(1-\beta_c)\gamma_{c,r}}{w_r}$$

$$0 \leq p_c \quad \perp \quad \sum_r x_{c,r} \geq \frac{\alpha_{c,P} e_P w_p + \alpha_{c,L} e_L w_L + \alpha_{c,O} \sum_r w_r(a_r - i_r)}{p_c}. \qquad (8)$$

Given the intervention purchase, we can now add a lower bound on w_r to avoid division by zero errors. In our model we chose 10^{-5} since this will never be active at the solution and therefore will not affect the positive homogeneity. After this reformulation, PATH 4.0 solves the problem. Furthermore, the gradient of the Fischer function, although slightly larger than the other residuals, is quite small, and can be made even smaller by reducing the convergence tolerance of PATH. Inspecting the listing file, the only difficulties mentioned are division by zero errors in the market clearance condition (8), that can be avoided a posteori by imposing an artificial (inactive) lower bound on these prices. We chose not to do this however.

6.3 Nested Production and Maintenance

Another observation that can be used to overcome the land price going to zero is the fact that land typically requires some maintenance labour input to keep it usable for crop growth. Traditionally, in economics, this is carried out by providing a nested CES function as technology input to the model. The idea is that commodity c in region r is made from labour and an intermediate good. The intermediate good is "maintained land". Essentially, the following production problem replaces (1):

$$\begin{array}{rl} \min_{y_M, y_L, y_r, g} & w_L(y_M + y_L) + w_r y_r \\ \text{subject to} & y_r \geq (1 - \beta_c - \epsilon) g \\ & y_M \geq \epsilon g \\ & \phi_c y_L^{\beta_c} g^{1-\beta_c} \geq 1, \\ & y_M, y_L, y_r, g \geq 0. \end{array}$$

Note that the variable y_M represents "maintenance labour" and g represents the amount of "maintained land" produced, an intermediate good. The process of generating maintained land uses a Leontieff production function, namely

$$\min(\lambda_r y_r, \lambda_M y_M) \geq g.$$

Here $\lambda_M = \frac{1}{\epsilon}$, ϵ small, corresponds to small amounts of maintenance labour, while $\lambda_r = \frac{1}{1-\beta_c-\epsilon}$ is chosen to calibrate the model correctly. A simple calculus exercise then generates appropriate demand and cost expressions. The resulting complementarity problem comprises (3), (6), (7) and

$$\gamma_{c,r} = \frac{w_L^{\beta_c}}{\phi_c}\left(\frac{w_L\epsilon + w_r(1-\beta_c-\epsilon)}{1-\beta_c}\right)^{1-\beta_c}$$

$$0 \leq w_L \quad \perp \quad e_L \geq \sum_{r,c} x_{c,r}\gamma_{c,r}\left(\frac{\beta_c}{w_L} + \frac{\epsilon(1-\beta_c)}{w_L\epsilon + w_r(1-\beta_c-\epsilon)}\right)$$

$$0 \leq w_r \quad \perp \quad a_r \geq \sum_{c} \frac{x_{c,r}\gamma_{c,r}(1-\beta_c)(1-\beta_c-\epsilon)}{w_L\epsilon + w_r(1-\beta_c-\epsilon)}$$

After making the appropriate modifications to the model file, PATH 4.0 solved the problem on defaults without any difficulties. All indicators showed he problem and solution found to be well-posed.

7 Conclusion

Models which are ill-defined, poorly scaled, or singular can cause algorithms to have extreme difficulty. Identifying the problem is a non-trivial matter for the modeler. An algorithm can provide additional statistics about the model to the modeler. This information can help the modeler to quickly identify and correct the problem.

Further algorithmic work would be useful; in particular methods to automatically identify implied lower bounds may help to avoid many of the domain violations that occur in ill-defined models.

Acknowledgement. The authors wish to thank Thomas Rutherford for numerous enlightening discussions of the von Thunen model and suggestions for improvements to the output of PATH that are relevant to modelers.

References

[1] A. Brooke, D. Kendrick, and A. Meeraus. *GAMS: A User's Guide*. The Scientific Press, South San Francisco, CA, 1988.

[2] S. P. Dirkse and M. C. Ferris. MCPLIB: A collection of nonlinear mixed complementarity problems. *Optimization Methods and Software*, 5:319–345, 1995.

[3] S. P. Dirkse and M. C. Ferris. The PATH solver: A non-monotone stabilization scheme for mixed complementarity problems. *Optimization Methods and Software*, 5:123–156, 1995.

[4] S. P. Dirkse and M. C. Ferris. A pathsearch damped Newton method for computing general equilibria. *Annals of Operations Research*, pages 211–232, 1996.

[5] S. P. Dirkse and M. C. Ferris. Crash techniques for large-scale complementarity problems. In Ferris and Pang [11], pages 40–61.

[6] M. C. Ferris, R. Fourer, and D. M. Gay. Expressing complementarity problems and communicating them to solvers. *SIAM Journal on Optimization, forthcoming*, 1999.

[7] M. C. Ferris, C. Kanzow, and T. S. Munson. Feasible descent algorithms for mixed complementarity problems. *Mathematical Programming, forthcoming*, 1999.

[8] M. C. Ferris and S. Lucidi. Nonmonotone stabilization methods for nonlinear equations. *Journal of Optimization Theory and Applications*, 81:53–71, 1994.

[9] M. C. Ferris and T. S. Munson. Complementarity problems in GAMS and the PATH solver. *Journal of Economic Dynamics and Control, forthcoming*, 1999.

[10] M. C. Ferris and T. S. Munson. Interfaces to PATH 3.0: Design, implementation and usage. *Computational Optimization and Applications*, 12:207–227, 1999.

[11] M. C. Ferris and J. S. Pang, editors. *Complementarity and Variational Problems: State of the Art*, Philadelphia, Pennsylvania, 1997. SIAM Publications.

[12] M. C. Ferris and J. S. Pang. Engineering and economic applications of complementarity problems. *SIAM Review*, 39:669–713, 1997.

[13] R. Fourer, D.M. Gay, and B.W. Kernighan. *AMPL: A Modeling Language for Mathematical Programming*. Duxbury Press, 1993.

[14] T. F. Rutherford. Applied general equilibrium modeling with MPSGE as a GAMS subsystem: An overview of the modeling framework and syntax. *Computational Economics, forthcoming*, 1999.

[15] P. Tseng. Growth behavior of a class of merit functions for the nonlinear complementarity problem. *Journal of Optimization Theory and Applications*, 89:17–37, 1996.

Convergence of a Smoothing Continuation Method for Mathematical Progams with Complementarity Constraints

Masao Fukushima * Jong-Shi Pang **

Abstract

We study the behavior of a sequence generated by a smoothing continuation method for mathematical programs with equilibrium constraints (MPEC). In particular, we show that under the linear independence constraint qualification and an additional condition called the asymptotic weak nondegeneracy, the limit of KKT points satisfying the second-order necessary conditions for the perturbed problems is a B-stationary point of the original MPEC. The notable fact is that this result does not rely on the strict complementarity assumption at the limit point.

Keywords: Mathematical programs with equilibrium constraints, stationary point, constraint qualification, smoothing continuation method.

AMS subject classification: 90C30, 90C33

1 Introduction

We consider the following mathematical programming problem with equilibrium constraints (MPEC):

$$\begin{array}{lll} \text{P:} & \text{minimize} & f(z) \\ & \text{subject to} & G(z) \geq 0, \quad H(z) \geq 0, \quad G(z)^T H(z) = 0, \\ & & g(z) \leq 0, \quad h(z) = 0, \end{array}$$

where $f : \Re^n \to \Re$, $G : \Re^n \to \Re^m$, $H : \Re^n \to \Re^m$, $g : \Re^n \to \Re^p$ and $h : \Re^n \to \Re^q$ are twice continuously differentiable functions. Although this formulation looks

*Department of Applied Mathematics and Physics, Graduate School of Informatics, Kyoto University, Kyoto 606-8501, Japan, Email: fuku@kuamp.kyoto-u.ac.jp

**Department of Mathematical Sciences, Whiting School of Engineering, The Johns Hopkins University, Baltimore, Maryland 21218-2692, U.S.A., Email: jsp@vicp1.mts.jhu.edu somewhat different from the conventional MPECs that involve the first-level and the second-level variables, the latter problems can be transformed into the form of P in a straightforward manner.

The major cause of difficulty in dealing with MPECs lies in the unusual nature of the constraints. For example, it is known [1, 12] that a standard constraint qualifications in nonlinear programming such as Mangasarian-Fromovitz constraint qualification (MFCQ) never holds for problem P. Nevertheless, serious attention has recently been paid to clarifying stationarity concepts for various types of MPECs under novel constraint qualifications pertaining to the peculiar structure of equilibrium constraints [7, 8, 11, 12, 13]. On the other hand, the development of efficient and robust algorithms for MPECs seems somewhat behind the theoretical study, although attempts have been made quite intensively in recent years [2, 4, 5, 7, 9, 10]. The algorithms proposed so far are only guaranteed to converge to a point satisfying some weak stationarity conditions such as Clarke stationarity condition, or require the nondegeneracy (strict complementarity) assumption at a limit point so as to ensure a stronger stationarity concept such as B-stationarity. The main theme of this paper is to establish convergence to a B-stationary point for a smoothing continuation method without the nondegeneracy assumption. The basic assumptions we adopt here are the linear independence constraint qualification (LICQ) and an additional condition called the asymptotic weak nondegeneracy at a limit of the generated sequence. Based on the fact that, under LICQ, a KKT point for a relaxed problem associated with the given MPEC affords a B-stationary point of the MPEC, we will show that the generated sequence converges to a KKT point for the relaxed problem.

Notations: The component functions of G, H, g and h are denoted by $G_i, i = 1,\dots,m$, $H_i, i = 1,\dots,m$, $g_l, l = 1,\dots,p$, and $h_k, k = 1,\dots,q$, respectively. For a vector $v \in \Re^m$ and index set $\alpha \in \{1,\dots,m\}$, v_α denotes the subvector of v with components $v_i, i \in \alpha$. Similar notations are used for vector valued functions as well. The index sets $\{1,\dots,p\}$ and $\{1,\dots,q\}$ are denoted by $\mathcal{I}$ and $\mathcal{J}$, respectively.

2 Preliminaries

Let $\bar{z}$ be a given feasible point of problem P. We define the index sets

$$\begin{aligned} \bar{\alpha} &\equiv \{\, i \,|\, G_i(\bar{z}) = 0 < H_i(\bar{z}) \,\} \\ \bar{\beta} &\equiv \{\, i \,|\, G_i(\bar{z}) = 0 = H_i(\bar{z}) \,\} \\ \bar{\gamma} &\equiv \{\, i \,|\, G_i(\bar{z}) > 0 = H_i(\bar{z}) \,\} \\ \bar{\mathcal{I}} &\equiv \{\, l \,|\, g_l(\bar{z}) = 0 \,\}. \end{aligned} \tag{1}$$

In particular, we say that $\bar{z}$ is *nondegenerate* [7] if $\bar{\beta} = \emptyset$, i.e., $\bar{z}$ satisfies the *strict complementarity* condition $G_i(\bar{z}) + H_i(\bar{z}) > 0$ for all $i = 1,\dots,m$. Associated with the index sets $\bar{\alpha}, \bar{\beta}$ and $\bar{\gamma}$, we define the *relaxed problem* of P at $\bar{z}$:

$$
\begin{array}{lll}
\mathrm{R}(\bar{z}): & \text{minimize} & f(z) \\
& \text{subject to} & G_i(z) = 0, \quad i \in \bar{\alpha} \\
& & G_i(z) \geq 0, \quad i \in \bar{\beta} \\
& & H_i(z) \geq 0, \quad i \in \bar{\beta} \\
& & H_i(z) = 0, \quad i \in \bar{\gamma} \\
& & g_l(z) \leq 0, \quad l \in \mathcal{I} \\
& & h_k(z) = 0, \quad k \in \mathcal{J}.
\end{array}
$$

We say that the *linear independence constraint qualification (LICQ)* holds at $\bar{z}$ if the vectors

$$
\begin{aligned}
&\{\nabla G_i(\bar{z}) \mid i \in \bar{\alpha} \cup \bar{\beta}\} \cup \{\nabla H_i(\bar{z}) \mid i \in \bar{\beta} \cup \bar{\gamma}\} \\
&\cup \{\nabla g_l(\bar{z}) \mid l \in \bar{\mathcal{I}}\} \cup \{\nabla h_k(\bar{z}) \mid k \in \mathcal{J}\}
\end{aligned}
$$

are linearly independent.
The *Karush-Kuhn-Tucker (KKT) conditions* for the relaxed problem $\mathrm{R}(\bar{z})$ are represented as

$$
\begin{aligned}
\nabla f(z) - \sum_{i \in \bar{\alpha}} v_i \nabla G_i(z) - \sum_{i \in \bar{\beta}} v_i \nabla G_i(z) - \sum_{i \in \bar{\beta}} w_i \nabla H_i(z) - \sum_{i \in \bar{\gamma}} w_i \nabla H_i(z) & \\
+ \sum_{l \in \mathcal{I}} \lambda_l \nabla g_l(z) + \sum_{k \in \mathcal{J}} \mu_k \nabla h_k(z) \ = \ 0, & \\
v_{\bar{\beta}} \geq 0, \quad G_{\bar{\beta}}(z) \geq 0, \quad v_{\bar{\beta}}^T G_{\bar{\beta}}(z) = 0, & \\
w_{\bar{\beta}} \geq 0, \quad H_{\bar{\beta}}(z) \geq 0, \quad w_{\bar{\beta}}^T H_{\bar{\beta}}(z) = 0, & \\
\lambda \geq 0, \quad g(z) \leq 0, \quad \lambda^T g(z) = 0, & \\
G_{\bar{\alpha}}(z) = 0, \quad H_{\bar{\gamma}}(z) = 0, \quad h(z) = 0. &
\end{aligned} \tag{2}
$$

Let S be an arbitrary subset of $\Re^n$. We call $\bar{z} \in S$ a *B-stationary point* of the minimization problem

$$
\begin{array}{ll}
\text{minimize} & f(z) \\
\text{subject to} & z \in S
\end{array}
$$

if it satisfies

$$
\nabla f(\bar{z})^T d \geq 0, \quad \forall d \in \mathcal{T}(\bar{z}; S),
$$

where $\mathcal{T}(\bar{z}; S)$ is the tangent cone of S at $\bar{z}$ that consists of all vectors d for which there exist a sequence $\{z^\nu\} \subset S$ converging to $\bar{z}$ and a sequence $\{\tau_\nu\}$ of positive numbers converging to zero such that $d = \lim_{\nu \to \infty}(z^\nu - \bar{z})/\tau_\nu$. Clearly any local optimal solution is a B-stationary point.

Under some constraint qualification, any B-stationary point of the relaxed problem R($\bar{z}$) satisfies the KKT conditions (2) along with some Lagrange multipliers $v_{\bar{\alpha}\cup\bar{\beta}}$, $w_{\bar{\beta}\cup\bar{\gamma}}$, λ and μ. Furthermore, if the LICQ holds at $\bar{z}$, then a B-stationary point of R($\bar{z}$) is also a B-stationary point for problem P [7, Proposition 4.3.7].

Proposition 2.1 *If a feasible point $\bar{z}$ of problem P satisfies the KKT conditions (2) for the relaxed problem R($\bar{z}$), and if the LICQ holds at $\bar{z}$, then $\bar{z}$ is a B-stationary point of problem P.*

3 Main Results

We consider a smoothing continuation method for solving problem P. This method uses the perturbed Fischer-Burmeister function $\phi : \Re^2 \times \Re_+ \to \Re$ defined by

$$\phi(a,b,\varepsilon) = a + b - \sqrt{a^2+b^2+\varepsilon}. \tag{3}$$

The function ϕ was first introduced by Kanzow [6] for linear complementarity problems, and used in particular to develop an SQP method for solving MPEC in [4]. When $\varepsilon = 0$, the function ϕ reduces to the Fischer-Burmeister function and satisfies

$$\phi(a,b,0) = 0 \quad \Longleftrightarrow \quad a = 0,\ b = 0,\ ab = 0. \tag{4}$$

When $\varepsilon > 0$, this function satisfies

$$\phi(a,b,\varepsilon) = 0 \quad \Longleftrightarrow \quad a > 0,\ b > 0,\ ab = \varepsilon/2 \tag{5}$$

and is smooth with respect to (a,b). More specifically, its first and second derivatives are given by

$$\begin{aligned}
\frac{\partial}{\partial a}\phi(a,b,\varepsilon) &= 1 - \frac{a}{\sqrt{a^2+b^2+\varepsilon}},\\
\frac{\partial}{\partial b}\phi(a,b,\varepsilon) &= 1 - \frac{b}{\sqrt{a^2+b^2+\varepsilon}},\\
\frac{\partial^2}{\partial a^2}\phi(a,b,\varepsilon) &= -\frac{b^2+\varepsilon}{(a^2+b^2+\varepsilon)^{3/2}},\\
\frac{\partial^2}{\partial b^2}\phi(a,b,\varepsilon) &= -\frac{a^2+\varepsilon}{(a^2+b^2+\varepsilon)^{3/2}},\\
\frac{\partial^2}{\partial a\partial b}\phi(a,b,\varepsilon) &= \frac{ab}{(a^2+b^2+\varepsilon)^{3/2}}.
\end{aligned} \tag{6}$$

With the perturbed Fischer-Burmeister function ϕ, we define the function $\Phi^\varepsilon : \Re^n \to \Re^m$ by

$$\Phi^\varepsilon(z) = \begin{pmatrix} \phi(G_1(z), H_1(z), \varepsilon) \\ \vdots \\ \phi(G_m(z), H_m(z), \varepsilon) \end{pmatrix}. \tag{7}$$

For $\varepsilon > 0$, the function Φ^ε is differentiable everywhere and the gradient of its component function Φ_i^ε at z is written as

$$\begin{aligned}\nabla\Phi_i^\varepsilon(z) &= \left(1 - \frac{G_i(z)}{\sqrt{G_i(z)^2 + H_i(z)^2 + \varepsilon}}\right)\nabla G_i(z) \\ &\quad + \left(1 - \frac{H_i(z)}{\sqrt{G_i(z)^2 + H_i(z)^2 + \varepsilon}}\right)\nabla H_i(z). \qquad (8)\end{aligned}$$

In particular, we have the following lemma that will be useful in subsequent discussions.

Lemma 3.1 *Let z satisfy $\Phi^\varepsilon(z) = 0$ with $\varepsilon > 0$. Then we have*

$$G_i(z) > 0, \quad H_i(z) > 0, \quad G_i(z)H_i(z) = \varepsilon/2$$

for all $i = 1, \ldots, m$. Moreover we have

$$\begin{aligned}\frac{\partial}{\partial a}\phi(G_i(z), H_i(z), \varepsilon) &= \frac{H_i(z)}{G_i(z) + H_i(z)}, \\ \frac{\partial}{\partial b}\phi(G_i(z), H_i(z), \varepsilon) &= \frac{G_i(z)}{G_i(z) + H_i(z)}, \\ \frac{\partial^2}{\partial a^2}\phi(G_i(z), H_i(z), \varepsilon) &= -\frac{H_i(z)(H_i(z) + 2G_i(z))}{(G_i(z) + H_i(z))^3}, \\ \frac{\partial^2}{\partial b^2}\phi(G_i(z), H_i(z), \varepsilon) &= -\frac{G_i(z)(G_i(z) + 2H_i(z))}{(G_i(z) + H_i(z))^3}, \\ \frac{\partial^2}{\partial a \partial b}\phi(G_i(z), H_i(z), \varepsilon) &= \frac{G_i(z)H_i(z)}{(G_i(z) + H_i(z))^3}\end{aligned}$$

for all $i = 1, \ldots, m$.

Proof. The first part is immediate from (5). Then we have

$$\begin{aligned}\sqrt{G_i(z)^2 + H_i(z)^2 + \varepsilon} &= \sqrt{G_i(z)^2 + H_i(z)^2 + 2G_i(z)H_i(z)} \\ &= |G_i(z) + H_i(z)| \\ &= G_i(z) + H_i(z).\end{aligned}$$

This together with (6) yields the latter part of the lemma. □

By Lemma 3.1, the gradient of Φ_i^ε at a point z such that $\Phi^\varepsilon(z) = 0$ is particularly given by

$$\nabla\Phi_i^\varepsilon(z) = \frac{H_i(z)}{G_i(z) + H_i(z)}\nabla G_i(z) + \frac{G_i(z)}{G_i(z) + H_i(z)}\nabla H_i(z). \qquad (9)$$

For $\varepsilon = 0$, the function Φ^0 is not differentiable in general. Specifically, when $G_i(\bar{z}) = H_i(\bar{z}) = 0$ for some $\bar{z}$ and i, i.e., $i \in \bar{\beta}$, the function Φ_i^0 is not differentiable at $\bar{z}$ and the Clarke generalized gradient

$$\partial\Phi_i^0(\bar{z}) \equiv \{r \in \Re^n \mid r = \lim_{k\to\infty} \nabla\Phi_i^0(z^k) \text{ with } z^k \to \bar{z} \text{ and } \nabla\Phi_i^0(z^k) \text{ exist}\}$$

is contained in the set

$$\partial_C\Phi_i^0(\bar{z}) \equiv \{r \in \Re^n \mid r = \bar{\xi}_i\nabla G_i(\bar{z}) + \bar{\eta}_i\nabla H_i(\bar{z}) \text{ for some } (\bar{\xi}_i, \bar{\eta}_i) \in B\}, \quad (10)$$

where $B \equiv \{(\bar{\xi}_i, \bar{\eta}_i) \in \Re^2 \mid (1-\bar{\xi}_i)^2 + (1-\bar{\eta}_i)^2 \le 1\}$.
Now consider the following perturbed problem for P with $\varepsilon > 0$:

$$\begin{aligned} \mathrm{P}^\varepsilon: \quad & \text{minimize} && f(z) \\ & \text{subject to} && \Phi^\varepsilon(z) = 0, \\ &&& g(z) \le 0, \quad h(z) = 0. \end{aligned}$$

We have the following lemma.

Lemma 3.2 *For each $\varepsilon > 0$, let z^ε be a feasible point of P^ε. Suppose that $z^\varepsilon \to \bar{z}$ as $\varepsilon \to 0$. Then $\bar{z}$ is a feasible point of P. Moreover, if the LICQ holds at $\bar{z}$, then the gradients*

$$\{\nabla\Phi_i^\varepsilon(z^\varepsilon) \mid i = 1,\ldots,m\} \cup \{\nabla g_l(z^\varepsilon) \mid l \in \mathcal{I}^\varepsilon\} \cup \{\nabla h_k(z^\varepsilon) \mid k \in \mathcal{J}\}$$

are linearly independent for all $\varepsilon > 0$ small enough, where $\mathcal{I}^\varepsilon \equiv \{l \mid g_l(z^\varepsilon) = 0\}$.

Proof. By (4), (5) and the continuity of ϕ, it is easy to see that $\bar{z}$ is feasible to P. Let index sets $\bar{\alpha}, \bar{\beta}, \bar{\gamma}, \bar{\mathcal{I}}$ be defined by (1). Then it follows from (8) that as $\varepsilon \to 0$

$$\begin{aligned} \|\nabla\Phi_i^\varepsilon(z^\varepsilon) - \nabla G_i(\bar{z})\| &\to 0 \quad \text{for } i \in \bar{\alpha}, \\ \|\nabla\Phi_i^\varepsilon(z^\varepsilon) - \nabla H_i(\bar{z})\| &\to 0 \quad \text{for } i \in \bar{\gamma}, \\ \mathrm{dist}[\nabla\Phi_i^\varepsilon(z^\varepsilon), \partial_C\Phi_i^0(\bar{z})] &\to 0 \quad \text{for } i \in \bar{\beta}, \end{aligned}$$

where $\mathrm{dist}[y, S] \equiv \min\{\|y - y'\| \mid y' \in S\}$ for $y \in \Re^n$ and $S \subset \Re^n$. Note that the set $\partial_C\Phi_i^0(\bar{z})$ contains the Clarke generalized gradient $\partial\Phi_i^0(\bar{z})$ of Φ_i^0 at $\bar{z}$. Moreover, by continuity, it is clear that

$$\begin{aligned} \|\nabla g_l(z^\varepsilon) - \nabla g_l(\bar{z})\| &\to 0 \quad \text{for } l \in \mathcal{I}, \\ \|\nabla h_k(z^\varepsilon) - \nabla h_k(\bar{z})\| &\to 0 \quad \text{for } k \in \mathcal{J}. \end{aligned}$$

Observe that $\mathcal{I}^\varepsilon \subseteq \bar{\mathcal{I}}$ for all $\varepsilon > 0$ sufficiently small. Then it is not difficult to deduce that the desired result holds under the LICQ at $\bar{z}$. □

Now let z^ε be a local optimal solution of problem P^ε. Then, since the perturbed problem P^ε is an ordinary nonlinear programming problem, under some constraint qualification, there exist Lagrange multipliers $u^\varepsilon \in \Re^m$, $\lambda^\varepsilon \in \Re^p$, $\mu^\varepsilon \in \Re^q$

such that the vector $(z, u, \lambda, \mu) = (z^\varepsilon, u^\varepsilon, \lambda^\varepsilon, \mu^\varepsilon)$ satisfies the following KKT conditions for problem P^ε:

$$\begin{aligned}
&\nabla f(z) - \sum_{i=1}^{m} u_i \nabla \Phi_i^\varepsilon(z) + \sum_{l \in \mathcal{I}} \lambda_l \nabla g_l(z) + \sum_{k \in \mathcal{J}} \mu_k \nabla h_k(z) = 0,\\
&\Phi^\varepsilon(z) = 0,\\
&\lambda \geq 0, \quad g(z) \leq 0, \quad \lambda^T g(z) = 0,\\
&h(z) = 0.
\end{aligned} \tag{11}$$

Moreover $(z, u, \lambda, \mu) = (z^\varepsilon, u^\varepsilon, \lambda^\varepsilon, \mu^\varepsilon)$ satisfies the inequalities

$$d^T \nabla_z^2 \mathcal{L}^\varepsilon(z, u, \lambda, \mu) d \geq 0 \quad \text{for all} \quad d \in \mathcal{T}^\varepsilon(z), \tag{12}$$

where

$$\mathcal{L}^\varepsilon(z, u, \lambda, \mu) \equiv f(z) - \sum_{i=1}^{m} u_i \Phi_i^\varepsilon(z) + \sum_{l \in \mathcal{I}} \lambda_l g_l(z) + \sum_{k \in \mathcal{J}} \mu_k h_k(z)$$

and

$$\mathcal{T}^\varepsilon(z) \equiv \{\, d \in \Re^n \mid \nabla \Phi^\varepsilon(z)^T d = 0, \ \nabla g_{\mathcal{I}_\varepsilon}(z)^T d = 0, \ \nabla h(z)^T d = 0 \,\}$$

with $\mathcal{I}_\varepsilon \equiv \{\, l \mid g_l(z^\varepsilon) = 0 \,\}$.

To establish our main result, we introduce the concept of asymptotic weak nondegeneracy. Suppose that $z^\varepsilon \to \bar{z}$ as $\varepsilon \to 0$. It then follows from (9) that, for each $i \in \bar{\beta}$, any accumulation point $\bar{r}$ of $\{\nabla \Phi_i^\varepsilon(z^\varepsilon)\}$ belongs to the set $\partial_C \Phi_i^0(\bar{z})$ defined by (10) and hence is represented as

$$\bar{r} = \bar{\xi}_i \nabla G_i(\bar{z}) + \bar{\eta}_i \nabla H_i(\bar{z})$$

for some $(\bar{\xi}_i, \bar{\eta}_i)$ such that $(1 - \bar{\xi}_i)^2 + (1 - \bar{\eta}_i)^2 \leq 1$. We say that $\{z^\varepsilon\}$ is *asymptotically weakly nondegenerate* if neither $\bar{\xi}_i$ nor $\bar{\eta}_i$ vanishes for any accumulation point $\bar{r}$. Roughly speaking, in view of (9), this condition means that $G_i(z^\varepsilon)$ and $H_i(z^\varepsilon)$ approach zero in the same order of magnitude. Obviously $\{z^\varepsilon\}$ is asymptotically weakly nondegenerate when $\bar{z}$ satisfies the nondegeneracy condition $\bar{\beta} = \emptyset$.

Now we state the main theorem.

Theorem 3.1 *Let $(z^\varepsilon, u^\varepsilon, \lambda^\varepsilon, \mu^\varepsilon)$ satisfy the second-order necessary conditions (11) and (12) for problem P^ε. Suppose that $(z^\varepsilon, u^\varepsilon, \lambda^\varepsilon, \mu^\varepsilon) \to (\bar{z}, \bar{u}, \bar{\lambda}, \bar{\mu})$ as $\varepsilon \to 0$. If the LICQ holds at $\bar{z}$ and $\{z^\varepsilon\}$ is asymptotically weakly nondegenerate, then $\bar{z}$ satisfies the KKT conditions (2) for the relaxed problem $R(\bar{z})$; moreover $\bar{z}$ is a B-stationary point for P.*

Proof. For each $i \in \bar{\beta}$, let $\bar{r}$ be an arbitrary accumulation point of $\{\nabla\Phi_i^\varepsilon(z^\varepsilon)\}$. Then, by the asymptotic weak nondegeneracy assumption, we have

$$\bar{r} = \bar{\xi}_i \nabla G_i(\bar{z}) + \bar{\eta}_i \nabla H_i(\bar{z})$$

with $(1-\bar{\xi}_i)^2 + (1-\bar{\eta}_i)^2 \leq 1$ and $\bar{\xi}_i > 0$ and $\bar{\eta}_i > 0$. Since

$$\begin{aligned} \nabla\Phi_i^\varepsilon(z^\varepsilon) &\to \nabla G_i(\bar{z}) \quad \text{for } i \in \bar{\alpha}, \\ \nabla\Phi_i^\varepsilon(z^\varepsilon) &\to \nabla H_i(\bar{z}) \quad \text{for } i \in \bar{\gamma}, \end{aligned}$$

it follows from the first equation in the KKT conditions (11) for problem P^ε that

$$\nabla f(\bar{z}) - \sum_{i\in\bar{\alpha}} \bar{u}_i \nabla G_i(\bar{z}) - \sum_{i\in\bar{\beta}} \bar{u}_i\bar{\xi}_i \nabla G_i(\bar{z}) - \sum_{i\in\bar{\beta}} \bar{u}_i\bar{\eta}_i \nabla H_i(\bar{z}) - \sum_{i\in\bar{\gamma}} \bar{u}_i \nabla H_i(\bar{z})$$
$$+ \sum_{l\in\mathcal{I}} \bar{\lambda}_l \nabla g_l(\bar{z}) + \sum_{k\in\mathcal{J}} \bar{\mu}_k \nabla h_k(\bar{z}) = 0. \tag{13}$$

It is easy to see that

$$\begin{aligned} \bar{\lambda}_l &\geq 0 \quad \text{for } l \in \bar{\mathcal{I}}, \\ \bar{\lambda}_l &= 0 \quad \text{for } l \notin \bar{\mathcal{I}}. \end{aligned}$$

Hence it suffices to show that

$$\begin{aligned} \bar{v}_i &\equiv \bar{u}_i\bar{\xi}_i \geq 0 \quad \text{for } i \in \bar{\beta}, \\ \bar{w}_i &\equiv \bar{u}_i\bar{\eta}_i \geq 0 \quad \text{for } i \in \bar{\beta}. \end{aligned} \tag{14}$$

We will derive a contradiction by assuming that $\bar{v}_j < 0$ for some $j \in \bar{\beta}$. We can argue the case where $\bar{w}_j < 0$ for some $j \in \bar{\beta}$ similarly. First note that, since $\bar{\xi}_i > 0$, $\bar{v}_j < 0$ implies $\bar{u}_j < 0$. Now let $\varepsilon > 0$ be sufficiently small and choose a vector $d^\varepsilon \in \Re^n$ such that

$$\begin{aligned} \nabla G_j(z^\varepsilon)^T d^\varepsilon &= \frac{\partial\phi_j^\varepsilon}{\partial b} \equiv \frac{\partial}{\partial b}\phi(G_j(z^\varepsilon), H_j(z^\varepsilon), \varepsilon), \\ \nabla H_j(z^\varepsilon)^T d^\varepsilon &= -\frac{\partial\phi_j^\varepsilon}{\partial a} \equiv -\frac{\partial}{\partial a}\phi(G_j(z^\varepsilon), H_j(z^\varepsilon), \varepsilon), \\ \nabla G_i(z^\varepsilon)^T d^\varepsilon &= 0 \quad \text{for } i \in \bar{\alpha} \cup \bar{\beta} \setminus \{j\}, \\ \nabla H_i(z^\varepsilon)^T d^\varepsilon &= 0 \quad \text{for } i \in \bar{\gamma} \cup \bar{\beta} \setminus \{j\}, \\ \nabla g_l(z^\varepsilon)^T d^\varepsilon &= 0 \quad \text{for } l \in \mathcal{I}^\varepsilon, \\ \nabla h_k(z^\varepsilon)^T d^\varepsilon &= 0 \quad \text{for } k \in \mathcal{J}. \end{aligned} \tag{15}$$

We can choose such d^ε because the LICQ at $\bar{z}$ ensures the linear independence of vectors

$$\{\, \nabla G_i(z^\varepsilon) \,|\, i \in \bar{\alpha} \cup \bar{\beta} \,\} \cup \{\, \nabla H_i(z^\varepsilon) \,|\, i \in \bar{\beta} \cup \bar{\gamma} \,\}$$
$$\cup \{\, \nabla g_l(z^\varepsilon) \,|\, l \in \mathcal{I}^\varepsilon \,\} \cup \{\, \nabla h_k(z^\varepsilon) \,|\, k \in \mathcal{J} \,\}$$

for any $\varepsilon > 0$ small enough. (Note that $\mathcal{I}^\varepsilon \subseteq \bar{\mathcal{I}}$ for all ε small enough.) Then since

$$\nabla\Phi_j^\varepsilon(z^\varepsilon) = \frac{\partial\phi_j^\varepsilon}{\partial a}\nabla G_j(z^\varepsilon) + \frac{\partial\phi_j^\varepsilon}{\partial b}\nabla H_j(z^\varepsilon),$$

it follows from (15) that

$$\begin{aligned}\nabla\Phi_j^\varepsilon(z^\varepsilon)^T d^\varepsilon &= \frac{\partial\phi_j^\varepsilon}{\partial a}\nabla G_j(z^\varepsilon)^T d^\varepsilon + \frac{\partial\phi_j^\varepsilon}{\partial b}\nabla H_j(z^\varepsilon)^T d^\varepsilon \\ &= \frac{\partial\phi_j^\varepsilon}{\partial a}\frac{\partial\phi_j^\varepsilon}{\partial b} - \frac{\partial\phi_j^\varepsilon}{\partial b}\frac{\partial\phi_j^\varepsilon}{\partial a} \\ &= 0.\end{aligned}$$

Moreover we have

$$\begin{aligned}\nabla\Phi_i^\varepsilon(z^\varepsilon)^T d^\varepsilon &= 0 \qquad \text{for } i = 1,\ldots,m \ (i \neq j), \\ \nabla g_l(z^\varepsilon)^T d^\varepsilon &= 0 \qquad \text{for } l \in \mathcal{I}^\varepsilon, \\ \nabla h_k(z^\varepsilon)^T d^\varepsilon &= 0 \qquad \text{for } k \in \mathcal{J}.\end{aligned}$$

Hence it follows that

$$d^\varepsilon \in \mathcal{T}^\varepsilon(z^\varepsilon).$$

Now we consider the limit

$$\begin{aligned}&\lim_{\varepsilon\to 0}\ (d^\varepsilon)^T\nabla_z^2\mathcal{L}(z^\varepsilon, u^\varepsilon, \lambda^\varepsilon, \mu^\varepsilon)d^\varepsilon \\ &= \lim_{\varepsilon\to 0}\ (d^\varepsilon)^T\Bigg[\nabla^2 f(z^\varepsilon) - \sum_{i=1}^m u_i^\varepsilon\nabla^2\Phi_i^\varepsilon(z^\varepsilon) \\ &\qquad + \sum_{l\in\mathcal{I}}\lambda_l^\varepsilon\nabla^2 g_l(z^\varepsilon) + \sum_{k\in\mathcal{J}}\mu_k^\varepsilon\nabla^2 h_k(z^\varepsilon)\Bigg] d^\varepsilon. \qquad (16)\end{aligned}$$

It follows from (15) that

$$\begin{aligned}&(d^\varepsilon)^T\nabla^2\Phi_j^\varepsilon(z^\varepsilon)d^\varepsilon \\ &= (d^\varepsilon)^T\Bigg(\frac{\partial\phi_j^\varepsilon}{\partial a}\nabla^2 G_j(z^\varepsilon) + \frac{\partial\phi_j^\varepsilon}{\partial b}\nabla^2 H_j(z^\varepsilon) \\ &\quad + \frac{\partial^2\phi_j^\varepsilon}{\partial a^2}\nabla G_j(z^\varepsilon)\nabla G_j(z^\varepsilon)^T + \frac{\partial^2\phi_j^\varepsilon}{\partial a\partial b}\nabla G_j(z^\varepsilon)\nabla H_j(z^\varepsilon)^T \\ &\quad + \frac{\partial^2\phi_j^\varepsilon}{\partial b\partial a}\nabla H_j(z^\varepsilon)\nabla G_j(z^\varepsilon)^T + \frac{\partial^2\phi_j^\varepsilon}{\partial b^2}\nabla H_j(z^\varepsilon)\nabla H_j(z^\varepsilon)^T\Bigg) d^\varepsilon\end{aligned}$$

$$= \frac{\partial \phi_j^\varepsilon}{\partial a}(d^\varepsilon)^T \nabla^2 G_j(z^\varepsilon) d^\varepsilon + \frac{\partial \phi_j^\varepsilon}{\partial b}(d^\varepsilon)^T \nabla^2 H_j(z^\varepsilon) d^\varepsilon$$
$$+ \frac{\partial^2 \phi_j^\varepsilon}{\partial a^2}\left(\frac{\partial \phi_j^\varepsilon}{\partial b}\right)^2 - 2\frac{\partial^2 \phi_j^\varepsilon}{\partial a \partial b}\left(\frac{\partial \phi_j^\varepsilon}{\partial a}\right)\left(\frac{\partial \phi_j^\varepsilon}{\partial b}\right) + \frac{\partial^2 \phi_j^\varepsilon}{\partial b^2}\left(\frac{\partial \phi_j^\varepsilon}{\partial a}\right)^2 \quad (17)$$

and, for all $i \neq j$,

$$(d^\varepsilon)^T \nabla^2 \Phi_i^\varepsilon(z^\varepsilon) d^\varepsilon = \frac{\partial \phi_i^\varepsilon}{\partial a}(d^\varepsilon)^T \nabla^2 G_i(z^\varepsilon) d^\varepsilon + \frac{\partial \phi_i^\varepsilon}{\partial b}(d^\varepsilon)^T \nabla^2 H_i(z^\varepsilon) d^\varepsilon,$$

where

$$\frac{\partial^2 \phi_j^\varepsilon}{\partial a^2} \equiv \frac{\partial^2}{\partial a^2}\phi_j(G_j(z^\varepsilon), H_j(z^\varepsilon), \varepsilon),$$
$$\frac{\partial^2 \phi_j^\varepsilon}{\partial a \partial b} \equiv \frac{\partial^2}{\partial a \partial b}\phi_j(G_j(z^\varepsilon), H_j(z^\varepsilon), \varepsilon),$$
$$\frac{\partial^2 \phi_j^\varepsilon}{\partial b^2} \equiv \frac{\partial^2}{\partial b^2}\phi_j(G_j(z^\varepsilon), H_j(z^\varepsilon), \varepsilon).$$

Note that $\partial \phi_i^\varepsilon/\partial a$, $\partial \phi_i^\varepsilon/\partial b$, $\nabla^2 G_i(z^\varepsilon)$, $\nabla^2 H_i(z^\varepsilon)$ are all bounded as $\varepsilon \to 0$, since, by assumption, $(z^\varepsilon, u^\varepsilon, \lambda^\varepsilon, \mu^\varepsilon) \to (\bar{z}, \bar{u}, \bar{\lambda}, \bar{\mu})$. Also note that the LICQ at $\bar{z}$ and the boundedness of $\partial \phi_i^\varepsilon/\partial a$, $\partial \phi_i^\varepsilon/\partial b$ imply that d^ε is also bounded as $\varepsilon \to 0$. Therefore $(d^\varepsilon)^T \nabla^2 f(z^\varepsilon) d^\varepsilon$, $u_i^\varepsilon (d^\varepsilon)^T \nabla^2 \Phi_i^\varepsilon(z^\varepsilon) d^\varepsilon$ $(i \neq j)$, $\lambda_l^\varepsilon (d^\varepsilon)^T \nabla^2 g_l(z^\varepsilon) d^\varepsilon$ and $\mu_k^\varepsilon (d^\varepsilon)^T \nabla^2 h_k(z^\varepsilon) d^\varepsilon$ are bounded as $\varepsilon \to 0$.
Let us show that $(d^\varepsilon)^T \nabla^2 \Phi_j^\varepsilon(z^\varepsilon) d^\varepsilon$ tends to $-\infty$ as $\varepsilon \to 0$. To simplify the notation, we denote $s^\varepsilon \equiv G_j(z^\varepsilon)$ and $t^\varepsilon \equiv H_j(z^\varepsilon)$. Note that $s^\varepsilon > 0$ and $t^\varepsilon > 0$ by Lemma 3.1, and $s^\varepsilon \to 0$ and $t^\varepsilon \to 0$ since $j \in \bar{\beta}$. Then, by Lemma 3.1 and (17), we can write

$$(d^\varepsilon)^T \nabla^2 \Phi_j^\varepsilon(z^\varepsilon) d^\varepsilon = \frac{\partial \phi_j^\varepsilon}{\partial a}(d^\varepsilon)^T \nabla^2 G_j(z^\varepsilon) d^\varepsilon + \frac{\partial \phi_j^\varepsilon}{\partial b}(d^\varepsilon)^T \nabla^2 H_j(z^\varepsilon) d^\varepsilon - R^\varepsilon, \quad (18)$$

where

$$R^\varepsilon \equiv \frac{t^\varepsilon(t^\varepsilon + 2s^\varepsilon)}{(s^\varepsilon + t^\varepsilon)^3}\left(\frac{s^\varepsilon}{s^\varepsilon + t^\varepsilon}\right)^2 + \frac{2s^\varepsilon t^\varepsilon}{(s^\varepsilon + t^\varepsilon)^3}\left(\frac{s^\varepsilon}{s^\varepsilon + t^\varepsilon}\right)\left(\frac{t^\varepsilon}{s^\varepsilon + t^\varepsilon}\right)$$
$$+ \frac{s^\varepsilon(s^\varepsilon + 2t^\varepsilon)}{(s^\varepsilon + t^\varepsilon)^3}\left(\frac{t^\varepsilon}{s^\varepsilon + t^\varepsilon}\right)^2$$
$$= \frac{2s^\varepsilon t^\varepsilon}{(s^\varepsilon + t^\varepsilon)^3}. \quad (19)$$

By the asymptotic weak nondegeneracy of z^ε, there exist $\rho_1, \rho_2 > 0$ such that $0 < \rho_1 \leq s^\varepsilon/t^\varepsilon \leq \rho_2 < \infty$ for all $\varepsilon > 0$. Therefore, as $\varepsilon \to 0$, we have

$$\frac{2s^\varepsilon t^\varepsilon}{(s^\varepsilon + t^\varepsilon)^3} = \frac{1}{t^\varepsilon} \cdot \frac{2(s^\varepsilon/t^\varepsilon)}{[(s^\varepsilon/t^\varepsilon) + 1]^3} \to \infty$$

and hence, by (18),

$$(d^\varepsilon)^T \nabla^2 \Phi_j^\varepsilon(z^\varepsilon) d^\varepsilon \to -\infty.$$

Because $(d^\varepsilon)^T \nabla^2 f(z^\varepsilon) d^\varepsilon$, $u_i^\varepsilon (d^\varepsilon)^T \nabla^2 \Phi_i^\varepsilon(z^\varepsilon) d^\varepsilon$ $(i \neq j)$, $\lambda_l^\varepsilon (d^\varepsilon)^T \nabla^2 g_l(z^\varepsilon) d^\varepsilon$ and $\mu_k^\varepsilon (d^\varepsilon)^T \nabla^2 h_k(z^\varepsilon) d^\varepsilon$ are bounded as $\varepsilon \to 0$, it follows from (16) that $\bar{u}_j < 0$ implies

$$(d^\varepsilon)^T \nabla_z^2 \mathcal{L}(z^\varepsilon, u^\varepsilon, \lambda^\varepsilon, \mu^\varepsilon) d^\varepsilon \to -\infty. \tag{20}$$

However, since $(z^\varepsilon, u^\varepsilon, \lambda^\varepsilon, \mu^\varepsilon)$ satisfies, by assumption, the second-order necessary conditions (12), $d^\varepsilon \in \mathcal{T}^\varepsilon(z^\varepsilon)$ implies

$$(d^\varepsilon)^T \nabla_z^2 \mathcal{L}(z^\varepsilon, u^\varepsilon, \lambda^\varepsilon, \mu^\varepsilon) d^\varepsilon \geq 0.$$

Since this inequality holds for each $\varepsilon > 0$ and d^ε is bounded, we have a contradiction with (20). Hence we have (14).

Consequently, by putting

$$\begin{aligned}
v_i &= \begin{cases} \bar{u}_i, & i \in \bar{\alpha} \\ \bar{u}_i \bar{\xi}_i, & i \in \bar{\beta} \end{cases} \\
w_i &= \begin{cases} \bar{u}_i \bar{\eta}_i, & i \in \bar{\beta} \\ \bar{u}_i, & i \in \bar{\gamma} \end{cases} \\
\lambda_l &= \bar{\lambda}_l, \quad l \in \bar{\mathcal{I}} \\
\mu_k &= \bar{\mu}_k, \quad k \in \mathcal{J},
\end{aligned}$$

we deduce that (13) yields the KKT conditions (2) for the relaxed problem $\mathrm{R}(\bar{z})$. The last part of the theorem follows from Proposition 2.1. □

Acknowledgments. We thank Nobuo Yamashita and Junji Imai for their comments on an earlier version of the paper.

References

[1] Y. Chen and M. Florian, "The nonlinear bilevel programming problem: Formulations, regularity and optimality conditions," *Optimization* 32 (1995) 193–209.

[2] F. Facchinei, H. Jiang and L. Qi, "A smoothing method for mathematical programs with equilibrium constraints," *Mathematical Programming*, forthcoming.

[3] A. Fischer, "A special Newton-type optimization method," *Optimization* 24 (1992) 269–284.

[4] M. Fukushima, Z.-Q. Luo and J.-S. Pang, "A globally convergent sequential programming algorithm for mathematical programs with linear complementarity constraints," *Computational Optimization and Applications* 10 (1998) 5–30.

[5] H. Jiang and D. Ralph, "Smooth SQP methods for mathematical programs with nonlinear complementarity constraints," Working paper, Department of Mathematics and Statistics, University of Melbourne, Parkville, Australia, 1997.

[6] C. Kanzow, "Some noninterior continuation methods for linear complementarity problems," *SIAM Journal on Matrix Analysis and Applications* 17 (1996) 851–868.

[7] Z.-Q. Luo, J.-S. Pang and D. Ralph, *Mathematical Programs with Equilibrium Constraints*, (Cambridge University Press, Cambridge, 1996).

[8] J.V. Outrata, "Optimality conditions for a class of mathematical programs with equilibrium constraints," *Mathematics of Operations Research*, forthcoming.

[9] J. Outrata, M. Kočvara and J. Zowe, *Nonsmooth Approach to Optimization Problems with Equilibrium Constraints*, (Kluwer Academic Publishers, Dordrecht, 1998).

[10] J. Outrata and J. Zowe, A numerical approach to optimization problems with variational inequality constraints, *Mathematical Programming* 68 (1995) 105–130.

[11] J.-S. Pang and M. Fukushima, "Complementarity constraint qualifications and simplified B-stationarity conditions for mathematical programs with equilibrium constraints," *Computational Optimization and Applications*, forthcoming.

[12] H. Scheel and S. Scholtes, "Mathematical programs with equilibrium constraints: Stationarity, optimality, and sensitivity," Working paper, The Judge Institute of Management Studies, University of Cambridge, Cambridge, 1997.

[13] J.J. Ye, "Optimality conditions for optimization problems with complementarity constraints," *SIAM Journal on Optimization*, forthcoming.

Stable Solution of Variational Inequalities with Composed Monotone Operators

A. Kaplan * R. Tichatschke **

Abstract

Convergence of proximal-based methods is analysed for variational inequalities with operators of the type $\mathcal{T}^0 + \partial F$, where $\mathcal{T}^0$ is a single-valued, hemicontinuous and monotone operator and ∂F is the subdifferential of a proper convex lower semicontinuous functional. The analysis is oriented to methods coupling successive approximation of the variational inequality with the proximal point algorithm as well as to related methods using regularization on a subspace and weak regularization. Conditions ensuring linear convergence are established. Finally, we observe briefly some classes of problems which can be solved by means of the methods considered.

Keywords: Variational inequalities with monotone operators, ill-posed problems, proximal point methods.

AMS subject classification: 47H05, 47H19, 65J20.

1 Introduction

Let V be a Hilbert space, V' the dual space of V and $< \cdot, \cdot >$ the duality pairing between V and V'. Let K be a convex closed subset of V.
The problem under consideration is:

$$\text{find } \; u \in K \; \text{ such that } \langle \mathcal{T}^0 u + \partial F(u), v - u \rangle \geq 0 \;\; \forall v \in K, \tag{1}$$

where $\mathcal{T}^0 : V \to V'$ is a single-valued monotone operator, $D(\mathcal{T}^0) \supset K$ and $\mathcal{T}^0$ is hemicontinuous on K; ∂F is the subdifferential of a convex, lower semicontinuous functional $F : V \to \bar{\mathbb{R}} \equiv \mathbb{R} \cup \{+\infty\}$, $D(\partial F) \supset K$. The variational inequality (1) means that

$$\exists \eta \in \partial F(u) : \; \langle \mathcal{T}^0 u + \eta, v - u \rangle \geq 0 \;\; \forall v \in K. \tag{2}$$

*University of Trier, D-54286 Trier
**University of Trier, D-54286, e-mail: tichat@uni-trier.de Trier

We recall that an operator $\mathcal{A} : V \to 2^{V'}$ is said to be *monotone* if

$$\langle w - z, u - v \rangle \geq 0 \quad \text{whenever } w \in \mathcal{A}u, \ z \in \mathcal{A}v$$

and *strongly monotone* if

$$\langle w - z, u - v \rangle \geq \gamma \|u - v\|^2 \quad \text{with some } \gamma > 0.$$

A monotone operator $\mathcal{A}$ is called *maximal monotone* if its graph is not properly contained in the graph of any other monotone operator $\mathcal{A}^1 : V \to 2^{V'}$.
Let $\mathcal{N}_K : V \to 2^{V'}$ be the normality operator for K:

$$\mathcal{N}_K(u) = \begin{cases} \{w \in V' : \langle w, u - v \rangle \geq 0 \ \ \forall v \in K\} & \text{if } u \in K \\ \emptyset & \text{if } u \notin K, \end{cases}$$

and put $\tilde{\mathcal{T}} = \mathcal{T}^0 + \mathcal{N}_K$.
According to Theorem 3 in [29], $\tilde{\mathcal{T}}$ is maximal monotone. It is well-known that u solves Problem (1) if and only if $0 \in \tilde{\mathcal{T}}u + \partial F(u)$.
In the sequel, Problem (1) is assumed to be solvable, and U^* denotes its solution set. Sufficient conditions for solvability of variational inequalities with monotone operators can be found, for instance, in [6], [29].

The proximal point method, originally introduced by MARTINET [24] to solve convex variational problems, has initiated a couple of new algorithms for various classes of variational inequalities. Proximal-based decomposition methods are developed in [3], [7] - [10], and the papers [20], [22], [25] are dealing with proximal-based bundle methods for non-smooth convex optimization problems. By the use of non-quadratic proximal mappings, new interior point methods for solving finite-dimensional variational inequalities on a polyhedron are constructed in [1], [2]. Solving ill-posed elliptic variational inequalities on a sequence of grids, in [14], [16] proximal iterations are inserted into the finite element method (FEM) to ensure stability of the approximation process. Here, we refer only to some recent publications, for earlier papers see the chapter "Comments" in [14].

The fundamental results of ROCKAFELLAR [30] on convergence of the proximal point method for finding a zero of a maximal monotone operator in a Hilbert space, have been generalized by LUQUE [23] concerning the rate of convergence, and a similar analysis was performed by HA [12] for methods with proximal regularization on a subspace. In these papers it is allowed that the proximal iterations can be calculated inexactly, however, an approximation of the problem data is not considered.

In the present paper we deal with methods coupling a successive approximation of the variational inequality and the proximal point approach, as well as with

related methods using a regularization on a subspace or a weak regularization. Throughout the whole paper it is supposed that H is a Hilbert space such that V can be continuously embedded into H, V_1 is a given closed subspace of V and $\mathcal{P} : V \to V_1$ is an orthoprojector. If V_1 is also closed in H then $\mathcal{P}$ can be defined as orthoprojector according to the norm in H. For the choice of V_1, H see the Assumptions 1 and 2(d).
In the framework of the approach considered, a solution of Problem (1) is obtained by solving approximately a sequence of well-posed variational inequalities

$$u \in K_i : \quad \langle \mathcal{T}^0 u + \nabla F_i(u), v - u\rangle + \chi_i \left(\mathcal{P}u - \mathcal{P}u^{i,s-1}, \mathcal{P}v - \mathcal{P}u\right)_H \geq 0 \ \ \forall v \in K_i$$

$$s = 1, ..., s(i); \ i = 1, 2, ...$$

Here, $F_i : V \to \bar{I\!R}$ and K_i are certain approximations for F and K, respectively, F_i is Gâteaux differentiable on K_i; $(\cdot,\cdot)_H$ is the inner product in H, $u^{i,s-1}$ is a solution of the previous problem ($u^{i,0} := u^{i-1,s(i-1)}$) and the regularization parameter χ_i satisfies $0 < \chi_i \leq \bar{\chi}$.
In the sequel, we refer to this scheme as *Multi-Step Regularization method (MSR-method)*. This notion reflects the presence of an inner cycle (w.r.t. s) of proximal iterations. MSR-methods for convex variational problems have been developed in [14] and adapted for some problems of elasticity theory and of optimal control with PDE's in [16], [13], [17]. The case $\mathcal{T}^0 = 0$, $H = V = V_1$, $s(i) = 1$ (for all i), includes several "diagonal" variants of the proximal point method for convex optimization (cf. [14], Sect. 9 and 12, [21] and references therein).
In [18], [19] the MSR-approach is studied for a more general variational inequality, namely with an arbitrary maximal monotone operator $\mathcal{T} : V \to 2^{V'}$ instead of $\mathcal{T}^0 + \partial F(\cdot)$ in(1). The special structure of the operator $\mathcal{T}^0 + \partial F(\cdot)$ allows much more convenient conditions for a successive approximation of the problem, based on an uniform approximation of F by $\{F_i\}$ (compare Assumption 2 below and Assumption 2 in [18]). However, the application of these conditions requires a completely different convergence analysis.

The paper is organized as follows: After a strong formulation of the MSR-methods in Section 2, their convergence is studied in Section 3, and estimates of the rate of convergence for the basic variant (with $H = V = V_1$) are given in Section 4. In Section 5 we observe briefly some classes of problems, which can be solved by means of the considered methods.

2 Multi-Step Proximal Regularization Scheme

We denote by $\|\cdot\|, \|\cdot\|_{V'}$ and $\|\cdot\|_H$ the norms in V, V' and H, respectively. With z, C, D from V, let

$$dist(z, D) = \inf_{w \in D} \|z - w\|, \quad e(C, D) = \sup_{z \in C} dist(z, D).$$

The following assumption concerns the chosen triple H, V_1 and $\mathcal{P}$.

Assumption 1 *For a given linear, continuous and monotone operator $\mathcal{B} : V \to V'$ with the symmetry property*

$$\langle \mathcal{B}u, v\rangle = \langle \mathcal{B}v, u\rangle \quad \forall u, v \in V$$

and a given $\tilde{\chi} > 0$ there exists $\beta_0 > 0$ such that

$$\tilde{\chi}\langle \mathcal{B}u, u\rangle + \|\mathcal{P}u\|_H^2 \geq \beta_0 \|u\|^2 \quad \forall u \in V. \tag{3}$$

If this assumption is valid, one can introduce in V a new norm $|||\cdot|||$ by

$$|||u|||^2 = \tilde{\chi}\langle \mathcal{B}u, u\rangle + \|\mathcal{P}u\|_H^2, \tag{4}$$

which is equivalent to the original one $\|\cdot\|$ because of

$$\beta_0\|u\|^2 \leq |||u|||^2 \leq (\beta_1\tilde{\chi} + \beta_2^2)\|u\|^2, \tag{5}$$

with $\beta_1 = \sup_{u\neq 0} \frac{\|\mathcal{B}u\|_{V'}}{\|u\|}$ and $\beta_2 : \|u\|_H \leq \beta_2\|u\| \ \ \forall u \in V$.

Let $\{F_i\}, F_i : V \to \bar{I\!R}$, be a sequence of convex functionals, approximating F, and $\{K_i\}, K_i \subset V$, be a sequence of closed, convex sets, approximating K. Moreover, let $K_i \subset D(\partial F) \cap D(\mathcal{T}^0)$ and F_i be Gâteaux-differentiable on K_i, with ∇F_i hemicontinuous on K_i.

MSR-method (Multi-step regularization)
Let $\{\chi_i\}, \{\epsilon_i\}, \{\delta_i\}$ be positive controlling sequences with

$$\sup_i \ \chi_i \leq \bar{\chi} < \infty, \quad \lim_{i\to\infty} \ \epsilon_i = 0.$$

Step i: *Given u^{i-1}.*

a) Set $u^{i,0} := u^{i-1}$, $s := 1$.

b) Given $u^{i,s-1}$, define $u^{i,s}$ satisfying

$$\|u^{i,s} - \bar{u}^{i,s}\| \leq \epsilon_i, \tag{6}$$

where $\bar{u}^{i,s}$ is the exact solution of the variational inequality

$$u \in K_i : \langle \mathcal{T}^0 u + \nabla F_i(u), v - u\rangle + \chi_i(\mathcal{P}u - \mathcal{P}u^{i,s-1}, \mathcal{P}v - \mathcal{P}u)_H \geq 0 \ \ \forall v \in K_i. \tag{7}$$

c) If $\|\mathcal{P}u^{i,s} - \mathcal{P}u^{i,s-1}\|_H > \delta_i$, then set $s := s + 1$ and repeat b).
Otherwise, set $s(i) := s$, $u^i := u^{i,s}$, $i := i + 1$ and repeat Step i.

The choice $H = V = V_1$ corresponds to the *basic variant of the MSR-method* [14], and $V_1 \neq V$ reflects the (MSR-) *method with regularization on a subspace* (cf. [14], [13], [17]). If $V_1 = V$ and the norm $\|\cdot\|_H$ is weaker than $\|\cdot\|$, we deal with a (MSR-) *method with weak regularization* (cf. [14], [16], [32]). For a diagonal method with regularization on a subspace we refer to [12] and to Rockafellar's interpretation of the multiplier method in [31].

3 Convergence Analysis

According to Riesz's representation theorem, the variational inequality (7) can be rewritten in the form

$$u \in K_i : \langle \mathcal{T}^0 u + \nabla F_i(u) + \chi_i \mathcal{M}(u - u^{i,s-1}), v - u\rangle \geq 0 \quad \forall v \in K_i, \tag{8}$$

where the linear operator $\mathcal{M} : V \to V'$ is defined by

$$\langle \mathcal{M}u, v\rangle = (\mathcal{P}u, \mathcal{P}v)_H \quad \forall u, v \in V.$$

In order to prove convergence of the MSR-method, we need some auxiliary statements.

Lemma 1 *Let $G \subset V$ be a convex, closed set, $\mathcal{A} : V \to V'$ be a single-valued operator with $D(\mathcal{A}) \supset G$, and $f : V \to \bar{\mathbb{R}}$ be a convex functional, Gâteaux-differentiable on G. Suppose that the inequality*

$$\langle \mathcal{A}u - \mathcal{A}v, u - v\rangle + f(v) - f(u) - \langle \nabla f(u), v - u\rangle \geq \langle \mathcal{B}u - \mathcal{B}v, u - v\rangle \quad \forall u, v \in G \tag{9}$$

is valid with $\mathcal{B}$ satisfying Assumption 1 (for given $\tilde{\chi}$).
Moreover, for arbitrary $a^0 \in V$ and $\chi \in (0, 2\tilde{\chi}^{-1}]$, let a^1 be a solution of the variational inequality

$$u \in G : \quad \langle \mathcal{A}u + \nabla f(u), v - u\rangle + \chi(\mathcal{P}u - \mathcal{P}a^0, \mathcal{P}v - \mathcal{P}u)_H \geq 0 \quad \forall v \in G. \tag{10}$$

Then the inequality

$$|||a^1 - v|||^2 - |||a^0 - v|||^2 \leq -\|\mathcal{P}a^1 - \mathcal{P}a^0\|_H^2 + \frac{2}{\chi}[\langle \mathcal{A}v, v - a^1\rangle + f(v) - f(a^1)] \tag{11}$$

is true for each $v \in G$.

Proof Using the definition of $|||\cdot|||$ and a^1, we obtain

$$\begin{aligned} &|||a^1 - v|||^2 - |||a^0 - v|||^2 \\ &\quad = -\|\mathcal{P}a^1 - \mathcal{P}a^0\|_H^2 - 2(\mathcal{P}a^1 - \mathcal{P}a^0, \mathcal{P}v - \mathcal{P}a^1)_H \\ &\quad + \tilde{\chi}\langle \mathcal{B}(a^1 - v), a^1 - v\rangle - \tilde{\chi}\langle \mathcal{B}(a^0 - v), a^0 - v\rangle \end{aligned}$$

$$
\begin{aligned}
&\leq -\|\mathcal{P}a^1 - \mathcal{P}a^0\|_H^2 + \frac{2}{\chi}\langle \mathcal{A}a^1 + \nabla f(a^1), v - a^1\rangle \\
&\quad + \tilde{\chi}\langle \mathcal{B}(a^1 - v), a^1 - v\rangle - \tilde{\chi}\langle \mathcal{B}(a^0 - v), a^0 - v\rangle \\
&= -\|\mathcal{P}a^1 - \mathcal{P}a^0\|_H^2 \\
&\quad + \frac{2}{\chi}\left[\langle \mathcal{A}a^1 - \mathcal{A}v, v - a^1\rangle - f(v) + f(a^1) + \langle \nabla f(a^1), v - a^1\rangle\right] \\
&\quad + \frac{2}{\chi}\left[\langle \mathcal{A}v, v - a^1\rangle + f(v) - f(a^1)\right] \\
&\quad + \tilde{\chi}\langle \mathcal{B}(a^1 - v), a^1 - v\rangle - \tilde{\chi}\langle \mathcal{B}(a^0 - v), a^0 - v\rangle.
\end{aligned}
$$

Due to $0 < \tilde{\chi} \leq 2\chi^{-1}$ and (9), this immediately yields (11). □

Lemma 2 *Let $G \subset V$, $G_1 \subset V$ be convex, closed sets with $intG_1 \neq \emptyset$, and let $\mathcal{A} : V \to 2^{V'}$ be a monotone operator with $D(\mathcal{A}) \supset G$. Assume that a solution $\bar{u}$ of the variational inequality*

$$u \in G \cap G_1 \ : \ \langle \mathcal{A}u, v - u\rangle \geq 0 \ \ \forall v \in G \cap G_1 \tag{12}$$

belongs to $intG_1$. Then $\bar{u}$ is also a solution of the variational inequality

$$u \in G \ : \ \langle \mathcal{A}u, v - u\rangle \geq 0 \ \ \forall v \in G. \tag{13}$$

Proof By definition of $\bar{u}$ there exists a point $y \in \mathcal{A}\bar{u}$ such that

$$\langle y, v - \bar{u}\rangle \geq 0 \ \ \forall v \in G \cap G_1. \tag{14}$$

Let $w \in G \setminus G_1$. Then there exists $\lambda \in (0,1)$ such that $\bar{u} + \lambda(w - \bar{u}) \in G_1$, hence $\bar{u} + \lambda(w - \bar{u}) \in G \cap G_1$. So, we get from (14) that

$$\langle y, \bar{u} + \lambda(w - \bar{u}) - \bar{u}\rangle \geq 0,$$

i.e.,

$$\langle y, w - \bar{u}\rangle \geq 0,$$

proving (13). □

Remark 1 *If, additionally, $\mathcal{A}$ is strictly monotone on $G \cap G_1$, then, of course, $\bar{u}$ is the unique solution of (13).* ◇

Lemma 3 *Let $G \subset V$ be a closed, convex set and $\mathcal{A} : V \to 2^{V'}$ be a monotone operator such that $D(\mathcal{A}) \supset G$. Let $\tilde{\mathcal{A}} = \mathcal{A} + \mathcal{N}_G$ be a maximal monotone operator. Moreover, assume that*

$$\sup_{v \in G} \ \inf_{y \in \mathcal{A}v} \|y\|_{V'} < \infty \tag{15}$$

and that for some $u \in G$ and any $v \in G, y(v) \in \mathcal{A}v$ it holds

$$\langle y(v), v - u\rangle \geq 0. \tag{16}$$

Then, with some $y \in \mathcal{A}u$, the inequality

$$\langle y, v - u \rangle \geq 0$$

is valid for all $v \in G$.

Proof Denote $\mathcal{A}^1 v = \tilde{\mathcal{A}}v + \mathcal{I}(v-u)$, with $\mathcal{I} : V \to V'$ the operator of canonical injection. The operator $\mathcal{A}^1$ is strongly monotone, $D(\mathcal{A}^1) = G$, and due to Theorem 1 in [29], $\mathcal{A}^1$ is maximal monotone. Therefore, there exists $w \in G$ such that $0 \in \mathcal{A}^1 w$. With other words, for some $y_1(w) \in \mathcal{A}w + \mathcal{I}(w-u)$ it holds

$$\langle y_1(w), v - w \rangle \geq 0 \quad \forall v \in G. \tag{17}$$

If $w = u$, then, of course, $y_1(w) \in \mathcal{A}u$, hence, the statement of the lemma is true. Otherwise, we use the relation

$$\langle \bar{y}(v), v - u \rangle \geq 0 \quad \forall v \in G, \tag{18}$$

which follows from (16) with $\bar{y}(v) = y(v) + \mathcal{I}(v-u)$, $y(v)$ an arbitrary element in $\mathcal{A}v$.
Let $w_\lambda = u + \lambda(w-u)$ for $\lambda \in (0,1]$. Obviously, $w_\lambda \in G$, and according to (18),

$$\langle \bar{y}(w_\lambda), w - u \rangle \geq 0 \tag{19}$$

is valid for any $\bar{y}(w_\lambda) \in \mathcal{A}w_\lambda + \mathcal{I}(w_\lambda - u)$.

For the sequel, choose $\bar{y}(w_\lambda) \in \mathcal{A}w_\lambda + \mathcal{I}(w_\lambda - u)$ such that the set $\{\bar{y}(w_\lambda) : \lambda \in (0,1]\}$ is bounded in V'. This is possible due to (15). Then, if λ tends to 0 in a suitable manner, the corresponding sequence $\{\bar{y}(w_\lambda)\}$ converges weakly in V' to some $\bar{y}$, and we obtain from (19) that

$$0 \leq \lim_{\lambda \to 0} \langle \bar{y}(w_\lambda), w - u \rangle = \langle \bar{y}, w - u \rangle. \tag{20}$$

But, since $0 \in \mathcal{N}_G(w_\lambda)$ for each w_λ, the inclusions $\bar{y}(w_\lambda) \in \mathcal{A}^1 w_\lambda$, $y_1(w) \in \mathcal{A}^1 w$ are obvious, and the maximal monotonicity of $\mathcal{A}^1$ together with $w_\lambda \to u$ in V yields $\bar{y} \in \mathcal{A}^1 u$.
Now, (20) and inequality (17), given with $v = u$, lead to

$$\langle \bar{y} - y_1(w), u - w \rangle \leq 0,$$

and, due to $u \neq w$, this is a contradiction to the strong monotonicity of $\mathcal{A}^1$. □

Remark 2 *Assuming that inequality (16) is valid for each $v \in G$ and some $y(v) \in \mathcal{A}v$ (not necessary for each $y(v) \in \mathcal{A}v$), then the statement of Lemma 3 remains true (with evident modifications in the proof) if we replace (15) by*

$$\sup_{v \in G} \sup_{y \in \mathcal{A}v} \|y\|_{V'} < \infty,$$

or by the condition that $\mathcal{A}$ is locally bounded for all $v \in G$. ◇

Throughout the remaining part of the paper we suppose that

- Assumption 1 is fulfilled,
- two radii r^* and r are chosen such that

$$U^* \cap S_{r^*/8} \neq \emptyset \quad \text{and} \quad r \geq r^*, \tag{21}$$

with $S_\tau = \{u \in V : |||u||| \leq \tau\}$,
- the operator $\mathcal{T}^0 + \partial F + \mathcal{N}_Q$ is maximal monotone with $Q = K \cap S_r$,
- the starting point u^0 of the MSR-method belongs to $S_{r^*/4}$.

In view of (21) and $K \subset D(\partial F)$, maximal monotonicity of $\mathcal{T}^0 + \partial F + \mathcal{N}_Q$ is guaranteed if, for instance, $int D(\partial F) \cap K \neq \emptyset$ (see Theorem 1 in [29]).
Denote

$$Q_i = K_i \cap S_r, \quad Q^* = U^* \cap S_{r^*}.$$

To investigate the convergence of the MSR-method for Problem (1), we make the following general

Assumption 2

(a) $\sup_{u \in Q} ||\mathcal{T}^0 u||_{V'} \leq \mu_0;\ \ \sup_{u \in Q} \inf_{y \in \partial F(u)} ||y||_{V'} < \infty;$

(b) $||\mathcal{T}^0 u - \mathcal{T}^0 v||_{V'} \leq d_{\mathcal{T}^0} \sqrt{\beta_0} ||u - v|| \quad \forall u, v \in S_r \cap D(\mathcal{T}^0) \cap D(\partial F);$

(c) $|F(u) - F(v)| \leq d_F ||u - v|| \quad \forall u, v \in S_r \cap D(\mathcal{T}^0) \cap D(\partial F);$

(d) with $\mathcal{B}$ and $\tilde{\chi}$ from Assumption 1,
the relation

$$\langle \mathcal{T}^0 u - \mathcal{T}^0 v, u - v \rangle + F_i(v) - F_i(u) - \langle \nabla F_i(u), v - u \rangle$$
$$\geq \langle \mathcal{B}u - \mathcal{B}v, u - v \rangle \quad \forall u, v \in Q_i$$

is fulfilled for each i,
and $\bar{\chi} \leq 2\tilde{\chi}^{-1}$ is valid
($\bar{\chi}$ is an upper bound for χ_i in the MSR-method);

(e) $\lim_{i \to \infty} e(Q, Q_i) = 0.$

For given sequences $\{\varphi_i\}, \{\sigma_i\}$ such that

$$\lim_{i \to \infty} \varphi_i = \lim_{i \to \infty} \sigma_i = 0$$

it holds

(f) $e(Q^*, Q_i) \leq \varphi_i, \quad e(Q_i, Q) \leq \varphi_i;$

(g) $\sup_{u\in Q_i} |F(u) - F_i(u)| \le \sigma_i$.

Remark 3

(i) In view of the convexity of F_i *and Assumption 2(d), the operator* $\mathcal{T}^0 + \nabla F_i + \chi_i \mathcal{M}$ *is strongly monotone on* Q_i.

(ii) If $V_1 = V = H$*, then Assumption 1 is fulfilled with* $\mathcal{B} = 0$*. Therefore,* $|||u||| = ||u||\ \forall u \in V$*, and Assumption 2(d) follows from the monotonicity of* $\mathcal{T}^0$ *and the convexity of* F_i. ◇

Lemma 4 *Let the Assumptions 2 (a) - (d), (f) and (g) be fulfilled, and assume that, for some counter* i_0*, the relations*

$$\frac{1}{4r}\left[\frac{4}{\chi_{i_0}}\left((\mu_0 + d_{\mathcal{T}^0} r + d_F)\varphi_{i_0} + \sigma_{i_0}\right) - \tilde{\epsilon}_{i_0}^2\right] + \beta_3 \epsilon_{i_0} < 0, \tag{22}$$

$$\tilde{\epsilon}_{i_0} = \delta_{i_0} - \beta_2 \epsilon_{i_0} > 0$$

hold with $\beta_3 = (2\beta_1 \bar{\chi}^{-1} + \beta_2^2)^{1/2}$*. Moreover, assume that in the MSR-method the relations* $s(i) < \infty$ *for* $i < i_0$*, and* $|||u^{i_0,s}||| < r^*$*,* $|||\bar{u}^{i_0,s}||| < r^*$ *for each* s *are valid. Then* $s(i_0) < \infty$ *is true.*

Proof Fix some $u^{**} \in Q^*$. Due to Assumption 2 (f), one can choose $v^{i_0} \in Q_{i_0}$, $v^{i_0,s} \in Q$ such that

$$||v^{i_0,s} - \bar{u}^{i_0,s}|| \le \varphi_{i_0}, \quad ||u^{**} - v^{i_0}|| \le \varphi_{i_0}.$$

Using the identity

$$\begin{aligned}
&\langle \mathcal{T}^0 v^{i_0}, v^{i_0} - \bar{u}^{i_0,s}\rangle + F_{i_0}(v^{i_0}) - F_{i_0}(\bar{u}^{i_0,s}) \\
&\quad = \langle \mathcal{T}^0 u^{**}, u^{**} - v^{i_0,s}\rangle + \langle \mathcal{T}^0 u^{**}, v^{i_0} - u^{**}\rangle + \langle \mathcal{T}^0 u^{**}, v^{i_0,s} - \bar{u}^{i_0,s}\rangle \\
&\quad + \langle \mathcal{T}^0 v^{i_0} - \mathcal{T}^0 u^{**}, v^{i_0} - \bar{u}^{i_0,s}\rangle + F(u^{**}) - F(v^{i_0,s}) - F(u^{**}) \\
&\quad + F(v^{i_0}) + F(v^{i_0,s}) - F(\bar{u}^{i_0,s}) + F_{i_0}(v^{i_0}) - F(v^{i_0}) + F(\bar{u}^{i_0,s}) - F_{i_0}(\bar{u}^{i_0,s}),
\end{aligned}$$

Assumptions 2 (a) - (c), (f) and (g) lead to

$$\begin{aligned}
&\langle \mathcal{T}^0 v^{i_0}, v^{i_0} - \bar{u}^{i_0,s}\rangle + F_{i_0}(v^{i_0}) - F_{i_0}(\bar{u}^{i_0,s}) \\
&\quad \le \langle \mathcal{T}^0 u^{**}, u^{**} - v^{i_0,s}\rangle + F(u^{**}) - F(v^{i_0,s}) \\
&\quad + 2\mu_0 \varphi_{i_0} + 2 d_{\mathcal{T}^0} r \varphi_{i_0} + 2 d_F \varphi_{i_0} + 2\sigma_{i_0}.
\end{aligned}$$

But, due to to the convexity of F, the inclusion $v^{i_0,s} \in Q$ and the definition of u^{**},

$$\langle \mathcal{T}^0 u^{**}, u^{**} - v^{i_0,s}\rangle + F(u^{**}) - F(v^{i_0,s}) \le 0$$

is valid, hence,

$$\langle \mathcal{T}^0 v^{i_0}, v^{i_0} - \bar{u}^{i_0,s}\rangle + F_{i_0}(v^{i_0}) - F_{i_0}(\bar{u}^{i_0,s}) \leq 2\left(\mu_0 + d_{\mathcal{T}^0} r + d_F\right)\varphi_{i_0} + 2\sigma_{i_0}. \quad (23)$$

Now, from Assumption 2(d) and Lemma 1 with

$$\mathcal{A} = \mathcal{T}^0,\ f = F_{i_0},\ G = K_{i_0},\ \chi = \chi_{i_0},\ v = v^{i_0} \text{ and } a^0 = u^{i_0,s-1},$$

in view of (23), we obtain

$$\begin{aligned} &|||\bar{u}^{i_0,s} - v^{i_0}|||^2 - |||u^{i_0,s-1} - v^{i_0}|||^2 \\ &\quad \leq -\|\mathcal{P}\bar{u}^{i_0,s} - \mathcal{P}u^{i_0,s-1}\|_H^2 + \frac{2}{\chi_{i_0}}\left[\langle \mathcal{T}^0 v^{i_0}, v^{i_0} - \bar{u}^{i_0,s}\rangle + F_{i_0}(v^{i_0}) - F_{i_0}(\bar{u}^{i_0,s})\right] \\ &\quad \leq -\|\mathcal{P}\bar{u}^{i_0,s} - \mathcal{P}u^{i_0,s-1}\|_H^2 + \frac{4}{\chi_{i_0}}\left[(\mu_0 + d_{\mathcal{T}^0} r + d_F)\varphi_{i_0} + \sigma_{i_0}\right]. \end{aligned} \quad (24)$$

Taking into account substep step c) of the method, (6) and $\|\mathcal{P}u\|_H \leq \beta_2\|u\|\ \forall u \in V$, as well as the second inequality in (22), then the inequality

$$\begin{aligned} \|\mathcal{P}\bar{u}^{i_0,s} - \mathcal{P}u^{i_0,s-1}\|_H &\geq \|\mathcal{P}u^{i_0,s} - \mathcal{P}u^{i_0,s-1}\|_H - \|\mathcal{P}u^{i_0,s} - \mathcal{P}\bar{u}^{i_0,s}\|_H \\ &\geq \|\mathcal{P}u^{i_0,s} - \mathcal{P}u^{i_0,s-1}\|_H - \beta_2\|u^{i_0,s} - \bar{u}^{i_0,s}\| > \delta_{i_0} - \beta_2\epsilon_{i_0} > 0 \end{aligned}$$

is true for $1 \leq s < s(i_0)$. Together with (24) this yields

$$\begin{aligned} &|||\bar{u}^{i_0,s} - v^{i_0}|||^2 - |||u^{i_0,s-1} - v^{i_0}|||^2 \\ &\quad < -\tilde{\epsilon}_{i_0}^2 + \frac{4}{\chi_{i_0}}\left[(\mu_0 + d_{\mathcal{T}^0} r + d_F)\varphi_{i_0} + \sigma_{i_0}\right], \end{aligned} \quad (25)$$

and using the first inequality in (22), we have

$$|||\bar{u}^{i_0,s} - v^{i_0}||| < |||u^{i_0,s-1} - v^{i_0}|||. \quad (26)$$

Moreover, due to $|||u^{i_0,s}||| < r^*$, $|||\bar{u}^{i_0,s}||| < r^*$, $|||v^{i_0}||| \leq r$, $\tilde{\chi} \leq 2\underline{\chi}^{-1}$ and the relations (5), (6), (25) and (26), we obtain for $1 \leq s < s(i_0)$ that

$$\begin{aligned} &|||u^{i_0,s} - v^{i_0}||| - |||u^{i_0,s-1} - v^{i_0}||| \\ &\quad < \frac{1}{4r}\left[\frac{4}{\chi_{i_0}}\left[(\mu_0 + d_{\mathcal{T}^0} r + d_F)\varphi_{i_0} + \sigma_{i_0}\right] - \tilde{\epsilon}_{i_0}^2\right] + \beta_3\epsilon_{i_0} < 0. \end{aligned} \quad (27)$$

Summing up the inequalities (27) w.r.t. $s = 1, ..., \bar{s}$, with $\bar{s} < s(i_0)$ arbitrarily chosen, one gets

$$\begin{aligned} &|||u^{i_0,\bar{s}} - v^{i_0}||| < |||u^{i_0,0} - v^{i_0}||| \\ &\quad + \bar{s}\left[\frac{1}{4r}\left(\frac{4}{\chi_{i_0}}\left[(\mu_0 + d_{\mathcal{T}^0} r + d_F)\varphi_{i_0} + \sigma_{i_0}\right] - \tilde{\epsilon}_{i_0}^2\right) + \beta_3\epsilon_{i_0}\right], \end{aligned}$$

therefore,

$$\bar{s} < |||u^{i_0,0} - v^{i_0}||| \left[\frac{1}{4r} \left[\tilde{\epsilon}_{i_0}^2 - \frac{4}{\chi_{i_0}} ((\mu_0 + d_{\mathcal{T}} \circ r + d_F)\varphi_{i_0} + \sigma_{i_0}) \right] - \beta_3 \epsilon_{i_0} \right]^{-1},$$

and

$$s(i_0) < 2r \left[\frac{1}{4r} \left[\tilde{\epsilon}_{i_0}^2 - \frac{4}{\chi_{i_0}} ((\mu_0 + d_{\mathcal{T}} \circ r + d_F)\varphi_{i_0} + \sigma_{i_0}) \right] - \beta_3 \epsilon_{i_0} \right]^{-1} + 1 \quad (28)$$

proves the lemma. □

Corollary 1 *Let the Assumptions 2 (a) - (d), (f), (g) and (for each i) the relations (22) be satisfied. Moreover, assume that the inequalities $|||\bar{u}^{i,s}||| < r^*$, $|||u^{i,s}||| < r^*$ are fulfilled in each step. Then the relation $s(i) < \infty$ holds for each i, i.e. the inner cycles are finite.*

Lemma 5 *Let the Assumptions 2 (a)-(d), (f) and (g) be fulfilled, and suppose that the controlling parameters $\epsilon_i, \chi_i, \delta_i, \sigma_i$ and φ_i satisfy the conditions*

$$\frac{1}{4r} \left[\frac{4}{\chi_i} ((\mu_0 + d_{\mathcal{T}} \circ r + d_F)\varphi_i + \sigma_i) - (\delta_i - \beta_2 \epsilon_i)^2 \right] + \beta_3 \epsilon_i < 0 \quad (29)$$

and

$$\sum_{i=1}^{\infty} \left[2 \left(\frac{(\mu_0 + d_{\mathcal{T}} \circ r + d_F)\varphi_i + \sigma_i}{\chi_i} \right)^{1/2} + \beta_3(\epsilon_i + 2\varphi_i) \right] < \frac{1}{2} r^*. \quad (30)$$

Then, in the MSR-method $s(i) < \infty$ is valid for each i, variational inequality (7) is uniquely solvable for all (i, s), and $|||u^{i,s}||| < r^*$, $|||\bar{u}^{i,s}||| < r^*$ are true.

Proof The inequality $\delta_i - \beta_2 \epsilon_i > 0$ follows immediately from (29), (30). Indeed, (30) yields $\beta_3 \epsilon_i < \frac{1}{2} r^*$, and due to (29), $\beta_3 \epsilon_i < \frac{1}{4r} (\delta_i - \beta_2 \epsilon_i)^2$. If $\delta_i \leq \beta_2 \epsilon_i$, then $(\delta_i - \beta_2 \epsilon_i)^2 \leq (\beta_2 \epsilon_i)^2$, hence,

$$\beta_3 \epsilon_i < \frac{1}{4r} (\beta_2 \epsilon_i)^2.$$

In view of $\beta_3 = \left(2\beta_1 \bar{\chi}^{-1} + \beta_2^2\right)^{1/2} \geq \beta_2$, we have $\beta_2 \epsilon_i \leq \beta_3 \epsilon_i < \frac{1}{2} r^*$, therefore

$$\beta_3 \epsilon_i < \frac{1}{4r} (\beta_2 \epsilon_i)^2 < \frac{r^*}{8r} \beta_2 \epsilon_i.$$

But this contradicts $\beta_3 \geq \beta_2, r^* < r$.
Suppose that i_0 and s_0 are kept fixed, $s(i) < \infty$ for $i < i_0$, and let $0 \leq s_0 < s(i_0)$. Denote

$$\Theta_0 = \{(i', s') \ : \ i' < i_0, \ 0 < s' \leq s(i') \text{ and } \ i' = i_0, \ 0 < s' \leq s_0\}.$$

Assume, moreover, that for all $(i,s) \in \Theta_0$ the variational inequality (7) is uniquely solvable[1] and the inequalities $|||u^{i,s}||| < r^*$, $|||\bar{u}^{i,s}||| < r^*$ are true.
For fixed $u^{**} \in S_{r^*/8} \cap U^*$ and each i, due to Assumption 2 (f), one can choose $v^i \in Q_i$ such that

$$\|v^i - u^{**}\| \leq \varphi_i. \tag{31}$$

Then, as in the proof of Lemma 4, we obtain (cf. (26) - (27)) for $i < i_0$, $0 \leq s < s(i) - 1$ and $i = i_0$, $0 \leq s < s_0$ that

$$|||u^{i,s+1} - v^i||| - |||u^{i,s} - v^i||| \leq \frac{1}{4r}\left(\tau_i - \tilde{\epsilon}_i^2\right) + \beta_3\epsilon_i < 0, \tag{32}$$

with $\tau_i = \frac{4}{\chi_i}\left[(\mu_0 + d_{\mathcal{T}^0} r + d_F)\varphi_i + \sigma_i\right]$, $\tilde{\epsilon}_i = \delta_i - \beta_2\epsilon_i$.
Also for $i < i_0$, $s = s(i) - 1$, similarly to (24), one can conclude that

$$|||\bar{u}^{i,s(i)} - v^i|||^2 - |||u^{i,s(i)-1} - v^i|||^2 \leq \tau_i,$$

and regarding (5) and (6),

$$|||u^{i,s(i)} - v^i||| - |||u^{i,s(i)-1} - v^i||| \leq \sqrt{\tau_i} + \beta_3\epsilon_i. \tag{33}$$

Summing up the inequalities (32), where $i < i_0$ is fixed and $0 \leq s < s(i) - 1$, together with (33), we get

$$|||u^{i,s(i)} - v^i||| - |||u^{i,0} - v^i||| \leq \sqrt{\tau_i} + \beta_3\epsilon_i.$$

Due to (5) and (31), this yields

$$|||u^{i,s(i)} - u^{**}||| - |||u^{i,0} - u^{**}||| \leq \sqrt{\tau_i} + \beta_3(\epsilon_i + 2\varphi_i). \tag{34}$$

By Theorem 5 in [29] and Remark 3 (i) the variational inequality

$$u \in Q_{i_0} \ : \ \langle \mathcal{T}^0 u + \nabla F_{i_0}(u), v - u\rangle + \chi_{i_0}\left(\mathcal{P}u - \mathcal{P}u^{i_0,s_0}, \mathcal{P}v - \mathcal{P}u\right)_H \geq 0 \ \ \forall v \in Q_{i_0} \tag{35}$$

is uniquely solvable. We denote its solution by $\hat{u}^{i_0,s_0+1}$. The use of Lemma 1 leads to

$$|||\hat{u}^{i_0,s_0+1} - v^{i_0}||| \leq |||u^{i_0,s_0} - v^{i_0}||| + \sqrt{\tau_{i_0}}. \tag{36}$$

Observing (31), (32) (taken with $i = i_0$, $0 \leq s < s_0$), and (36), the inequality

$$|||\hat{u}^{i_0,s_0+1} - u^{**}||| \leq |||u^{i_0,0} - u^{**}||| + \sqrt{\tau_{i_0}} + 2\beta_3\varphi_{i_0}. \tag{37}$$

can be obtained similarly to (34). Regarding that $u^{i+1,0} := u^{i,s(i)}$ (cf. Step c) of the MSR-method), the inequalities (34) and (37) imply

$$|||\hat{u}^{i_0,s_0+1} - u^{**}||| \leq |||u^{1,0} - u^{**}||| + \sum_{k=1}^{i_0-1}\left(\sqrt{\tau_k} + \beta_3(\epsilon_k + 2\varphi_k)\right) + \sqrt{\tau_{i_0}} + 2\beta_3\varphi_{i_0}. \tag{38}$$

[1]This can be proved directly. We conclude solvability of (7) from the solvability of (35) below.

Due to (30) and the choice of u^{**} and $u^{1,0}$, estimate (38) yields

$$|||\hat{u}^{i_0,s_0+1}||| < r^* - \beta_3 \epsilon_{i_0}. \tag{39}$$

So, we are in the situation of Lemma 2 and Remark 1, and hence, the variational inequality (7) with $i = i_0$ and $s = s_0 + 1$ is uniquely solvable, $\bar{u}^{i_0,s_0+1} = \hat{u}^{i_0,s_0+1}$, and

$$|||u^{i_0,s_0+1}||| < r^*. \tag{40}$$

This enables to conclude that

$$|||\bar{u}^{i_0,s}||| < r^*, \ |||u^{i_0,s}||| < r^* \text{ for all } s,$$

and Lemma 4 ensures that $s(i_0) < \infty$. To complete the induction, note that the unique solvability of (7) with $i = 1, s = 1$, as well as the inequalities $|||\bar{u}^{1,1}||| < r^*$, $|||u^{1,1}||| < r^*$ and the finiteness of $s(1)$, can be established quite analogously. □

For further considerations it should be noted that, due to (30) and (38), the relation $\sup_{(i,s)} |||\bar{u}^{i,s}||| < r^*$ is also true.

Theorem 1 *Let Assumption 2 be fulfilled, and suppose that the points $u^{i,s}$, generated by the MSR-method for Problem (1), as well as the points $\bar{u}^{i,s}$ which solve the variational inequalities (7), satisfy the relations*

$$|||\bar{u}^{i,s}||| < r^* \text{ and } |||u^{i,s}||| < r^* \text{ for all } i, s.$$

Moreover, assume that the controlling parameters satisfy (29) and

$$\sum_{i=1}^{\infty} \sqrt{\frac{\sigma_i}{\chi_i}} < \infty, \quad \sum_{i=1}^{\infty} \sqrt{\frac{\varphi_i}{\chi_i}} < \infty, \quad \sum_{i=1}^{\infty} \epsilon_i < \infty. \tag{41}$$

Then, $s(i) < \infty$ holds for each i and the sequences $\{u^{i,s}\}$, $\{\bar{u}^{i,s}\}$ converge weakly in V to an element $u^* \in U^*$.

Proof Corollary 1 ensures that $s(i) < \infty$ for each i. With an arbitrary $w \in U^* \cap S_{r^*}$, the inequality

$$|||u^{i+1,0} - w||| - |||u^{i,0} - w||| \le \sqrt{\tau_i} + \beta_3(\epsilon_i + 2\varphi_i)$$

can be established similarly to (34). Therefore, condition (41) and Lemma 2.2.2 in [28] imply the convergence of the sequence $\{|||u^{i,0} - w|||\}$.
Note that the inequalities (32), (33) remain true with a "new" $v^i \in Q_i$ which is chosen according to Assumption 2 (f) such that

$$\|v^i - w\| \le \varphi_i.$$

For $1 \le s < s(i)$, this leads to the relation

$$-(\sqrt{\tau_i} + \beta_3(\epsilon_i + 2\varphi_i)) + |||u^{i+1,0} - w||| \le |||u^{i,s} - w||| < |||u^{i,0} - w||| + 2\beta_3\varphi_i. \tag{42}$$

Taking into account the convergence of the sequence $\{|||u^{i,0} - w|||\}$ and conditions (41) and (5), (6), the sequences $\{|||u^{i,s} - w|||\}$ and $\{|||\bar{u}^{i,s} - w|||\}$ converge to the same limit as $\{|||u^{i,0} - w|||\}$.
Now, from inequality (24), which can be extended to each i and $1 \le s \le s(i)$, one can conclude that

$$\|\mathcal{P}\bar{u}^{i,s} - \mathcal{P}u^{i,s-1}\|_H^2 \le 4r\left(\left|-|||\bar{u}^{i,s} - u^{**}||| + |||u^{i,s-1} - u^{**}|||\right| + 2\beta_3\varphi_i\right) + \tau_i.$$

Therefore,

$$\lim_{i\to\infty} \max_{1\le s\le s(i)} \|\mathcal{P}\bar{u}^{i,s} - \mathcal{P}u^{i,s-1}\|_H = 0. \tag{43}$$

Using the definition of $\bar{u}^{i,s}$ and that $|||\bar{u}^{i,s}||| < r^*$, we get

$$\langle \mathcal{T}^0\bar{u}^{i,s} + \nabla F_i(\bar{u}^{i,s}), v - \bar{u}^{i,s}\rangle + \chi_i\left(\mathcal{P}\bar{u}^{i,s} - \mathcal{P}u^{i,s-1}, \mathcal{P}v - \mathcal{P}\bar{u}^{i,s}\right)_H \ge 0 \;\; \forall v \in Q_i, \tag{44}$$

and hence,

$$\langle \mathcal{T}^0 v, v - \bar{u}^{i,s}\rangle + F_i(v) - F_i(\bar{u}^{i,s}) + \chi_i\left(\mathcal{P}\bar{u}^{i,s} - \mathcal{P}u^{i,s-1}, \mathcal{P}v - \mathcal{P}\bar{u}^{i,s}\right)_H \ge 0 \;\; \forall v \in Q_i. \tag{45}$$

For an arbitrary $z \in Q$, due to Assumption 2 (e), there exists $v^i(z) \in Q_i$ such that $\lim_{i\to\infty} \|v^i(z) - z\| = 0$. With these elements z and $v^i(z)$ we obtain

$$\begin{aligned}
&\langle \mathcal{T}^0 z, z - \bar{u}^{i,s}\rangle + F(z) - F(\bar{u}^{i,s}) + \chi_i\left(\mathcal{P}\bar{u}^{i,s} - \mathcal{P}u^{i,s-1}, \mathcal{P}z - \mathcal{P}\bar{u}^{i,s}\right)_H \\
&\quad = \langle \mathcal{T}^0 z, z - v^i(z)\rangle + F(z) - F(v^i(z)) + \chi_i\left(\mathcal{P}\bar{u}^{i,s} - \mathcal{P}u^{i,s-1}, \mathcal{P}z - \mathcal{P}v^i(z)\right)_H \\
&\quad + \langle \mathcal{T}^0 z - \mathcal{T}^0 v^i(z), v^i(z) - \bar{u}^{i,s}\rangle + F(v^i(z)) - F_i(v^i(z)) - F(\bar{u}^{i,s}) + F_i(\bar{u}^{i,s}) \\
&\quad + \big[\langle \mathcal{T}^0 v^i(z), v^i(z) - \bar{u}^{i,s}\rangle + F_i(v^i(z)) - F_i(\bar{u}^{i,s}) \\
&\quad + \chi_i\left(\mathcal{P}\bar{u}^{i,s} - \mathcal{P}u^{i,s-1}, \mathcal{P}v^i(z) - \mathcal{P}\bar{u}^{i,s}\right)_H\big],
\end{aligned} \tag{46}$$

and, due to (45), the term in the square brackets is nonnegative.
Let $\{\bar{u}^{i,s}\}_{(i,s)\in\Gamma}$ be a subsequence of $\{\bar{u}^{i,s}\}$, which converges weakly in V to some u^*. Observing that $\sup_{(i,s)} |||\bar{u}^{i,s}||| < r^*$, one can show that $u^* \in K \cap int S_{r^*}$. Take the limit inferior in (46) w.r.t. $(i,s) \in \Gamma$. Then, using (43), the relation $\lim_{i\to\infty} \|v^i(z) - z\| = 0$, the Assumptions 2 (b), (c), (g), the boundedness of $\{\chi_i\}$ and the weak lower semi-continuity of F (which follows from the convexity and the lower semi-continuity of F), we obtain

$$\langle \mathcal{T}^0 z, z - u^*\rangle + F(z) - F(u^*) \ge 0,$$

therefore,

$$\langle \mathcal{T}^0 z, z - u^*\rangle + \langle \zeta, z - u^*\rangle \ge 0 \;\; \forall \zeta \in \partial F(z). \tag{47}$$

Taking into account (47), Assumption 2 (b) as well as the second Assumption 2 (a), Lemma 3 (given with $G = Q$, $\mathcal{A} = \mathcal{T}^0 + \partial F$) permits to conclude that, for some $y^* \in \partial F(u^*)$,

$$\langle \mathcal{T}^0 u^*, z - u^*\rangle + \langle y^*, z - u^*\rangle \ge 0 \;\; \forall z \in Q.$$

Because $u^* \in int S_{r^*}$, Lemma 2 (with $G_1 = S_{r*}$, $G = K$ and $\mathcal{A} = \mathcal{T}^0 + \partial F$) guarantees that $u^* \in U^*$. Now, due to Lemma 1 in [27], both sequences $\{\bar{u}^{i,s}\}$ and $\{u^{i,s}\}$ converge weakly in V to u^*. □

Remark 4 *Theorem 1 establishes convergence of the MSR-method for a bounded set K. Indeed, if the radius r^* is chosen such that $K \subset S_\tau$ holds with $\tau < r^* - \beta_3(\sup_i \varphi_i + \sup_i \epsilon_i)$, then, due to Assumption 2 (f) and inequality (6), we infer that*

$$|||\bar{u}^{i,s}||| < r^*, \quad |||u^{i,s}||| < r^* \quad \forall (i, s).$$

◇

If boundedness of K is not supposed, compiling Lemma 5 and Theorem 1, the following statement on convergence of the MSR-method can be obtained immediately.

Theorem 2 *Let Assumption 2 as well as the conditions (29), (30) be fulfilled. Then, the MSR-method for Problem (1) has the following properties:*

(i) $s(i) < \infty$ *for all* i;

(ii) $|||u^{i,s}||| < r^*, \quad |||\bar{u}^{i,s}||| < r^*$ *for each* (i, s);

(iii) both sequences $\{u^{i,s}\}$ *and* $\{\bar{u}^{i,s}\}$ *converge weakly in* V *to an element* $u^* \in U^*$.

Remark 5 *If δ_i is chosen sufficiently large (for instance, $\delta_i > 2\beta_0^{-1/2}\beta_2 r^*$ for all i), then from Theorem 2 (also from Remark 4) it follows that $s(i) = 1$ for each i. This means that the MSR-method passes over to the usual diagonal method. Obviously, in this case condition (29) can be omitted in Lemma 5.* ◇

Theorem 3 *Let the conditions of Theorem 1 or Theorem 2 be fulfilled and let*

$$\lim_{i\to\infty} \max_{1\le s\le s(i)} \|\mathcal{P}u^{i,s} - \mathcal{P}u^*\|_H = 0, \tag{48}$$

with u^* a weak limit of $\{u^{i,s}\}$. Then both sequences $\{u^{i,s}\}$ and $\{\bar{u}^{i,s}\}$ converge to u^* (strongly) in V.

Proof Due to (6) and the continuous embedding of V into H, relation (48) leads to

$$\lim_{i\to\infty} \max_{1\le s\le s(i)} \|\mathcal{P}\bar{u}^{i,s} - \mathcal{P}u^*\|_H = 0. \tag{49}$$

Assumption 2 (f) permits to choose $v^i \in Q_i$ and $v^{i,s} \in Q$ such that

$$\|u^* - v^i\| \le \varphi_i \quad \text{for } i = 1, 2, ..., \tag{50}$$

and

$$\|\bar{u}^{i,s} - v^{i,s}\| \le \varphi_i \quad \text{for } 1 \le s \le s(i), \; i = 1, 2, ... \tag{51}$$

On account of Assumption 2 (d) and the symmetry of the operator $\mathcal{B}$, we get

$$\begin{aligned}\langle \mathcal{B}\bar{u}^{i,s} - \mathcal{B}u^*, \bar{u}^{i,s} - u^*\rangle &\leq \langle \mathcal{T}^0\bar{u}^{i,s} - \mathcal{T}^0 v^i, \bar{u}^{i,s} - v^i\rangle \\ &+ F_i(v^i) - F_i(\bar{u}^{i,s}) - \langle \nabla F_i(\bar{u}^{i,s}), v^i - \bar{u}^{i,s}\rangle \\ &- \langle \mathcal{B}\bar{u}^{i,s} - \mathcal{B}u^*, u^* - v^i\rangle - \langle \mathcal{B}\bar{u}^{i,s} - \mathcal{B}v^i, u^* - v^i\rangle,\end{aligned}$$

and using (44), with $v = v^i$, this implies

$$\begin{aligned}\langle \mathcal{B}\bar{u}^{i,s} - \mathcal{B}u^*, \bar{u}^{i,s} - u^*\rangle &\leq \langle \mathcal{T}^0 v^i, v^i - \bar{u}^{i,s}\rangle + F_i(v^i) - F_i(\bar{u}^{i,s}) \\ &+ \chi_i \left(\mathcal{P}\bar{u}^{i,s} - \mathcal{P}u^{i,s-1}, \mathcal{P}v^i - \mathcal{P}\bar{u}^{i,s}\right)_H + \langle \mathcal{B}u^* + \mathcal{B}v^i - 2\mathcal{B}\bar{u}^{i,s}, u^* - v^i\rangle \\ &= \langle \mathcal{T}^0 v^i - \mathcal{T}^0 u^*, v^i - \bar{u}^{i,s}\rangle + \langle \mathcal{T}^0 u^*, v^i - u^*\rangle + \langle \mathcal{T}^0 u^*, v^{i,s} - \bar{u}^{i,s}\rangle \\ &+ F_i(v^i) - F(v^i) - F_i(\bar{u}^{i,s}) + F(\bar{u}^{i,s}) + F(v^i) - F(u^*) - F(\bar{u}^{i,s}) + F(v^{i,s}) \\ &+ \langle \mathcal{T}^0 u^*, u^* - v^{i,s}\rangle + F(u^*) - F(v^{i,s}) + \chi_i \left(\mathcal{P}\bar{u}^{i,s} - \mathcal{P}u^{i,s-1}, \mathcal{P}v^i - \mathcal{P}\bar{u}^{i,s}\right)_H \\ &+ \langle \mathcal{B}u^* + \mathcal{B}v^i - 2\mathcal{B}\bar{u}^{i,s}, u^* - v^i\rangle. \end{aligned} \tag{52}$$

But

$$\langle \mathcal{T}^0 u^*, u^* - v^{i,s}\rangle + F(u^*) - F(v^{i,s}) \leq 0,$$

because u^* is a solution of Problem (1) and $v^{i,s} \in Q$. Combining this inequality with the relations (49) - (52) and (30) or (41), and using the Assumptions 2 (b), (c) , (g) and the monotonicity of B, we obtain

$$\lim_{i\to\infty} \max_{1\leq s\leq s(i)} \langle \mathcal{B}\bar{u}^{i,s} - \mathcal{B}u^*, \bar{u}^{i,s} - u^*\rangle = 0.$$

The last relation together with (49) leads to

$$\lim_{i\to\infty} \max_{1\leq s\leq s(i)} |||\bar{u}^{i,s} - u^*||| = 0,$$

and all we have to do is to use (5) and (6). □

4 Rate of Convergence

In this section we investigate the rate of convergence of the basic variant of the MSR-method that corresponds to $H = V = V_1, \mathcal{B} = 0$, and hence, the norms $|||\cdot|||$ and $||\cdot||$ coincide.
Let $\delta \in (0, 2r)$, $l > 0$ be fixed, and denote $U_\delta = \{u \in Q : dist(u, Q^*) \leq \delta\}$. We need the following additional

Assumption 3 *There exists a constant $d_0 > 0$ such that, for each $u \in U_\delta$ the inequality*

$$\inf_{v\in Q^*} \left[\langle \mathcal{T}^0 u, u - v\rangle + F(u) - F(v)\right] \geq d_0 ||u - u^*(u)||^l \tag{53}$$

is valid with $u^*(u) = arg\min_{w\in Q^*} ||u - w||$.

Lemma 6 *Let Assumption 3 be fulfilled. Then, for each $u \in Q$ the inequality*

$$\inf_{v \in Q^*} \left[\langle \mathcal{T}^0 u, u - v \rangle + F(u) - F(v) \right] \geq d \|u - u^*(u)\|^l \tag{54}$$

holds with $d = (\frac{\delta}{2r^*})^l d_0$.

The proof of this lemma is quite similar to the proof of Lemma 6 in [18].

Now, let

$$\begin{aligned}
\rho_{i,s} &= dist^2\left(u^{i,s}, Q^*\right), \\
\gamma_i &= 4r(\epsilon_i + \varphi_i) + \frac{4}{\chi_i}\left[(\mu_0 + d_{\mathcal{T}^0} r + d_F + 2dr)\varphi_i + \sigma_i + 2dr\epsilon_i\right], \\
c_i &= \rho_{1,0} \prod_{k=1}^{i} \left(1 + \frac{2d}{\chi_k}\right)^{-s(k)+\frac{1}{2}}, \quad c_0 = \rho_{1,0},
\end{aligned}$$

and we define by $\bar{s}(i)$ the largest integer of the value $\sqrt{c_{i-1}}\varphi_i^{-1} + 2$.

Theorem 4 *Let the Assumptions 2 (a) - (c), (f), (g) be fulfilled. Moreover, suppose that the controlling parameters of the MSR-method satisfy (30).*

(i) If Assumption 3 with $l = 2$ is valid, as well as, for each i,

$$\gamma_i \leq c_{i-1} \left(1 + \frac{2d}{\chi_i}\right)^{-\bar{s}(i)+1} \left[\left(1 + \frac{2d}{\chi_i}\right)^{1/2} - 1\right] \tag{55}$$

and

$$\begin{aligned}
-(\delta_i - \epsilon_i)^2 &+ \frac{4}{\chi_i}\left[(\mu_0 + d_{\mathcal{T}^0} r + d_F + 2dr)\varphi_i + \sigma_i\right] \\
&+ 4r\epsilon_i + \frac{8d}{\chi_i} r\epsilon_i + \left(4 + \frac{8d}{\chi_i}\right) r\varphi_i < 0,
\end{aligned} \tag{56}$$

then, for each i and $0 \leq s < s(i)$, the estimate

$$\begin{aligned}
\rho_{i,s} &\leq c_{i-1} \left(1 + \frac{2d}{\chi_i}\right)^{-s} \\
&= \rho_{1,0} \left(1 + \frac{2d}{\chi_i}\right)^{-s} \prod_{k=1}^{i-1} \left(1 + \frac{2d}{\chi_k}\right)^{-s(k)+\frac{1}{2}}
\end{aligned} \tag{57}$$

holds true.

(ii) If Assumption 3 with $l = 1$ and relation (29) are valid, then there exists i_0 such that the estimate

$$dist(u^{i,s}, Q^*) \le \left(3 + \frac{4r\bar{\chi}}{d}\right)\epsilon_i \\ +2\left(1 + \frac{2(\mu_0 + d_{\mathcal{T}^0}r + d_F + r\bar{\chi})}{d}\right)\varphi_i + \frac{4}{d}\sigma_i \tag{58}$$

holds for $i \ge i_0$ and $1 \le s \le s(i)$.

Sketch of the proof We make use of the inequality

$$\|\bar{u}^{i,s} - v\|^2 - \|u^{i,s-1} - v\|^2 \\ \le -\|\bar{u}^{i,s} - u^{i,s-1}\|^2 + \frac{2}{\chi_i}\langle\mathcal{T}^0\bar{u}^{i,s} + \nabla F_i(\bar{u}^{i,s}), v - \bar{u}^{i,s}\rangle, \tag{59}$$

which follows for $v \in K_i$ from the definition of $\bar{u}^{i,s}$.
Choose $v^{i,s} \in Q$, $q^{i,s-1} \in Q_i$ such that

$$\|v^{i,s} - \bar{u}^{i,s}\| \le \varphi_i, \qquad \|q^{i,s-1} - u^*(u^{i,s-1})\| \le \varphi_i.$$

Estimating $\langle\mathcal{T}^0\bar{u}^{i,s} + \nabla F_i(\bar{u}^{i,s}), q^{i,s-1} - \bar{u}^{i,s}\rangle$, from the Assumptions 2 (a) - (c), (f) and (g) we obtain

$$\langle\mathcal{T}^0\bar{u}^{i,s} + \nabla F_i(\bar{u}^{i,s}), q^{i,s-1} - \bar{u}^{i,s}\rangle \\ \le \langle\mathcal{T}^0\bar{u}^{i,s} - \mathcal{T}^0 v^{i,s}, q^{i,s-1} - \bar{u}^{i,s}\rangle + \langle\mathcal{T}^0 v^{i,s}, q^{i,s-1} - u^*(u^{i,s-1})\rangle \\ + \langle\mathcal{T}^0 v^{i,s}, u^*(u^{i,s-1}) - v^{i,s}\rangle + \langle\mathcal{T}^0 v^{i,s}, v^{i,s} - \bar{u}^{i,s}\rangle \\ + F_i(q^{i,s-1}) - F(q^{i,s-1}) - F_i(\bar{u}^{i,s}) + F(\bar{u}^{i,s}) + F(q^{i,s-1}) - F(u^*(u^{i,s-1})) \\ + F(v^{i,s}) - F(\bar{u}^{i,s}) + F(u^*(u^{i,s-1})) - F(v^{i,s}) \\ \le \langle\mathcal{T}^0 v^{i,s}, u^*(u^{i,s-1}) - v^{i,s}\rangle + F(u^*(u^{i,s-1})) - F(v^{i,s}) \\ + 2\left((\mu_0 + d_{\mathcal{T}^0}r + d_F)\varphi_i + \sigma_i\right). \tag{60}$$

Thus, (60) and (59) (given with $v = q^{i,s-1}$) yield

$$\|\bar{u}^{i,s} - q^{i,s-1}\|^2 - \|u^{i,s-1} - q^{i,s-1}\|^2 \le -\|\bar{u}^{i,s} - u^{i,s-1}\|^2 \\ +\frac{2}{\chi_i}\left[\langle\mathcal{T}^0 v^{i,s}, u^*(u^{i,s-1}) - v^{i,s}\rangle + F(u^*(u^{i,s-1})) - F(v^{i,s})\right] \\ +\frac{4}{\chi_i}\left((\mu_0 + d_{\mathcal{T}^0}r + d_F)\varphi_i + \sigma_i\right). \tag{61}$$

Using Assumption 3 and Lemma 6, we conclude from (61) that

$$\|\bar{u}^{i,s} - q^{i,s-1}\|^2 - \|u^{i,s-1} - q^{i,s-1}\|^2 \le -\|\bar{u}^{i,s} - u^{i,s-1}\|^2 \\ -\frac{2d}{\chi_i}\|v^{i,s} - u^*(u^{i,s-1})\|^l + \frac{4}{\chi_i}\left((\mu_0 + d_{\mathcal{T}^0}r + d_F)\varphi_i + \sigma_i\right). \tag{62}$$

The remaining part of the proof is based on inequality (62) and repeats the corresponding part of the proof of Theorem 4 in [18] (there starting from inequality (58), which differs from (62) above in the combination of the controlling parameters in the last term only). □

Remark 6 *Estimate (57) shows that the distance from $u^{i,s}$ to U^* tends to 0 not slower than a geometrical progression with the factor $(1+\frac{2d}{\bar{\chi}})^{-1/4}$. Using the proof of Theorem 14.6 in [14], linear convergence of $\{u^{i,s}\}$ to an element $u^* \in U^*$ can be established, too.*
Estimate (58), in particular, leads to the known result on "finite convergence" of the exact proximal point method (cf. [30], [23]). ◇

5 Applications

Let us briefly observe the application of the scheme described to some classes of problems. If an abstract formulation of the problem is considered, then we shall suppose that both the corresponding operator $\mathcal{T}^0$ and the corresponding functional F satisfy the Assumptions 2 (a) - (c).

5.1 Convex Variational Problems

$$\min\{f_1(u) + f_2(u) : \ u \in K\},$$

where K is given as in Problem (1), $f_1, \ f_2 : V \to \bar{I\!R}$ are proper convex, lsc. functionals, *int dom* $f_1 \cap D(\partial f_2) \supset K$, and f_1 is Gâteaux-differentiable on *int dom* f_1.
This problem is equivalent to Problem (1) by setting $\mathcal{T}^0 = \nabla f_1, \ F = f_2$, and the operator $\nabla f_1 + \partial f_2 + \mathcal{N}_K$ is maximal monotone.

In several applications, f_2 can be uniformly approximated on $dom f_2 \cap S_r$ by convex differentiable functionals F_i such that (with an appropriate approximation of K) Assumption 2 is fulfilled, and hence, the MSR-method is applicable. This concerns, in particular, the *problem of a laminar stationary flow of a Bingham fluid* in a cylindrical pipe of cross-section Ω, which has the following formulation [11]:

$$J(u) = \int_\Omega \left[\frac{1}{2}|\nabla u|^2 - fu + g|\nabla u|\right] d\Omega \to min, \quad u \in H_0^1(\Omega), \qquad (63)$$

with given $f \in L_2(\Omega)$ and a positive constant g. This problem has a unique solution u^*, and if Ω is a bounded convex domain with sufficiently smooth boundary, or Ω is a convex polygon, then $u^* \in H_0^1(\Omega) \cap H^2(\Omega)$ and

$$||u^*||_{H^2(\Omega)} \leq \gamma(\Omega)||f||_{L_2(\Omega)}$$

with $\gamma(\Omega)$ a certain constant (see [5]).
Setting

$$f_1(u) = \int_\Omega \left[\frac{1}{2}|\nabla u|^2 - fu\right] d\Omega,$$
$$f_2(u) = g\int_\Omega |\nabla u| d\Omega,$$

and approximating the non-smooth functional f_2 by

$$F_i(u) = g\int_\Omega \sqrt{|\nabla u|^2 + \tau_i} d\Omega \quad \text{(with suitably chosen } \tau_i > 0)$$

and the set $K = H_0^1(\Omega) = V$ by means of the FEM (with piece-wise linear basic functions) on a sequence of triangulations, then Assumption 2 can be satisfied for $V_1 = V$ and $H = V$ or $H = L_2(\Omega)$. Numerical results for the basic variant of the MSR-method and for the method with weak regularization are described in [32].
The application of the MSR-methods to the static problem of linear elasticity with friction and to the contact problem of two elastic bodies is studied in [16]. Both problems are not uniquely solvable, in general. The MSR-method for semi-infinite problems is developed in [15].

5.2 Saddle-Point Problems

The analysis described in Sections 2 - 4 can be extended to some special saddle point problems for convex-concave functions on the product of convex, closed subsets in Hilbert spaces.
First of all, this concerns the saddle-point problem corresponding to a convex, non-differentiable optimization problem:

$$\min\{f(x) : g_j(x) \le 0, \ j = 1, ..., m, \ x \in K_0\},$$

with K_0 a convex, closed subset of a Hilbert space X, $f : X \to \bar{I\!R}$ a proper convex, lsc. functional, $D(\partial f) \supset K_0$, and g_j convex, Gâteaux-differentiable functionals on X.
Let us suppose that the set of saddle-points $U^* = \{(x^*, y^*)\}$ of the Lagrangian

$$L(x,y) = f(x) + \sum_{j=1}^m y_j g_j(x)$$

on $K_0 \times I\!R_+^m$ is non-empty. Now we introduce $u = (x, y)$, the space

$$V = X \times I\!R^m \ \text{ normed by } \|u\| = \left(\|x\|_X^2 + \|y\|_{I\!R^m}^2\right)^{1/2}, \quad K = K_0 \times I\!R_+^m,$$

and

$$\bar{L}(x,y) = \begin{cases} L(x,y) & \text{if } x \in K_0,\ y \in \mathbb{R}^m_+, \\ -\infty & \text{if } x \in K_0,\ y \notin \mathbb{R}^m_+, \\ +\infty & \text{if } x \notin K_0, \end{cases}$$

$$\mathcal{T} : u \to \partial_x \bar{L}(x,y) \times \partial_y(-\bar{L}(x,y)) \quad \text{for} \quad u \in K,$$

with $\partial_x \bar{L}(\cdot, y), \partial_y(-\bar{L}(x,\cdot))$ subdifferentials of $\bar{L}(\cdot,y), -\bar{L}(x,\cdot)$, correspondingly. Then

$$u \in U^* \Longleftrightarrow 0 \in \mathcal{T}u. \tag{64}$$

Relation (64) can be rewritten as

$$u \in K : \quad \langle \mathcal{T}^0 u, v-u \rangle + \langle \partial F(u), v-u \rangle \geq 0 \ \ \forall v \in K, \tag{65}$$

with

$$\mathcal{T}^0 u = \begin{cases} \left[\sum_{j=1}^m y_j \nabla_x g_j(x), -g(x)\right] \in X' \times \mathbb{R}^m & \text{if } u \in X \times \mathbb{R}^m_+, \\ \emptyset & \text{if } u \notin X \times \mathbb{R}^m_+, \end{cases}$$

and

$$F(u) = f(x), \quad g(x) = [g_1(x), ..., g_m(x)]\,;$$

the dual space of $(X \times \mathbb{R}^m)$ is identified with $X' \times \mathbb{R}^m$. The monotonicity of $\mathcal{T}^0$ follows from the convexity of the functionals g_j. Hence, the operator $\mathcal{T}^0 + \mathcal{N}_K$ is maximal monotone, and the maximal monotonicity of $\mathcal{T}^0 + \partial F + \mathcal{N}_K$ is guaranteed if $K_0 \cap int D(\partial f) \neq \emptyset$ ($int D(\partial f)$ in X).
Thus, the conditions of Section 1 are valid for Problem (65), but to satisfy Assumption 2 the problem has to be specialized, first of all concerning the choice of the space X and the kind of the non-smoothness of f.

As an example, we observe the saddle-point problem corresponding to the following *maximum entropy reconstruction problem.* An unknown spectral density function x on a domain $\Omega \subset \mathbb{R}^n$ has to be restored from a finite number of known measurements

$$\int_\Omega a_j(t)x(t)d\Omega = b_j, \ j = 1, ..., m, \tag{66}$$

with given weight functions a_j and constants b_j. According to the maximum entropy principle, the non-negative function x has to maximize a given measure of entropy $H(x)$, or equivalently, to minimize the information measure $I(x) = -H(x)$. Similar problems arise in various applications such as time series analysis, image reconstruction, models in crystallography, etc. (see [9], [26] and references therein).
Here we restrict ourselves to the consideration of the averaged Fisher information measure, i.e.

$$I(x) = \int_\Omega \varphi(x(t), \nabla x(t))d\Omega, \tag{67}$$

with

$$\varphi(p,q) = \begin{cases} q^2/p & \text{if } p > 0, \\ 0 & \text{if } p = 0, q = 0, \\ +\infty & \text{otherwise} \end{cases}$$

The problem now is [2]

$$\begin{aligned} &\textit{minimize } I(x) = \int_\Omega \varphi(x(t), \nabla x(t)) d\Omega \\ &\quad s.t. \int_\Omega a_j(t)x(t)d\Omega = b_j, \; j = 1, ..., m, \\ &\qquad x(t) \geq 0, \; x \in H^1(\Omega). \end{aligned} \tag{68}$$

The choice of the space $H^1(\Omega)$ is based on the analysis of the problem in [26]; in case $n = 1$, Ω is simply an interval, and if $n \geq 2$, Ω is assumed to be bounded, with a sufficiently smooth boundary. Moreover it is supposed that $\int_\Omega a_1(t)d\Omega \neq 0$, $a_1, ..., a_m$ are sufficiently smooth and form a Pseudo-Haar system (it means that $a_1, ..., a_m$ are linearly independent on any subset of Ω of positive measure, see [4]). This guarantees the unique solvability of the problem and a sufficient smoothness of the solution x^*. There is a more exact result in [26], Prop. 2.2, and the analyticity of the solution (under analytic data) is a partial case.
To simplify the consideration, we don't go beyond the case when

$$x^* > 0, \; x^* \in C(\bar{\Omega}). \tag{69}$$

The conditions in [26], Corollary 3.2, which ensure this property for a relaxed model (including the random noise), remain sufficient for our problem under the following constraint qualification:

$$\text{there exists } \; \tilde{x} \in C^1(\bar{\Omega}), \; \tilde{x} > 0, \; \text{ satisfying (66).}$$

We define the Lagrangian

$$L(x, y) = I(x) + \sum_{j=1}^m y_j \left(\int_\Omega a_j(t)x(t)d\Omega - b_j \right) \tag{70}$$

with

$$x \in H^1(\Omega), \; x \geq 0, \; y = (y_1, ..., y_m) \in I\!R^m. \tag{71}$$

Defining

$$\begin{aligned} Ax = \left(\int_\Omega a_1 x d\Omega, ..., \int_\Omega a_m x d\Omega \right), \; b = (b_1, ..., b_m), \\ K_0 = \{x \in H^1(\Omega) : \; x \geq 0\}, \; u = (x, y), \end{aligned}$$

[2]Taking into account the definition of the Fisher function I, the restriction $x(t) \geq 0$ could be omitted. But anyway, we shall use it to define the set K.

$$V = H^1(\Omega) \times \mathbb{R}^m, \quad K = K_0 \times \mathbb{R}^m,$$

$$F(u) = I(x), \quad \mathcal{T}^0 : u \to \left(\sum_{j=1}^{m} y_j a_j, \ -Ax + b \right),$$

the variational inequality associated with the saddle point problem for (70), (71) can be written in the form (1), with these data $\mathcal{T}^0$, F, K and V.
Applying the basic variant of the MSR-method, it is convenient to define the approximate sets K_i as follows: for given constants $\kappa_i > 0$, $h_i > 0$ ($\lim \kappa_i = \lim h_i = 0$) we put $K_{0,\kappa_i} = \{x \in K_0 : x \geq \kappa_i\}$ and construct $K_{0,i}$ approximating K_{0,κ_i} by means of the FEM with piece-wise linear basic functions on a h_i-triangulation of Ω. The set K_i has the structure $K_{0,i} \times \mathbb{R}^m$.
Under an appropriate tuning between κ_i and h_i this construction of K_i ensures the inclusion $K_i \subset K$ and permits to avoid an approximation of F. So we can choose $F_i = F$.
To satisfy the conditions of Theorem 2, the main difficulties are connected with the calculation of the constants r (see (21)) and d_F (see Ass. 2 (c)). For the time being, we are ready to give the answer only if

$$n = 1 \text{ and } 1 \in span(a_1, ..., a_m). \tag{72}$$

In this case $\|u^*\|$ can be estimated in the following way. Using the substitution $x(t) = z^2(t)$, the objective functional (67) takes the form

$$\int_\Omega \frac{(2z(t)z'(t))^2}{z(t)^2} d\Omega = 4 \int_\Omega |z'(t)|^2 d\Omega$$

and, with an arbitrary feasible $\tilde{z}$ we obtain an upper bound for $\|(z^*)'\|_{L_2(\Omega)}$:

$$\|(z^*)'\|_{L_2(\Omega)} \leq \|(\tilde{z})'\|_{L_2(\Omega)} \equiv \alpha_0.$$

The conditions (66) and (72) deliver an α_1 such that $\|z^*\|_{L_2(\Omega)} \leq \alpha_1$. Indeed, due to $1 \in span(a_1, ..., a_m)$, we have $\sum_{j=1}^m \xi_j a_j = 1$ for some $\xi \in \mathbb{R}^m$. Multiplying the constraints in (66) by $\xi_1, ..., \xi_m$ and summing up, one gets

$$\int_\Omega x(t) d\Omega = \sum_{j=1}^{m} \xi_j b_j \equiv \tilde{\alpha}_1.$$

Therefore,

$$\|z^*\|^2_{L_2(\Omega)} = \int_\Omega (z^*)^2 d\Omega = \tilde{\alpha}_1, \quad \text{hence,} \quad \|z^*\|_{L_2(\Omega)} = \tilde{\alpha}_1^{1/2} \leq \alpha_1.$$

Due to the continuous embedding $H^1(\Omega) \hookrightarrow C(\bar{\Omega})$, we obtain $\|z\|_{C(\bar{\Omega})} \leq \bar{\alpha}\|z\|_{H^1(\Omega)}$ with a known constant $\bar{\alpha}$, thus

$$\|z^*\|_{C(\bar{\Omega})} \leq \bar{\alpha}\|z\|_{H^1(\Omega)} \leq \bar{\alpha}(\alpha_1^2 + \alpha_0^2)^{1/2} \equiv \alpha_2.$$

Now, using the relation

$$\|x\|^2_{H^1(\Omega)} = \int_\Omega z^4(t)d\Omega + 4\int_\Omega (z(t)z'(t))^2 d\Omega$$

with $x = x^*$, $z = z^*$, we obtain from the mean value theorem

$$\|x^*\|^2_{H^1(\Omega)} \leq \|z^*\|^2_{C(\bar{\Omega})} \left(\int_\Omega (z^*(t))^2 d\Omega + 4\int_\Omega (z^*(t)')^2 d\Omega\right) \leq \alpha_2^2(\alpha_1^2 + 4\alpha_0^2).$$

Hence, $\|x^*\|_{H^1(\Omega)} \leq \alpha_2(\alpha_1^2+4\alpha_0^2)^{1/2}$. Now, the linear independence of $(a_1, ..., a_m)$ permits to estimate $\|y^*\|_{\mathbb{R}^m}$, and hence $\|u^*\|$, in a standard manner. Thus, we can choose the radii r^* and r via (21).
The function F is not Lipschitz continuous on $K \cap S_r$, but it is Lipschitz continuous on $K^\kappa = \{u \in K : x \geq \kappa > 0\} \cap S_r$ with arbitrary $\kappa > 0$. It is not difficult to calculate the corresponding Lipschitz constant d_F^κ, using $H^1(\Omega) \hookrightarrow C(\bar{\Omega})$ and the Hoelder inequality only. With regard to (69), we may handle with a variable d_F^κ. However, because $d_F^\kappa \to \infty$ as $\kappa \to 0$, this may enforce that the controlling parameters of the method decrease too fast. Therefore, an alternative is to add to K the constraint $\|u\|^2 \leq r_1$, with $r^* < r_1 < r$. Then the controlling parameters can be defined by means of the weaker conditions of Theorem 1 (see also Remark 4).

References

[1] A. Auslender and M. Haddou, *An interior-proximal method for convex linearly constrained problems and its extension to variational inequalities*, Math. Programming 71 (1995) 77-100.

[2] A. Auslender, M. Teboulle and S. Ben-Tiba, *Interior proximal and Multiplier Methods based on second order homogeneous kernels*, MOR (to appear).

[3] D.P. Bertsekas and P. Tseng, *Partial proximal minimization algorithms for convex programming*, SIAM J. Optimization 4 (1994) 551-572.

[4] J.M. Borwein and A.S. Lewis, *Duality relationship for entropy-like minimization problems*, SIAM J. Control and Optim. 29 (1991) 325-338.

[5] H. Bresis, *Monotonicity in Hilbert spaces and some applications to nonlinear partial differential equations*, in: Contributions to Nonlinear Functional Analysis, Zarantonello (ed.), Academic, New York, 1971, 101-116.

[6] F.E. Browder, *Nonlinear monotone operators and convex sets in Banach spaces*, Bull. Amer. Math. Soc. 71 (1965) 780-785.

[7] G. Chen and M. Teboulle, *A proximal-based decomposition method for convex minimization problems*, Math. Programming 64 (1994) 81-101.

[8] J. Eckstein and D.P. Bertsekas, *On the Douglas-Rachford splitting method and the proximal point algorithm for maximal monotone operators*, Math. Programming 55 (1992) 293-318.

[9] G.J. Erickson and C.R. Smith (Eds.) *Maximum-entropy and Bayesian Methods in Science and Engeneering*, Kluwer, Dordrecht, The Netherlands, 1988.

[10] M. Fukushima, *The primal Douglas-Rachford splitting algorithm for a class of monotone mappings with applications to the traffic equilibrium problem*, Math. Programming 72 (1996) 1-15.

[11] R. Glowinski, J.L. Lions and R. Trémolières, *Numerical Analysis of Variational Inequalities*, North-Holland, Amsterdam, 1981.

[12] Cu D. Ha, *A generalization of the proximal point algorithm*, SIAM J. Control and Optimization 28 (1990) 503-512.

[13] R. Hettich, A. Kaplan and R. Tichatschke, *Regularized penalty methods for ill-posed optimal control problems with elliptic equations*, Part I and II, Control and Cybernetics 26 (1997) 5-27, 29-43.

[14] A. Kaplan and R. Tichatschke, *Stable Methods for Ill-posed Variational Problems - Prox-Regularization of Elliptic Variational Inequalities and Semi-Infinite Problems*, Akademie Verlag, Berlin, 1994.

[15] A. Kaplan and R. Tichatschke, *Path-following proximal approach for solving ill-posed convex semi-infinite programming problems*, JOTA 90 (1996) 113-137.

[16] A. Kaplan and R. Tichatschke, *Prox-Regularization and solution of ill-posed variational inequalities*, Applications of Mathematics 42 (1997), No.2, 111-145.

[17] A. Kaplan and R. Tichatschke, *Regularized penalty method for non-coercive parabolic optimal control problems*, Control and Cybernetics 27 (1998) 1, 5-27.

[18] A. Kaplan and R. Tichatschke, *Proximal Point Approach and Approximation of Variational Inequalities*, SIAM J. Control and Optimization, (submitted)

[19] A. Kaplan and R. Tichatschke, *Multi-step proximal method for variational inequalities with monotone operators*, in: Recent Advances in Optimization, P. Gritzmann, R. Horst, E. Sachs, R. Tichatschke (eds.), Lecture Notes in Economics and Mathematical Systems, Springer 1997, pp. 138-153.

[20] K. C. Kiwiel, *Proximal level bundle methods for convex nondifferentiable optimization, saddle-point problems and variational inequalities*, Math. Programming, 69 (1995) 89-109.

[21] B. Lemaire, *Coupling optimization methods and variational convergence*, ISNM 84 (1988) 163-179.

[22] C. Lemaréchal and C.A. Sagastizábal, *An approach to variable metric bundle methods*, Lecture Notes in Control and Information Sciences, No. 197, Springer, Berlin, 1994, 144-162.

[23] F. J. Luque, *Asymptotic convergence analysis of the proximal point algorithms*, SIAM J. Control and Optimization 22 (1984) 277-293.

[24] B. Martinet, *Régularisation d'inéquations variationelles par approximations successives*, RAIRO 4 (1970) 154-159.

[25] R. Mifflin, *A quasi-second-order proximal bundle algorithm*, Math. Programming 73 (1996) 51-72.

[26] D. Noll, *Reconstruction with noisy data: an approach via eigenvalue optimization*, SIAM J. Optimization 8 (1997) 82-104.

[27] Z. Opial, *Weak convergence of the successive approximaions for nonexpansive mappings in Banach spaces*, Bull. Amer. Math. Soc. 73 (1967) 591-597.

[28] B. T. Polyak, *Introduction to Optimization*, Optimization Software, Inc. Publ. Division, New York, 1987.

[29] R. T. Rockafellar, *On the maximality of sums of nonlinear monotone operators*, Trans. Amer. Math. Soc. 149 (1970) 75-88.

[30] R. T. Rockafellar, *Monotone operators and the proximal point algorithm*, SIAM J. Control and Optim. 14 (1976) 877-898.

[31] R. T. Rockafellar, *Augmented Lagrange multiplier functions and applications of the proximal point algorithm in convex programming*, Math. Oper. Res. 1 (1976) 97-116.

[32] H. Schmitt, *Normschwächere Prox-Regularisierungen*, PhD-Thesis, University of Trier, 1996.

A Projection-Proximal Bundle Method for Convex Nondifferentiable Minimization *

Krzysztof C. Kiwiel **

Abstract

We consider the problem $\inf f_X := f + \imath_X$, where f is a closed proper convex function and $\imath_X$ is the indicator function of a closed convex set X in a real Hilbert space. We study an inexact proximal point algorithm under mild conditions on accuracy tolerances that ensure weak convergence of its iterates to a solution, or strong convergence when f_X is boundedly well-set. We also give a bundle method for minimizing f_X. At each iteration, it finds a trial point by minimizing a simple "unconstrained" polyhedral model of f_X augmented with a proximal term, and projecting a certain point onto X. We establish convergence of descent and nondescent versions. Finally, we discuss an application in Tikhonov's regularization of ill-posed minimization problems.

Keywords: Convex programming, nondifferentiable optimization, proximal methods, bundle methods, well-set problems, ill-posed problems

AMS subject classification: 90C25, 65K05

1 Introduction

Let $\mathcal{H}$ be a real Hilbert space with inner product $\langle\cdot,\cdot\rangle$ and norm $|\cdot|$. We consider the problem

$$f_* = \inf\{ f(x) : x \in X \}, \tag{1}$$

where $f : \mathcal{H} \to (-\infty, \infty]$ is closed proper convex and $\emptyset \neq X \subset \mathcal{H}$ is closed convex. The proximal point algorithm for (1), starting from any $x^0 \in \mathcal{H}$, generates a sequence

$$x^k = \arg\min\{ f(x) + \tfrac{1}{2t_k}|x - x^{k-1}|^2 : x \in X \}, \tag{2}$$

where $t_k > 0$; approximate versions allow inexact minimization in (2) (cf., e.g., [1, 2, 3, 4, 5, 7, 9, 10, 17, 18, 19, 20, 21, 24]).

The first part of this paper studies convergence of an approximate form of (2) in an infinite-dimensional extension of the framework of [16]. Our results improve the corresponding ones in [2, 3, 20].

*Research supported by State Committee for Scientific Research Grant 8T11A00115.

**Systems Research Institute, Newelska 6, 01–447 Warsaw, Poland (kiwiel@ibspan.waw.pl)

The second part of this paper gives a bundle projection-proximal method. At each iteration, it finds a trial point by minimizing a simple "unconstrained" polyhedral model of f_X augmented with a proximal term, and projecting a certain point onto X. (Standard bundle methods [11, 13] require minimizing the model over X, i.e., a more difficult subproblem.) We show that this basic procedure can implement an approximate prox iteration in a finite number of steps. We establish convergence of descent and nondescent versions. Our results subsume the corresponding ones in [2, 3]. Finally, we give an application in Tikhonov's [26] regularization of ill-posed minimization problems.

The paper is organized as follows. In §2 we analyze convergence of an approximate form of (2). In §3 we present a descent bundle method. Its convergence is established in §4. A nondescent bundle method is discussed in §5, and Tikhonov's regularization in §6.

Our notation is fairly standard. $B(x, \epsilon) = \{y : |x - y| \leq \epsilon\}$ is the ball with center x and radius ϵ. The *indicator* function of a set C is denoted by $\imath_C$ ($\imath_C(x) = 0$ if $x \in C$, ∞ otherwise), its *distance* function by $d_C(x) = \inf_C |\cdot - x|$, and its *projector* by $P_X x = \arg\min_C |\cdot - x|$. $\mathcal{D}_f = \{x : f(x) < \infty\}$ denotes the *effective domain* of f, $\partial_\epsilon f(\cdot) = \{p : f(y) \geq f(\cdot) + \langle p, y - \cdot\rangle - \epsilon \ \forall y\}$ its ϵ-*subdifferential*, and $\partial f = \partial_0 f$ its *subdifferential*. "$\to$" and "$\rightharpoonup$" stand for norm and weak convergence, respectively. "s.t." abbreviates "such that".

2 Convergence of a General Prox Method

We now recall the inexact prox method and improve results on its convergence. First, we list various assumptions about our problem (1) and the algorithm.

A_1: f is a closed proper convex function.

A_2: X is a nonempty closed convex set.

A_3: $\mathcal{D}_{f_X} = \mathcal{D}_f \cap X \neq \emptyset$, where $f_X = f + \imath_X$ is the *essential objective* of (1).

A_4: $\{t_k\}$ is a sequence of positive numbers s.t. $s_l := \sum_{k=1}^{l} t_k \to \infty$ as $l \to \infty$.

A_5: $\{\epsilon_k\}$ is a sequence of nonnegative numbers s.t. $\sum_{k=1}^{l} t_k \epsilon_k / s_l \to 0$.

At iteration $k \geq 1$ of the method, having $x^{k-1} \in \mathcal{D}_{f_X}$, one has to find x^k s.t.

$$(x^{k-1} - x^k)/t_k \in \partial_{\epsilon_k} f_X(x^k). \tag{3}$$

Remarks 2.1. (i) Let $\phi_k(\cdot) = f_X(\cdot) + \frac{1}{2t_k}|\cdot - x^{k-1}|^2$. By (3), $0 \in \partial_{\epsilon_k}\phi_k(x^k)$, so $\phi_k(x^k) \leq \min \phi_k + \epsilon_k \leq \phi_k(x^{k-1}) + \epsilon_k$ and hence

$$f_X(x^k) + \tfrac{1}{2t_k}|x^k - x^{k-1}|^2 \leq f_X(x^{k-1}) + \epsilon_k. \tag{4}$$

(ii) The *exact* version of (2) employs $\epsilon_k = 0$ and $x^k = \arg\min \phi_k$.

(iii) Since $f_X(x^k) > f_X(x^{k-1})$ is possible, the method is a *nondescent* one, but

$$f_X(x^k) \leq f_X(x^{k-1}) + \left(\epsilon_k - \tfrac{1}{2t_k}|x^k - x^{k-1}|^2\right) \leq f_X(x^{k-1}) + \epsilon_k \tag{5}$$

(cf. (4)). In contrast, the *descent* versions of [18, §3], [15, §3] demand that

$$f_X(x^k) \le f_X(x^{k-1}) \qquad \forall k. \tag{6}$$

It will be useful to express (5) in terms of

$$\hat\epsilon_k := \epsilon_k - \tfrac{1}{2t_k}|x^k - x^{k-1}|^2 \le \epsilon_k \quad \text{and} \quad \hat\epsilon_k^+ := \max\{\hat\epsilon_k, 0\}. \tag{7}$$

Following [10], we now derive *global convergence rate estimates.*

Lemma 2.2. *For all $x \in \mathcal{H}$ and $k \le l$, we have*

$$\tfrac12|x - x^k|^2 \le \tfrac12|x - x^{k-1}|^2 + t_k\hat\epsilon_k \quad \text{if} \quad f_X(x) \le f_X(x^k), \tag{8}$$

$$\sum_{k=1}^{l} t_k f_X(x^k)/s_l - f_X(x) \le \tfrac12|x - x^0|^2/s_l + \sum_{k=1}^{l} t_k\hat\epsilon_k/s_l, \tag{9}$$

$$f_X(x^l) - f_X(x) \le \tfrac12|x - x^0|^2/s_l + \sum_{k=1}^{l} s_k\hat\epsilon_k/s_l, \tag{10}$$

where $s_l = \sum_{k=1}^{l} t_k$; moreover, if the descent condition (6) *holds then*

$$f_X(x^l) - f_X(x) \le \tfrac12|x - x^0|^2/s_l + \sum_{k=1}^{l} t_k\hat\epsilon_k/s_l. \tag{11}$$

Proof. By (3), $f_X(x) - f_X(x^k) \ge \langle x^{k-1} - x^k, x - x^k\rangle/t_k - \epsilon_k$, so

$$\begin{aligned} t_k[f_X(x) - f_X(x^k) + \epsilon_k] &\ge \langle x^{k-1} - x^k, x - x^k\rangle \\ &= \tfrac12|x^k - x^{k-1}|^2 + \tfrac12|x - x^k|^2 - \tfrac12|x - x^{k-1}|^2 \end{aligned}$$

yields (8) by (7), whereas summing this inequality over $k = 1{:}l$ gives

$$s_l f_X(x) - \sum_{k=1}^{l} t_k f_X(x^k) \ge -\sum_{k=1}^{l} t_k\hat\epsilon_k + \tfrac12|x - x^l|^2 - \tfrac12|x - x^0|^2 \tag{12}$$

and hence (9). Multiply (cf. (5)) $f_X(x^{k-1}) - f_X(x^k) \ge -\hat\epsilon_k$ by $s_{k-1} = s_k - t_k$ ($s_0 = 0$) to get $s_{k-1}f_X(x^{k-1}) - s_k f_X(x^k) + t_k f_X(x^k) \ge -s_{k-1}\hat\epsilon_k$ and sum over $k = 1{:}l$ to obtain $-s_l f_X(x^l) + \sum_{k=1}^{l} t_k f_X(x^k) \ge -\sum_{k=1}^{l} s_{k-1}\hat\epsilon_k$. Add this inequality to (12) and rearrange, using $s_k = s_{k-1} + t_k$, to get (10). Use $f_X(x^{k-1}) - f_X(x^k) \ge 0$ similarly for (11). □

Our first convergence result below employs the *aggregates* $\bar x^l := \sum_{k=1}^{l} t_k x^k/s_l$ (convex combinations of $\{x^k\}_{k=1}^{l}$) and *records* $x_{\rm rec}^l \in \operatorname{Arg\,min}_{\{x^k\}_{k=1}^{l}} f_X(x^k)$. By convexity,

$$f_X(\bar x^l) \le \sum_{k=1}^{l} t_k f_X(x^k)/s_l, \tag{13a}$$

$$f_X(x^l_{\rm rec}) = \min_{k=1:l} f_X(x^k) \le \sum_{k=1}^{l} t_k f_X(x^k)/s_l. \tag{13b}$$

It will be useful to employ the following weaker version of A_5

A_5': $\overline{\lim}_{l\to\infty} \sum_{k=1}^{l} t_k\hat\epsilon_k/s_l \le 0$.

Theorem 2.3. *If* A_1–A_4, A_5' *hold, then:*

(a) $f_X(x^l_{\rm rec})$, $f_X(\bar x^l)$ *and* $\sum_{k=1}^{l} t_k f_X(x^k)/s_l$ *converge to* f_* *as* $l\to\infty$, *and* $\underline{\lim}_{k\to\infty} f_X(x^k) = f_*$.

(b) *If* $\overline{\lim}_{l\to\infty} \sum_{k=1}^{l} s_k\hat\epsilon_k/s_l \le 0$, *then* $f_X(x^k)\to f_*$.

(c) *If the descent condition* (6) *holds, then* $f_X(x^k)\downarrow f_*$.

(d) *If* $\sum_{k=1}^{\infty}\hat\epsilon_k^+ < \infty$, *then* $f_X(x^k)\to f_*$.

Proof. (a) (1), (13) and (9) yield

$$f_* \le f_X(\bar x^l) \le \sum_{k=1}^{l} t_k f_X(x^k)/s_l \le f_X(x) + \tfrac12|x-x^0|^2/s_l + \sum_{k=1}^{l} t_k\hat\epsilon_k/s_l \quad \forall x, \tag{14}$$

also with $\bar x^l$ replaced by $x^l_{\rm rec}$ above. Take upper and lower limits in (14), using $s_l\to\infty$ and $\sum_{k=1}^{l} t_k\hat\epsilon_k/s_l \to 0$ (cf. A_4, A_5'). (b) The upper limit in (10) for any x yields $\overline{\lim}_l f_X(x^l) \le f_*$. (c) For any x, taking the limit in (11) yields $\lim_l f_X(x^l) \le f_X(x)$. (d) $\underline{\lim}_k f_X(x^k) = f_*$ by (a), so $\lim_k f_X(x^k)$ exists by summing (5) (cf. (7)). □

Having established convergence of their objective values, we now discuss convergence of $\{\bar x^k\}$, $\{x^k_{\rm rec}\}$ and $\{x^k\}$ to $X_* := \operatorname{Arg\,min} f_X$, the *solution set* of (1). Of course, we start with the case of $X_* = \emptyset$.

Theorem 2.4. *Let* A_1–A_3 *hold and* $X_* = \emptyset$. *If* $\{y^k\}$ *is s.t.* $f_X(y^k)\to f_*$, *then* $|y^k|\to\infty$. *Thus, under the additional assumptions of Thm* 2.3, *we have* $|\bar x^k|\to\infty$ *and* $|x^k_{\rm rec}|\to\infty$ *in Thm* 2.3(a), *and* $|x^k|\to\infty$ *in Thm* 2.3(b–d).

Proof. If $|y^k|\not\to\infty$, then $\{y^k\}$ has a weak limit point y with $f_X(y)\le f_*$ (since f_X is weakly closed, being closed convex by A_1–A_3), so $f_* = \inf f_X$ yields $y\in X_*\ne\emptyset$. □

The remaining case of $X_*\ne\emptyset$ analyzed in Thm 2.7 below needs the following

Lemma 2.5. *If* A_1–A_3 *hold,* $\sum_{k=1}^{\infty} t_k\hat\epsilon_k^+ < \infty$ *and* $X_*\ne\emptyset$, *then:*

(i) $|x^k-x|^2 \le |x^{k-1}-x|^2 + 2t_k\epsilon_k^+$ $\forall x\in X_*$, $d^2_{X_*}(x^k) \le d^2_{X_*}(x^{k-1}) + 2t_k\epsilon_k^+$ $\forall k\ge 1$.

(ii) $\{x^k\}$ *is bounded,* $\{d_{X_*}(x^k)\}$ *and* $\{|x^k-x|\}$ *converge* $\forall x\in X_*$.

(iii) $\{x^k\}$ *has at most one weak limit point* x^∞ *in* X_*.

(iv) *For any subsequence* $\{x^{k_j}\}$, *if* $f_X(x^{k_j})\to f_*$ *then* $x^{k_j}\rightharpoonup x^\infty\in X_*$.

(v) *If* $\underline{\lim}_{k\to\infty} d_{X_*}(x^k) = 0$, *then* $x^k\to x^\infty\in X_*$.

(vi) *If the interior* $\operatorname{int} X_*$ *of* X *is nonempty, then* $\{x^k\}$ *converges.*

Proof. (i) Recall (8). (ii) Use (i) and [23, Lem. 2.2.2]. (iii) If $x^{k_j} \rightharpoonup x \in X \ni x' \leftharpoonup x^{k'_j}$, $|x^k - x| \to a$, $|x^k - x'| \to a'$ (cf. (ii)), then $|x^{k_j} - x'|^2 = |x^{k_j} - x|^2 + |x' - x|^2 - 2\langle x^{k_j} - x, x' - x\rangle$ gives in the limit $a' = a + |x' - x|^2$, whereas by a symmetric argument $a = a' + |x' - x|^2$, so $x' = x$. (iv) $\{x^{k_j}\}$ is bounded by (ii), its weak limit points lie in X_* (f_X is weakly closed by A_1–A_3) and coincide by (iii). (v) By (ii), $d_{X_*}(x^k) \to 0$. For $m > l$, by (i), $\frac{1}{4}|x^m - x^l|^2 \le \frac{1}{2}|x^m - P_{X_*}x^l|^2 + \frac{1}{2}|x^l - P_{X_*}x^l|^2 \le d_{X_*}^2(x^l) + \sum_{k=l+1}^{\infty} t_k\epsilon_k^+ \to 0$ as $l \to \infty$ (using $(a+b)^2 \le 2(a^2+b^2)$), i.e., $\{x^k\}$ is a Cauchy sequence. Its limit lies in X_*, since d_{X_*} is continuous. (vi) Fix $x \in X_*$, $\epsilon > 0$ s.t. $B(x,\epsilon) \subset X_*$. Suppose $x^k \ne x^{k-1}$. Let $y = (x^{k-1} - x^k)/|x^{k-1} - x^k|$, so that $x + \epsilon y \in X_*$. By (i), $|x^k - x - \epsilon y|^2 \le |x^{k-1} - x - \epsilon y|^2 + 2t_k\epsilon_k^+$. Expanding yields $2\epsilon|x^k - x^{k-1}| \le |x^{k-1} - x|^2 - |x^k - x|^2 + 2t_k\epsilon_k^+$, so $\sum_k |x^k - x^{k-1}| < \infty$. □

The following property will ensure strong convergence.

Definition 2.6. We say f_X is *well-set* if $d_{X_*}(y^k) \to 0$ whenever $f_X(y^k) \to f_*$; f_X is *boundedly well-set* if $d_{X_*}(y^k) \to 0$ whenever $\{y^k\}$ is bounded, $f_X(y^k) \to f_*$.

Theorem 2.7. *If* A_1–A_4 *hold,* $\sum_{k=1}^{\infty} t_k\hat\epsilon_k^+ < \infty$ *and* $X_* \ne \emptyset$*, then:*
(a) $x_{\text{rec}}^k \rightharpoonup x^\infty \in X_*$.
(b) *If* $f_X(x^k) \to f_*$ *(as in Thm 2.3(b–d)), then* $x^k \rightharpoonup x^\infty$.
(c) *If* f_X *is boundedly well-set, then* $x^k \to x^\infty$.
(d) *If* $\operatorname{int} X_* \ne \emptyset$*, then* $x^k \to x^\infty$.
(e) $\{\bar x^k\}$ *is bounded and all its weak cluster points lie in* X_*.
(f) *If* $x^k \rightharpoonup x^\infty$ *(as in (b)), then* $\bar x^k \rightharpoonup x^\infty$*; if* $x^k \to x^\infty$ *(as in (c,d)), then* $\bar x^k \to x^\infty$.

Proof. (a) Pick $\{x^{k_j}\}$ s.t. $x^{k_j} = x_{\text{rec}}^j\ \forall j$. Then $f_X(x^{k_j}) = f_X(x_{\text{rec}}^j) \to f_*$ by Thm 2.3(a), so $x_{\text{rec}}^j = x^{k_j} \rightharpoonup x^\infty \in X_*$ by Lem. 2.5(iv). (b) Let $\{x^{k_j}\} = \{x^k\}$ in Lem. 2.5(iv). (c) For $\{x^{k_j}\}$ s.t. $f_X(x^{k_j}) \to f_*$ (cf. Thm 2.3(a)), $d_{X_*}(x^{k_j}) \to 0$ by Lem. 2.5(ii) and Def. 2.6, so $x^k \to x^\infty$ by Lem. 2.5(v). (d) This follows from $\underline{\lim}_k f_X(x^k) = f_*$ (Thm 2.3(a)) and Lem. 2.5(iv,vi). (e) Since $\{x^k\}$ is bounded by Lem. 2.5(ii), so is $\{\bar x^k = \sum_{j=1}^k t_j x^j/s_k\}$. Use Thm 2.3(a) and the weak closedness of f_X. (f) Use $s_l = \sum_{k=1}^l t_k \to \infty$ (cf. A_4), $\langle \bar x^l, y\rangle = \sum_{k=1}^l t_k\langle x^k, y\rangle/s_l\ \forall y$, $|\bar x^l - x^\infty| \le \sum_{k=1}^l t_k|x^k - x^\infty|/s_l$ and Lem. 2.8 below. □

Lemma 2.8 (Silverman-Toeplitz's theorem [8, p. 75]). *Let* $a_{lk} \in \mathbb{R}_+$ *for* $k = 1{:}l$, $l = 1, 2, \ldots$, *be s.t.* $\sum_{k=1}^l a_{lk} = 1\ \forall l$, $\lim_{l\to\infty} a_{lk} = 0\ \forall k$. *If* $\{u_k\}$ *is a sequence s.t.* $\lim_{k\to\infty} u_k = u$*, then* $\lim_{l\to\infty}\sum_{k=1}^l a_{lk}u_k = u$.

For choosing $\{\epsilon_k\}$ and $\{t_k\}$, we may use the following results based on A_4.

Lemma 2.9. (i) *If* $\epsilon_k \to 0$ *(e.g.,* $\sum_k \epsilon_k < \infty$*), then* $\sum_{k=1}^l t_k\epsilon_k/s_l \to 0$.

(ii) *If* $\lim_{k\to\infty} \epsilon_k s_k/t_k = 0$, *then* $\lim_{l\to\infty} \sum_{k=1}^{l} s_k\epsilon_k/s_l = 0$.
(iii) *If* $\{t_k\} \subset [t_{\min}, t_{\max}] \subset (0,\infty)$ *and* $k\epsilon_k \to 0$, *then* $\sum_{k=1}^{l} s_k\epsilon_k/s_l \to 0$.
(iv) *If* $\sum_k \epsilon_k < \infty$ *and* $\{t_k\} \subset (0, t_{\max}]$ *for some* $t_{\max} < \infty$, *then* $\sum_k t_k\epsilon_k < \infty$.

Proof. (i) Use Lem. 2.8 with $u_k = \epsilon_k$, $a_{lk} = t_k/s_l$, $s_l = \sum_{k=1}^{l} t_k \to \infty$ (cf. A_4). (ii) Use Lem. 2.8 with $a_{lk} = t_k/s_l$, $u_k = \epsilon_k s_k/t_k$. (iii) Use (ii) with $t_l/s_l \in [t_{\min}/lt_{\max}, t_{\max}/lt_{\min}]$ $\forall l$. (iv) We have $\sum_k t_k\epsilon_k \le t_{\max}\sum_k \epsilon_k < \infty$. □

Remarks 2.10. (i) In practice one is interested in weakest conditions on $\{\epsilon_k\}$ that still ensure convergence. Replacing $\hat\epsilon_k$ by ϵ_k and A_5' by A_5 yields *weaker* versions of Thms 2.3 and 2.7. The stronger ones are useful when ϵ_k and $\hat\epsilon_k$ are generated "on-line", as in §§4–5.

(ii) Theorems 2.3–2.4 generalize [16, Thms 3.3–3.4]. Thm 2.3(a) corresponds to [3, Thm 2.1] (with $\epsilon_k \to 0$, $x_{\rm rec}^l$ and $\bar x^l$ omitted), Thm 2.3(d) to [3, Rem. 2.2] (with $\sum_k \epsilon_k < \infty$), Lem. 2.5(vi) to [20, Lem. 5.1], Thm 2.7(b) to [3, Thm 2.2] (with $\sum_k t_k\epsilon_k < \infty$, $\sum_k \epsilon_k < \infty$), Thm 2.7(c) to [20, Cor. 5.1] (with f_X well-set, $\sum_k t_k\epsilon_k < \infty$, $\sum_k \epsilon_k < \infty$), Thm 2.7(d) to [20, Prop. 5.1] (with $\sum_k t_k\epsilon_k < \infty$, $\epsilon_k \to 0$). We may omit $f_X(x^k) \to f_*$ in Thm 2.7(b) if $\dim\mathcal{H} < \infty$ [16, Thm 3.6(b)].

(iii) Bounded well-setness required in Thm 2.7(c) is significantly weaker than well-setness; see [6, 22, 25, 27] and references therein.

3 A Descent Bundle Method

From now on, we adopt the following stronger version of assumptions A_1–A_3.
B_1: X is a nonempty closed convex set.
B_2: f is a closed proper convex function with $\mathcal{D}_f \supset X$.
B_3: We have a subgradient mapping $X \ni x \mapsto g_f(x) \in \partial f(x)$ that is *bounded on bounded subsets* of X; then f is Lipschitz continuous on bounded subsets of X (since $f(x) - f(y) \le \langle g_f(x), x - y\rangle \le |g_f(x)||x - y|$ $\forall x, y \in X$).

The rules of the following *bundle method* will be discussed below.

Algorithm 3.1.
Step 0 (*Initialization*). Choose a parameter $\kappa \in (0,1)$. Select initial $x^0 \in X$, $g^0 = g_f(x^0)$, $t_1 > 0$. Let $y^0 = x^0$, $\bar f_0(\cdot) = f_0(\cdot) = f(y^0) + \langle g^0, \cdot - y^0\rangle$, $\bar\imath_X^0(\cdot) = 0$. Set the counters $k = 1$, $l = 0$ and $k_0 = 0$ (k_l will denote the iteration number of the lth descent step).

Step 1 (*Model selection*). Choose $\check f_k : \mathcal{H} \to (-\infty, \infty]$ closed convex and s.t. $f_X \ge \check f_k \ge \max\{\bar f_{k-1}, f_{k-1}\}$.

Step 2 (*Trial point finding*). Let

$$y_f^k := \arg\min\left\{\phi_f^k(\cdot) := \check f_k(\cdot) + \bar\imath_X^{k-1}(\cdot) + \tfrac{1}{2t_k}|\cdot - x^{k_l}|^2\right\}, \qquad (15)$$

$$\bar{f}_k(\cdot) := \check{f}_k(y_f^k) + \langle (x^{k_l} - y_f^k)/t_k - \nabla \bar{\imath}_X^{k-1}, \cdot - y_f^k \rangle . \tag{16}$$

Step 3 (*Projection*). Set

$$y^k := \arg\min \left\{ \phi_{\imath_X}^k(\cdot) := \bar{f}_k(\cdot) + \imath_X(\cdot) + \tfrac{1}{2t_k} |\cdot - x^{k_l}|^2 \right\} = P_X(x^{k_l} - t_k \nabla \bar{f}_k), \tag{17}$$

$$\bar{\imath}_X^k(\cdot) := \langle (x^{k_l} - y^k)/t_k - \nabla \bar{f}_k, \cdot - y^k \rangle , \tag{18}$$

$$v_k := \bar{f}_k(y^k) - f(x^{k_l}). \tag{19}$$

Step 4 (*Descent test*). If

$$f(y^k) \le f(x^{k_l}) + \kappa v_k, \tag{20}$$

set $x^k = y^k$, $i_k = 1$, $k_{l+1} = k$ and increase l by 1 (*descent step*); otherwise, set $x^k = x^{k-1}$ and $i_k = 0$ (*null step*).

Step 5 (*Bundle updating*). Set $g^k = g_f(y^k)$, $f_k(\cdot) = f(y^k) + \langle g^k, \cdot - y^k \rangle$.

Step 6 (*Stepsize updating*). Choose $t_{k+1} > 0$ s.t. $t_{k+1} \le t_k$ if $x^k = x^{k-1}$.

Step 7. Increase k by 1 and go to Step 1.

Remarks 3.2. (i) Choices for $\check{f}_k$ at Step 1 include $\check{f}_k = \max\{\bar{f}_{k-1}, f_{k-1}\}$, $\check{f}_k = \max_{j=0}^{k-1} f_j$, $\check{f}_k = \max\{\bar{f}_{k-1}, f_j : j \in J^k\}$ with $k-1 \in J^k \subset \{0{:}k-1\}$. (In fact, after a descent step ($i_{k-1} = 1$), one may take any $\check{f}_k \le f_X$ without impairing our subsequent results.)

(ii) For $\check{f}_k$ polyhedral as in (i), subproblem (15) may be solved via simple dual QP methods [12]; in contrast, standard approaches [11, 13] employ a more difficult version of (15) with $\bar{\imath}_X^{k-1}$ replaced by $\imath_X$, thus requiring more sophisticated methods even for a polyhedral X [14]. The projection of (17) is easily found if X is "simple" (e.g., the Cartesian product of boxes, simplices and ellipsoids in $\mathcal{H} = \mathrm{IR}^n$).

(iii) (15)–(16) yield $0 \in \partial \phi_f^k(y_f^k) = \partial \check{f}_k(y_f^k) + \nabla \bar{\imath}_X^{k-1} + (y_f^k - x^{k_l})/t_k$, $\nabla \bar{f}_k \in \partial \check{f}_k(y_f^k)$, $\bar{f}_k(y_f^k) = \check{f}_k(y_f^k)$, so $\bar{f}_k \le \check{f}_k \le f_X$. Similarly, (17)–(18) give $0 \in \partial \phi_{\imath_X}^k(y^k)$, $\nabla \bar{\imath}_X^k \in \partial \imath_X(y^k)$, $\bar{\imath}_X^k(y^k) = \imath_X(y^k) = 0$, so $\bar{\imath}_X^k \le \imath_X$. Hence $\bar{f}_X^k := \bar{f}_k + \bar{\imath}_X^k \le f_X$, $\nabla \bar{f}_X^k = (x^{k_l} - y^k)/t_k$ (cf. (18)), $\bar{f}_X^k(x^{k_l}) \le f_X(x^{k_l}) = f(x^{k_l})$ ($x^{k_l} \in X$ by Steps 0 and 4), $\bar{f}_k(y^k) = \bar{f}_X^k(y^k) = \bar{f}_X^k(x^{k_l}) + \langle \nabla \bar{f}_X^k, y^k - x^{k_l} \rangle$ and (19) yield

$$-v_k = f(x^{k_l}) - \bar{f}_X^k(x^{k_l}) + |y^k - x^{k_l}|^2/t_k \ge 0, \tag{21}$$

$$(x^{k_l} - y^k)/t_k \in \partial_{\epsilon_k} f_X(y^k) \text{ with } \epsilon_k := f_X(y^k) - \bar{f}_X^k(y^k) = f(y^k) - \bar{f}_k(y^k). \tag{22}$$

(If $v_k = 0$ then $0 \in \partial f_X(x^{k_l})$ and the method may stop.) Further (cf. Steps 0 and 4), $x^{k-1} = x^{k_l}$. Thus a descent step with $x^{k_{l+1}} = y^k$ in (22) corresponds to the prox iteration (3) with k replaced by k_{l+1} and $k-1$ by k_l.

4 Convergence of the Bundle Prox Method

For convergence analysis, in addition to B_1–B_3, we impose the following requirements.

C_1: $t_k \downarrow t_\infty > 0$ if there are finitely many descent steps.
C_2: $\{\nabla \check{f}_k\}$ is bounded if there are finitely many descent steps.
C_3: $\sum_{k:i_k=1} t_k = \infty$ if there are infinitely many descent steps.

It will be seen that C_2 holds for the choices of $\check{f}_k$ specified in Rem. 3.2(i).

It is instructive to consider first an ideal bundle prox iteration (cf. [1, 3, 5, 7, 13]).

Lemma 4.1. *Suppose that, once Algorithm 3.1 reaches a certain prox-center x^{k_l}, the descent test (20) is omitted, i.e., only null steps are made. Then $\{y^k\} \subset X$ is bounded,*

$$\epsilon_k = f(y^k) - \bar{f}_k(y^k) \to 0, \tag{23}$$

$$y^k \to y^\infty := \arg\min\left\{ \phi_\infty(\cdot) := f_X(\cdot) + \tfrac{1}{2t_\infty}|\cdot - x^{k_l}|^2 \right\}. \tag{24}$$

Proof. Fix $k > k_l$. By (15)–(16), $\bar{f}_k(y_f^k) = \check{f}_k(y_f^k)$ and

$$y_f^k = \arg\min\left\{ \bar{\phi}_f^k(\cdot) := \bar{f}_k(\cdot) + \bar{\imath}_X^{k-1}(\cdot) + \tfrac{1}{2t_k}|\cdot - x^{k_l}|^2 \right\}, \tag{25}$$

since $\bar{\phi}_f^k$ is quadratic and $\nabla \bar{\phi}_f^k(y_f^k) = 0$, so

$$\bar{\phi}_f^k(\cdot) \geq \bar{\phi}_f^k(y_f^k) + \tfrac{1}{2t_k}|\cdot - y_f^k|^2 \quad \text{with} \quad \bar{\phi}_f^k(y_f^k) = \phi_f^k(y_f^k). \tag{26}$$

Similarly, by (17)–(18), we have $\bar{\imath}_X^k(y^k) = \imath_X(y^k) = 0$,

$$y^k = \arg\min\left\{ \bar{\phi}_{\imath_X}^k(\cdot) := \bar{f}_k(\cdot) + \bar{\imath}_X^k(\cdot) + \tfrac{1}{2t_k}|\cdot - x^{k_l}|^2 \right\}, \tag{27}$$

$$\bar{\phi}_{\imath_X}^k(\cdot) \geq \bar{\phi}_{\imath_X}^k(y^k) + \tfrac{1}{2t_k}|\cdot - y^k|^2 \quad \text{with} \quad \bar{\phi}_{\imath_X}^k(y^k) = \phi_{\imath_X}^k(y^k). \tag{28}$$

Now, $\imath_X \geq \bar{\imath}_X^{k-1}$ (cf. Step 0 and Rem. 3.2(iii)), (25) and (17) give $\phi_{\imath_X}^k \geq \bar{\phi}_f^k$, whereas $t_{k+1} \leq t_k$ (cf. Step 6), $\check{f}_{k+1} \geq \bar{f}_k$ (cf. Step 1), (27) and (15) yield $\phi_f^{k+1} \geq \bar{\phi}_{\imath_X}^k$. Hence by (26) and (28),

$$\phi_{\imath_X}^k(y^k) \geq \phi_f^k(y_f^k) + \tfrac{1}{2t_k}|y^k - y_f^k|^2 \quad \text{and} \quad \phi_f^{k+1}(y_f^{k+1}) \geq \phi_{\imath_X}^k(y^k) + \tfrac{1}{2t_k}|y_f^{k+1} - y^k|^2. \tag{29}$$

Next, (25), (27), $\bar{f}_k \leq f_X$ and $\bar{\imath}_X^{k-1}, \bar{\imath}_X^k \leq \imath_X$ (cf. Rem. 3.2(iii)) give

$$\bar{\phi}_f^k(y_f^k) \leq \bar{\phi}_f^k(x^{k_l}) \leq f(x^{k_l}) \quad \text{and} \quad \bar{\phi}_{\imath_X}^k(y^k) \leq \bar{\phi}_{\imath_X}^k(x^{k_l}) \leq f(x^{k_l}). \tag{30}$$

Since $t_k \leq t_{k_l+1}$ (cf. Step 6), by (26), (28), (29) and (30),

$$|y_f^k - x^{k_l}|^2 \leq 2t_k[\bar{\phi}_f^k(x^{k_l}) - \bar{\phi}_f^k(y_f^k)] \leq 2t_{k_l+1}[f(x^{k_l}) - \phi_f^{k_l+1}(y_f^{k_l+1})], \tag{31}$$

$$|y^k - x^{k_l}|^2 \le 2t_k[\bar\phi^k_{ix}(x^{k_l}) - \bar\phi^k_{ix}(y^k)] \le 2t_{k_l+1}[f(x^{k_l}) - \phi_f^{k_l+1}(y_f^{k_l+1})]. \quad (32)$$

By (32), $\{y^k\}$ is bounded, so by B_3 and C_2, there exists $L_f < \infty$ s.t. $|g^k|, |\nabla \bar f_k| \le L_f$ $\forall k$. Now, $\epsilon_k = f(y^k) - \bar f_k(y^k)$ (cf. (22)), where (cf. (16)) $\bar f_k(y^k) = \bar f_k(y_f^k) + \left\langle \nabla \bar f_k, y^k - y_f^k \right\rangle$ with (cf. Steps 1 and 5)

$$\bar f_k(y_f^k) = \check f_k(y_f^k) \ge f_{k-1}(y_f^k) = f(y^{k-1}) + \langle g^{k-1}, y_f^k - y^{k-1}\rangle$$

and $f(y^{k-1}) \ge f(y^k) + \langle g^k, y^{k-1} - y^k\rangle$, so using the Cauchy-Schwarz and triangle inequalities, we get

$$\begin{aligned}
\epsilon_k &\le \langle g^k, y^k - y^{k-1}\rangle - \langle g^{k-1}, y_f^k - y^{k-1}\rangle - \langle \nabla \bar f_k, y^k - y_f^k\rangle \\
&\le L_f(|y^k - y^{k-1}| + |y_f^k - y^{k-1}| + |y^k - y_f^k|) \\
&\le 2L_f(|y^k - y_f^k| + |y_f^k - y^{k-1}|)
\end{aligned}$$

and thus $\epsilon_k^2 \le 8L_f^2(|y^k - y_f^k|^2 + |y_f^k - y^{k-1}|^2)$. Hence by (29) with $t_k \le t_{k_l+1}$,

$$\epsilon_k^2 \le 8L_f^2 2t_{k_l+1}[\phi^k_{ix}(y^k) - \phi^k_f(y_f^k) + \phi^k_f(y_f^k) - \phi^{k-1}_{ix}(y^{k-1})] \quad \text{if} \quad k > k_l + 1,$$

$$\epsilon_k^2 \le 8L_f^2 2t_{k_l+1}[\phi^k_{ix}(y^k) - \phi^k_f(y_f^k) + f(x^{k_l}) - \phi_f^{k_l+1}(y_f^{k_l+1})] \quad \text{if} \quad k = k_l + 1,$$

using (31) with $y^{k-1} = x^{k_l}$ for $k = k_l + 1$. Sum up the above inequalities and use $\phi^k_{ix}(y^k) \le f(x^{k_l})$ (cf. (28), (30)) to get

$$\begin{aligned}
\sum_{j=k_l+1}^{k} \epsilon_j^2 &\le 16L_f^2 t_{k_l+1}[\phi^k_{ix}(y^k) - \phi_f^{k_l+1}(y_f^{k_l+1}) + f(x^{k_l}) - \phi_f^{k_l+1}(y_f^{k_l+1})] \\
&\le 32L_f^2 t_{k_l+1}[f(x^{k_l}) - \phi_f^{k_l+1}(y_f^{k_l+1})]. \qquad (33)
\end{aligned}$$

This yields $\epsilon_k \to 0$, i.e., (23). It remains to verify (24). By strong convexity, $\phi_\infty(y^k) \ge \phi_\infty(y^\infty) + \frac{1}{2t_\infty}|y^k - y^\infty|^2$, whereas $t_\infty \le t_k$, $\bar f_k \le f_X$ and (17) yield $\phi_\infty(y^\infty) \ge \phi^k_{ix}(y^\infty) \ge \phi^k_{ix}(y^k) = \phi_\infty(y^k) - \tilde\epsilon_k$ with $\tilde\epsilon_k := \epsilon_k + (\frac{1}{2t_\infty} - \frac{1}{2t_k})|y^k - x^{k_l}|^2$, so $|y^k - y^\infty| \le \sqrt{2t_\infty \tilde\epsilon_k} \to 0$, since $\epsilon_k \to 0$ by (23), $t_k \downarrow t_\infty > 0$ by C_1, and $|y^k - x^{k_l}|$ is bounded by (32). □

Remarks 4.2. (i) Let $\hat y^k := \arg\min f_X + \frac{1}{2t_k}|\cdot - x^{k_l}|^2$. Then $|y^k - \hat y^k| \le \sqrt{2t_k\epsilon_k}$ (set $t_\infty = t_k$ at the end of the proof of Lemma 4.1).

(ii) C_2 is not needed if $X = \mathcal{H}$: then $\bar\imath^k_X \equiv 0$ and $y^k \equiv y_f^k$ (cf. (25), (27)).

(iii) For $\check f_k$ chosen as in Rem. 3.2(i), $\bar f_k \in \text{co}\{f_j\}_{j=0}^{k-1}$ (e.g., $\check f_k = \max\{\bar f_{k-1}, f_{k-1}\} \Rightarrow \partial \check f_k(y_f^k) = \nu_f^k \nabla \bar f_{k-1} + (1-\nu_f^k)\nabla f_k$, $\nu_f^k \in [0,1]$, $\nu_f^k[\check f_k(y_f^k) - \bar f_{k-1}(y_f^k)] = 0$, $(1-\nu_f^k)[\check f_k(y_f^k) - f_{k-1}(y_f^k)] = 0 \Rightarrow \bar f_k = \nu_f^k \bar f_{k-1} + (1-\nu_f^k) f_{k-1}$). Hence $\nabla \bar f_k \in \text{co}\{g^j\}_{j=0}^{k-1}$ yields $|\nabla \bar f_k| \le \max_{j=0}^{k-1}|g^j|$, so that, by B_3 and the boundedness of $\{y^k\}$ (cf. (32)), C_2 holds *automatically*.

Next, we finish the case of infinitely many null steps.

Theorem 4.3. *If only $l < \infty$ descent steps occur then $x^{k_l} \in X_*$.*

Proof. By Lem. 4.1, $X \supset \{y^k\} \to y^\infty \in X$ yields $f(y^k) \to f(y^\infty)$, since f is continuous on X (cf. B$_3$). For $k > k_l$ (cf. (19), (20), (23))

$$f(y^k) - f(x^{k_l}) > \kappa v_k = \kappa[\bar{f}_k(y^k) - f(x^{k_l})] = \kappa[f(y^k) - \epsilon_k - f(x^{k_l})],$$

so $(1-\kappa)[f(y^k) - f(x^{k_l})] > -\kappa\epsilon_k \to 0$ (cf. (23)) and $\kappa \in (0,1)$ give $f(y^\infty) \geq f(x^{k_l})$. Hence by (24), $\phi_\infty(y^\infty) \geq f(x^{k_l}) = \phi_\infty(x^{k_l})$, $y^\infty = x^{k_l}$, $0 = (x^{k_l} - y^\infty)/t_\infty \in \partial f_X(x^{k_l})$. □

We now show that the descent steps fit the framework of §2.

Theorem 4.4. *Suppose infinitely many descent steps occur. Then:*
(a) $f(x^k) \downarrow f_*$.
(b) *If $X_* \neq \emptyset$ and $\sum_{k:i_k=1} t_k\epsilon_k < \infty$ (e.g., $t_k \leq t_{\max} < \infty$ $\forall k$), then $x^k \rightharpoonup x^\infty \in X_*$; moreover, $x^k \to x^\infty$ if f_X is boundedly well-set or* $\operatorname{int} X_* \neq \emptyset$.
(c) *If $X_* = \emptyset$, then $|x^k| \to \infty$.*

Proof. If $f(x^k) \downarrow -\infty$ then, since $\{x^k\} \subset X$, $f_* = -\infty$, $X_* = \emptyset$ and $|x^k| \to \infty$ by Thm 2.4. Suppose $\inf_k f(x^k) > -\infty$. At Step 4, $x^k = x^{k_l}$ if $i_k = 0$. Let $k \in K := \{k_{l+1}\}_{l=0}^\infty = \{k : i_k = 1\}$. With $k = k_{l+1}$ and $y^k = x^k$ in (19)–(22), we have

$$\epsilon_k = [f(y^k) - f(x^{k_l})] + [f(x^{k_l}) - \bar{f}_k(y^k)] \leq \kappa v_k - v_k \leq (1/\kappa - 1)[f(x^{k_l}) - f(x^{k_{l+1}})],$$

so $\sum_{l=1}^\infty \epsilon_{k_l} < \infty$. But $\check{s}_m := \sum_{l=1}^m t_{k_l} \to \infty$ as $m \to \infty$ (cf. C$_3$), so by Lem. 2.9(i,iv), $\sum_{l=1}^m t_{k_l}\epsilon_{k_l}/\check{s}_m \to 0$, and $\sum_{l=1}^\infty t_{k_l}\epsilon_{k_l} < \infty$ if $\sup_l t_{k_l} < \infty$. Moreover, for $l \geq 1$, $f(x^{k_l}) \leq f(x^{k_{l-1}})$, $(x^{k_{l-1}} - x^{k_l})/t_{k_l} \in \partial_{\epsilon_{k_l}} f_X(x^{k_l})$ by (19)–(22). These relations correspond to A$_4$, A$_5$, (3) and (6) with k replaced by k_l, $k-1$ by k_{l-1} and l by m. Further, $\{x^k\}$ and $\{x^{k_l}\}$ have the same cluster points (if any), whereas B$_1$–B$_2$ yield A$_1$–A$_3$. Hence the conclusion follows from Thms 2.3(c), 2.4 and 2.7(b–d) via suitable identifications. □

Remark 4.5. Consider replacing (20) by the *weaker* descent test

$$f(y^k) \leq f(x^{k_l}) - \kappa\delta_k, \tag{34}$$

where

$$\delta_k := f(x^{k_l}) - \phi_{ix}^k(y^k) = -v_k - \tfrac{1}{2t_k}|y^k - x^{k_l}|^2 = f(x^{k_l}) - \bar{f}_k(x^{k_l}) + \tfrac{1}{2t_k}|y^k - x^{k_l}|^2,$$

$0 \leq \delta_k \leq -v_k \leq 2\delta_k$ (cf. (21)). This test, which may reduce the number of null steps, is used in [11, §XV.3]. Since if (34) fails then so does (20), the proof of Thm 4.3 is not affected. In the proof of Thm 4.4, use $\epsilon_k \leq -\kappa\delta_k - v_k \leq (2-\kappa)\delta_k$ as before.

Remark 4.6. In the unconstrained case ($X = \mathcal{H}$), Lem. 4.1 corresponds to [3, Prop. 3.1] (with a constant t_k), Thm 4.3 to [3, Thm 4.1(i)] (with a constant t_k), Thm 4.4(a) to [3, Thm 4.1(ii)], Thm 4.4(b) to [3, Thm 4.1(ii)] (with only $x^k \rightharpoonup x^\infty$ if $\{t_k\} \subset [t_{\min}, t_{\max}] \subset (0, \infty)$).

5 A Nondescent Bundle Prox Method

Suppose Step 4 is replaced by

Step 4' (*Approximation test*). For a given approximation tolerance $\check{\epsilon}_k > 0$, if

$$\hat{\epsilon}_k := \epsilon_k - \tfrac{1}{2t_k}|y^k - x^{k_l}|^2 = [f(y^k) - \bar{f}_k(y^k)] - \tfrac{1}{2t_k}|y^k - x^{k_l}|^2 \le \check{\epsilon}_k, \tag{35}$$

set $x^k = y^k$, $i_k = 1$, $k_{l+1} = k$ and $l \leftarrow l + 1$; else, set $x^k = x^{k-1}$ and $i_k = 0$.

Remark 5.1. If $\check{\epsilon}_k$ is bounded away from 0 during a series of null steps, the test (35) is eventually passed (cf. Lem. 4.1). Further, (17) gives $\bar{f}_k(y^k) + \frac{1}{2t_k}|y^k - x^{k_l}|^2 \le \phi^k_{ix}(x^{k_l}) \le f(x^{k_l})$, so (35) with $x^{k_{l+1}} = y^{k_{l+1}}$ yields (cf. (5))

$$f(x^{k_{l+1}}) \le f(x^{k_l}) + \hat{\epsilon}_{k_{l+1}} \le f(x^{k_l}) + \check{\epsilon}_{k_{l+1}}.$$

We now exploit the nondescent prox framework of §2.

Theorem 5.2. *Suppose* $\check{\epsilon}_{k_l} \to 0$ *as* $l \to \infty$ *in Algorithm* 3.1 *with Step* 4'. *Let* $\check{s}_m := \sum_{l=1}^m t_{k_l}$ ($\check{s}_m \to \infty$ *as* $m \to \infty$; *cf.* C_3) *and* $x^k_{\mathrm{rec}} \in \operatorname{Arg\,min}_{\{x^j\}_{j=1}^k} f_X(x^j)$ *for all* k. *Then*:

(a) $f_X(x^k_{\mathrm{rec}}) \downarrow f_*$; *also* $f_X(x^k) \to f_*$ *if* $\sum_{l=1}^\infty \check{\epsilon}_{k_l} < \infty$ *or* $\sum_{l=1}^m \check{s}_l \check{\epsilon}_{k_l}/\check{s}_m \to 0$.

(b) *If* $X_* \neq \emptyset$ *and* $\sum_{l=1}^\infty t_{k_l}\check{\epsilon}_{k_l} < \infty$ (*e.g.*, $\sum_l \check{\epsilon}_{k_l} < \infty$ *and* $t_k \le t_{\max} < \infty$ $\forall k$), *then* $x^k_{\mathrm{rec}} \rightharpoonup x^\infty \in X_*$. *Morever*, $x^k \rightharpoonup x^\infty$ *if* $\sum_{l=1}^\infty \check{\epsilon}_{k_l} < \infty$ *or* $\sum_{l=1}^m \check{s}_l\check{\epsilon}_{k_l}/\check{s}_m \to 0$ (*e.g.*, $\{t_{k_l}\} \subset [t_{\min}, t_{\max}] \subset (0, \infty)$ *and* $k_l \check{\epsilon}_{k_l} \to 0$, *or* $\check{\epsilon}_{k_l}\check{s}_l/t_{k_l} \to 0$), *with* $x^k \to x^\infty$ *if additionally* f_X *is boundedly well-set or* $\operatorname{int} X_* \neq \emptyset$.

(c) *If* $X_* = \emptyset$, *then* $|x^k_{\mathrm{rec}}| \to \infty$; *also* $|x^k| \to \infty$ *if* $f_X(x^k) \to f_*$.

Proof. Since $\hat{\epsilon}^+_{k_{l+1}} \le \check{\epsilon}_{k_{l+1}}$ in (35) with $k = k_{l+1}$ and $y^k = x^{k_{l+1}}$ corresponds to (7), the conclusion follows from Lem. 2.9 and Thms 2.3, 2.4 and 2.7 as in the proof of Thm 4.4. □

Remark 5.3. In the unconstrained case ($X = \mathcal{H}$) with (35) replaced by the (stronger) test $\epsilon_k \le \check{\epsilon}_k$, Thm 5.2(a) and Thm 5.2(b) concerning $x^k \rightharpoonup x^\infty$ subsume [3, Thm 5.1].

6 Application to Tikhonov's Regularization

Assuming that $X_* \neq \emptyset$, we now give a modification of Algorithm 3.1 for finding the Tikhonov [26] *regularized solution* $\hat{x} := \arg\min_{X_*} |\cdot|$ as in [1].

Suppose Step 1 sets $x^{k_l} = 0$ and Steps 4–6 are replaced by

Step 4" (*Approximation test*). For a given approximation tolerance $\check{\epsilon}_k > 0$, if

$$\epsilon_k = f(y^k) - \bar{f}_k(y^k) \le \check{\epsilon}_k, \tag{36}$$

set $i_k = 1$, $k_{l+1} = k$ and increase l by 1; otherwise, set $i_k = 0$.

Step 5 (*Bundle updating*). Set $g^k = g_f(y^k)$, $f_k(\cdot) = f(y^k) + \langle g^k, \cdot - y^k\rangle$.

Step 6' (*Stepsize updating*). Choose $t_{k+1} \ge t_k$ s.t. $t_{k+1} = t_k$ if $i_k = 0$.

We say that a *serious* step occurs if $i_k = 1$, i.e., (36) holds. To show that $y^{k_l} \to \hat{x}$, in addition to B_1–B_3, we impose the following requirements.

D_1: $\{\check{\epsilon}_k\}$ is bounded away from 0 if there are finitely many serious steps.
D_2: $\{\nabla \bar{f}_k\}$ is bounded if there are finitely many serious steps.
D_3: $t_k \to \infty$ and $t_k\check{\epsilon}_k \to 0$ if there are infinitely many serious steps.

We start with the following modification of Lem. 4.1.

Lemma 6.1. *Suppose that, for some l, the approximation test* (36) *is omitted, i.e., only null steps are made. Let $\hat{t} = t_{k_l+1}$. Then $\{y^k\} \subset X$ is bounded,* (23) *holds and*

$$y^k \to \hat{y} := \arg\min\left\{ \hat{\phi}(\cdot) := f_X(\cdot) + \tfrac{1}{2\hat{t}}|\cdot|^2 \right\}. \tag{37}$$

Proof. In the proof of Lem. 4.1, set $x^{k_l} = 0$ in (25) and (27), and replace $\bar{\phi}_f^k(x^{k_l})$, $\bar{\phi}_{iX}^k(x^{k_l})$, $f(x^{k_l})$ by $\bar{\phi}_f^k(\hat{x})$, $\bar{\phi}_{iX}^k(\hat{x})$, $\hat{\phi}(\hat{x})$, respectively, in (30) and what follows. Finally, replace t_∞ by $\hat{t}$. □

Remark 6.2. Note the following consequences of Lem. 6.1 and its proof: $l \to \infty$ (cf. (23), (36) and D_1), D_2 is not needed if $X = \mathcal{H}$ or $\bar{f}_k$ is chosen as in Rem. 3.2(i) (cf. Rem. 4.2(iii)).

We may now prove our convergence result as in [1, Thm 3.2].

Theorem 6.3. $y^{k_l} \to \hat{x} := \arg\min_{X_*} |\cdot|$ *as* $l \to \infty$.

Proof. Let $k \in K := \{k_l\}_{l=1}^\infty$. Use $x^{k_l} = 0$ in (22) and argue as in Rem. 2.1(i) to get

$$f_X(y^k) + \tfrac{1}{2t_k}|y^k|^2 \le f_X(\hat{x}) + \tfrac{1}{2t_k}|\hat{x}|^2 + \epsilon_k \le f_X(\hat{x}) + \tfrac{1}{2t_k}|\hat{x}|^2 + \check{\epsilon}_k. \tag{38}$$

Since $f_X(y^k) \ge f_X(\hat{x})$, (38) gives $|y^k|^2 \le |\hat{x}|^2 + 2t_k\check{\epsilon}_k$, so by D_3, $\overline{\lim}_{k\in K} |y^k| \le |\hat{x}|$. In particular, $\{y^k\}_{k\in K}$ is bounded. Let $\{y^k\}_{k\in K'} \rightharpoonup y^\infty$, $K' \subset K$. Then $|y^\infty| \le |\hat{x}|$, since $|\cdot|$ is weakly closed and $\overline{\lim}_{k\in K'} |y^k| \le |\hat{x}|$. Since f_X is weakly closed (being closed convex by B_1–B_2), $t_k \to \infty$ and $\check{\epsilon}_k \to 0$ by D_3, (38) yields $f_X(y^\infty) \le f_X(\hat{x})$. Hence $y^\infty = \hat{x}$. Thus $\{y^k\}_{k\in K'} \rightharpoonup \hat{x}$ and $\overline{\lim}_{k\in K'} |y^k| \le |\hat{x}|$, so $\{y^k\}_{k\in K'} \to \hat{x}$. Therefore, $\{y^k\}_{k\in K} \to \hat{x}$. □

References

[1] A. Auslender, *Numerical methods for nondifferentiable convex optimization*, Math. Programming Stud. **30** (1986) 102–126.

[2] M. A. Bahraoui, *Suites diagonalement stationnaires en optimisation convexe*, Thése de doctorat, Université Montpellier II, Montpellier, France, 1994.

[3] M. A. Bahraoui and B. Lemaire, *Approximate proximal point algorithm and bundle methods for constrained convex minimization*, Tech. report, Mathématiques, Université Montpellier II, Montpellier, 1995.

[4] H. Brézis and P.-L. Lions, *Produits infinis de résolvantes*, Israel J. Math. **29** (1978) 329–345.

[5] R. Cominetti, *Coupling the proximal point algorithm with approximation methods*, J. Optim. Theory Appl. **95** (1997) 581–600.

[6] O. Cornejo, A. Jourani and C. Zălinescu, *Conditioning and upper-lipschitz inverse subdifferentials in nonsmooth optimization problems*, J. Optim. Theory Appl. **95** (1997) 127–148.

[7] R. Correa and C. Lemaréchal, *Convergence of some algorithms for convex minimization*, Math. Programming **62** (1993) 261–275.

[8] N. Dunford and J. T. Schwartz, *Linear Operators, Part I: General Theory*, Wiley-Interscience, New York, 1988.

[9] J. Eckstein and D. P. Bertsekas, *On the Douglas-Rachford splitting method and the proximal point algorithm for maximal monotone operators*, Math. Programming **55** (1992) 293–318.

[10] O. Güler, *On the convergence of the proximal point algorithm for convex minimization*, SIAM J. Control Optim. **29** (1991) 403–419.

[11] J.-B. Hiriart-Urruty and C. Lemaréchal, *Convex Analysis and Minimization Algorithms*, Springer-Verlag, Berlin, 1993.

[12] K. C. Kiwiel, *A method for solving certain quadratic programming problems arising in nonsmooth optimization*, IMA J. Numer. Anal. **6** (1986) 137–152.

[13] ———, *Proximity control in bundle methods for convex nondifferentiable minimization*, Math. Programming **46** (1990) 105–122.

[14] ———, *A Cholesky dual method for proximal piecewise linear programming*, Numer. Math. **68** (1994) 325–340.

[15] ———, *Proximal minimization methods with generalized Bregman functions*, SIAM J. Control Optim. **35** (1997) 1142–1168.

[16] ———, *A bundle Bregman proximal method for convex nondifferentiable minimization*, Math. Programming ? (1998). To appear.

[17] B. Lemaire, *The proximal algorithm*, in New Methods in Optimization and Their Industrial Uses, J. P. Penot, ed., International Series of Numerical Mathematics 87, Birkhäuser, Basel, 1989, pp. 73–87.

[18] ———, *About the convergence of the proximal method*, in Advances in Optimization, W. Oettli and D. Pallaschke, eds., Lecture Notes in Economics and Mathematical Systems 382, Springer-Verlag, Berlin, 1992, pp. 39–51.

[19] ———, *Bounded diagonally stationary sequences in convex optimization*, J. Convex Anal. **1** (1994) 75–86.

[20] ———, *On the convergence of some iterative methods for convex minimization*, in Recent Developments in Optimization, R. Durier and C. Michelot, eds., Lecture Notes in Economics and Mathematical Systems 429, Springer-Verlag, Berlin, 1995, pp. 252–268.

[21] B. Martinet, *Régularisation d'inéquations variationelles par approximations successives*, RAIRO Rech. Opér. **4(R3)** (1970) 154–158.

[22] J.-P. Penot, *Conditioning convex and nonconvex problems*, J. Optim. Theory Appl. **90** (1996) 535–554.

[23] B. T. Polyak, *Introduction to Optimization*, Nauka, Moscow, 1983. English transl., Optimization Software Inc., New York, 1987.

[24] R. T. Rockafellar, *Monotone operators and the proximal point algorithm*, SIAM J. Control Optim. **14** (1976) 877–898.

[25] P. Shunmugaraj, *Well-set and well-posed minimization problems*, Set-Valued Anal. **3** (1995) 281–294.

[26] A. N. Tikhonov and V. Ya. Arsenin, *Methods for Solving Ill-Posed Problems*, Nauka, Moscow, 1974 (Russian). English transl., *Solutions of Ill-Posed Problems*, Wiley, New York, 1977.

[27] T. Zolezzi, *Extended well-posedness of optimization problems*, J. Optim. Theory Appl. **91** (1996) 257–266.

Regularization of Fixed-Point Problems and Staircase Iteration

B. Lemaire *

Abstract

Let C be a nonempty closed convex subset of the Hilbert space $\mathcal{H}$ and P be a nonexpansive mapping from C into C. The Tikhonov regularization method is extended to the fixed point problem for P. This method generates a family of strongly contractive mappings P_r from $\mathcal{H}$ into $\mathcal{H}$ by composition of P with the projector onto C and with the resolvent of a given maximal and strongly monotone operator R on $\mathcal{H}$ with positive parameter r. If the fixed point set S of P is nonempty (for instance C bounded), then, as r tends to zero, u_r converges to $u^\star$ in S the unique solution to the variational inequality defined by R and the closed convex subset S. Moreover, the iteration method suitably combined, by a staircase technique, with approximation of P by a sequence of nonexpansive mappings P_n and with regularization generates a sequence that converges strongly to $u^\star$. Applications to some variational problems are considered.

Keywords: Approximation, fixed point, iteration, monotone, nonexpansive, regularization, staircase iteration, variational inequality, well-posed.

AMS subject classification: 65K10, 49M07, 90C25, 90C48.

1 Introduction

As this work deals with Tikhonov regularization, let us give some insights on Tikhonov well-posedness. First, let us recall this well known notion for minimization problems ([11]). Let X be a normed vector space. An extended real-valued function ϕ on X is minimization Tikhonov well-posed (TWP) iff ϕ has a unique minimizer $\bar{u}$ and $\phi(u_n) \to \inf \phi$ implies $u_n \to \bar{u}$.
The analogue notion for inclusion problems is as follows. A set valued operator T from X into another normed space Y is inclusion TWP iff T has a unique zero $\bar{u}$ and $v_n \in Tu_n$, $v_n \to 0$ implies $u_n \to \bar{u}$.
We can also consider the analogue for fixed point problems. I denoting the identity mapping, a self mapping P on X is fixed point TWP iff P has a unique fixed point $\bar{u}$ and $(I - P)\, u_n \to 0$ implies $u_n \to \bar{u}$.

*Université Montpellier II, Place E. Bataillon, 34 060 Montpellier cedex 05, France

See [19] for connexions between these notions. In particular, it is shown that if X is a Hilbert space and ϕ is a closed proper convex function, then ϕ is minimization TWP iff its subdifferential $\partial\phi$ is inclusion TWP iff its proximal mapping is fixed point TWP.
It is worth noting that, relatively to the above definitions, nonexistence or nonuniqueness of solutions implies Tikhonov ill-posedness (TIP) that is non TWP.
Actually, the purpose of Tikhonov regularization techniques is to approximate, by a suitable perturbation, a possibly TIP problem by a family (in practice a sequence) of TWP problems of same type (minimization, inclusion or fixed point in our cases). This is well known for minimization and inclusion ([30, 13, 29]). Extensions to fixed point problems has yet been considered ([6, 7, 8, 12, 1, 19, 22, 25, 31]). Here we present such a technique with a more general perturbation than the one of the standard Tikhonov regularization. This is done in section 2 in which we prove the main feature that is a selection principle. In section 3 we prove that the selection principle remains true for the iteration method suitably combined, by a staircase technique, with approximation and with regularization, akin to recent results ([9, 15, 23, 17, 18, 25]. Applications are considered to the degenerate Neumann problem in section 4 and to periodic solutions in section 5.

2 Regularization

As it is well known for minimization ([30, 16, 2]), the regularization (or viscosity) method consists in replacing a TIP problem by a family (in practice a sequence) of TWP ones of same type.
Let $\mathcal{H}$ be a real Hilbert space. First let us recall the regularization method for convex minimization. Let ϕ be a closed proper convex function on $\mathcal{H}$, g a continuous real valued strongly convex function on $\mathcal{H}$ and $r > 0$. The regularized minimization problem for ϕ is the minimization problem for

$$\phi_{g,r} := \phi + \frac{r}{2} g.$$

Tikhonov regularization corresponds to the particular case where g is the square norm. As $\phi_{g,r}$ is closed proper, strongly convex, it has a unique minimizer u_r. It is known (see for instance [16, 2]) that when r tends to 0 then $\{u_r\}$ is minimizing for ϕ and if $S :=$ Argmin ϕ is nonempty then u_r norm converges to the unique minimizer of g over S. Since u_r is characterized as the unique zero of the maximal and strongly monotone operator $\partial\phi + r\partial g$ we can extend the regularization method to the inclusion problem as follows.
Let T be a maximal monotone operator on $\mathcal{H}$, R a maximal and strongly monotone operator on $\mathcal{H}$ everywhere defined ($D(R) = \mathcal{H}$) and $r > 0$. The regularized

inclusion problem for T is the inclusion problem for

$$T_{R,r} := T + rR.$$

Tikhonov regularization corresponds to $R := I - \hat{u}$ (i.e. $Ru := u - \hat{u}$, $\forall u \in \mathcal{H}$) for some $\hat{u} \in \mathcal{H}$. As $T_{R,r}$ is maximal and strongly monotone, it has a unique zero u_r.

When T is the subdifferential of a closed proper saddle function ([28]) L on the product $X := X_1 \times X_2$ and R is the subdifferential of a closed real valued strongly convex-concave function G, then the inclusion problem for $T_{R,r}$ is equivalent to the saddle-point problem for

$$L_{G,r} := L + \frac{r}{2}G.$$

So, some convergence results for saddle-point regularization can be deduced from convergence for inclusion regularization (see Proposition 2.1 below).

Now let us consider $J_\lambda^T := (I + \lambda T)^{-1}$ the resolvent of T with positive parameter λ and the three equivalent problems

$$0 \in Tu \quad \Leftrightarrow \quad u = J_\lambda^T u \quad \Leftrightarrow \quad 0 = (I - J_\lambda^T)u$$

with common solution set S. Let $\hat{u} \in \mathcal{H}$ and $r > 0$ be given. Let us consider the Tikhonov regularizations respectively of the equation in the right handside above with regularizing parameter r and of the inclusion in the left handside with regularizing parameter $\dfrac{r}{\lambda(1+r)}$:

$$0 = (r(I - \hat{u}) + \underbrace{I - J_\lambda^T}) \, v_r, \qquad 0 \in (\frac{r}{\lambda(1+r)}(I - \hat{u}) + \underbrace{T}) \, w_r.$$

A simple calculation shows that these are respectively equivalent to

$$v_r = J_r^R \circ J_\lambda^T \, v_r, \qquad w_r = J_\lambda^T \circ J_r^R \, w_r, \tag{1}$$

where $R := I - \hat{u}$, because in this case, $J_r^R = \dfrac{I}{1+r} + \dfrac{r}{1+r}\hat{u}$. As it is known ([29]), if S is nonempty then, when r tends to 0, v_r and w_r norm converge to $\mathrm{proj}_S \hat{u}$ the projection of $\hat{u}$ onto the (closed convex set) S, i.e. the unique solution to the variational inequality $0 \in (R + N_S)\, u$, where N_S denotes the normal cone to S.

Now we can extend the regularization method to fixed point problems in a natural way replacing in (1) J_λ^T by a general nonexpansive mapping and $I - \hat{u}$ by a more general maximal and strongly monotone operator R. More precisely, let C be a nonempty closed convex subset of $\mathcal{H}$ and P be a nonexpansive mapping from C into C, R a maximal and α-strongly monotone operator on $\mathcal{H}$ and $r > 0$. Let us define the external regularized fixed point problem for P as the fixed point problem for

$$P_{R,r}^e := J_r^R \circ P \circ \Pi_C, \tag{2}$$

where Π_C denotes the projection operator onto C, and the internal regularized fixed point problem for P as the fixed point problem for

$$P^i_{R,r} := P \circ \Pi_C \circ J^R_r. \tag{3}$$

As J^R_r is $\dfrac{1}{1+\alpha r}$-strongly contractive, $P^e_{R,r}$ (resp. $P^i_{R,r}$) is strongly contractive and so has a unique fixed-point v_r (resp. w_r), and both regularized mappings are fixed point TWP ([19]). It should be noticed that the external regularized problem is nothing but the regularized inclusion problem for the maximal monotone operator $I - P\circ\Pi_C$. If the resolvent J^R_r maps C into C, in both definitions (2,3), Π_C can be omitted. For instance, $R := I - \hat{u}$ with $\hat{u} \in C$ ([7, 8, 19, 25]). Generally, $w_r \in C$ but v_r may be not in C. Obviously, we have $w_r = P \circ \Pi_C v_r$ and $v_r = J^R_r w_r$.

In the following, we show that the selection principle recalled above for convex minimization and Tikhonov regularization for maximal monotone inclusions holds true for the external and internal fixed point regularization methods. Actually this will be obtain as a corollary of a more general result. Indeed, in applications it is usual to deal with a sequence of approximations P_n of P rather than with P itself. So, if P_n maps C into C and is nonexpansive we can define the two sequences $\{v_n\}$ and $\{w_n\}$ where v_n (resp. w_n) is the unique fixed point of $J^R_{r_n} \circ P_n \circ \Pi_C$ (resp. $P_n \circ \Pi_C \circ J^R_{r_n}$), $\{r_n\}$ being a given sequence of positive reals.

Theorem 2.1 *Let C, P, P_n, v_n (resp. w_n) defined as above and S be the fixed point set of P. Let us assume*
(i) $r_n \to 0$,
(ii) R is the subdifferential of a continuous real valued strongly convex function g on $\mathcal{H}$ or is an everywhere defined maximal and α-strongly monotone bounded operator on $\mathcal{H}$.
(iii) $\forall u \in C,\ \ P_n u \to Pu$,
(iv) $\forall u \in S,\ \ (P_n u - u)/r_n \to 0$.
Then,
if $S \neq \emptyset$, both v_n and w_n norm converge to $u^\star$ the unique solution to the variational inequality

$$0 \in (R + N_S)\, u, \tag{4}$$

if $S = \emptyset$, $\|v_n\| \to +\infty$ and $\|w_n\| \to +\infty$.

Proof. First we note that, S being a closed convex subset and $D(R) = \mathcal{H}$, if S is nonempty then $R + N_S$ is maximal and strongly monotone with nonempty domain S and so $u^\star$ is well defined.
As $w_n = P_n \circ \Pi_C v_n$, the strong convergence of w_n results from the one of v_n thanks to the nonexpansiveness of $P_n \circ \Pi_C$ and assumption (iii). So, let

$$T := I - P \circ \Pi_C, \quad T_n := I - P_n \circ \Pi_C.$$

T, T_n are maximal monotone self mappings on $\mathcal{H}$ and the fixed points of P are the zeroes of T. Moreover, assumptions (iii) and (iv) amounts respectively to

$$\forall u \in \mathcal{H},\ T_n u \to Tu \ \text{ and } \ \forall u \in S,\ \frac{T_n u}{r_n} \to 0.$$

Finally, v_n is nothing but the unique zero of the regularized operator:

$$0 \in (r_n R + T_n)\, v_n \tag{5}$$

The proof is in four steps.
1. $\{v_n\}$ IS BOUNDED.
By definition of v_n, $z_n \in Rv_n$ where $-r_n z_n = T_n v_n$. So, thanks to the monotonicity of T_n, we get

$$\forall u \in \mathcal{H},\ \langle \frac{T_n u}{r_n} + z_n, u - v_n \rangle \geq 0. \tag{6}$$

Let $u \in \mathcal{H}$ and $z \in Ru$. Thanks to the α-strong monotonicity of R, we get $\langle z - z_n, u - v_n \rangle \geq \alpha \|u - v_n\|^2$. So, by (6),

$$\forall u \in \mathcal{H}, \forall z \in Ru,\ \langle z + \frac{T_n u}{r_n}, u - v_n \rangle \geq \alpha \|u - v_n\|^2. \tag{7}$$

Therefore, we have

$$\forall u \in \mathcal{H},\ \forall z \in Ru,\ \|v_n - u\| \leq \|z + \frac{T_n u}{r_n}\|/\alpha. \tag{8}$$

Taking $u \in S$ in (8), thanks to assumption (iv), we get the boundedness of $\{v_n\}$.

2. ANY WEAK LIMIT POINT OF $\{v_n\}$ IS IN S.
Let us consider a subsequence $\{v_{n'}\}$ that converges weakly to v_∞. Let $u \in \mathcal{H}$ and $z \in Ru$. Thanks to the monotonicity of $T_n + r_n R$, from (5) we get $\langle T_{n'} u + r_{n'} z, u - v_{n'} \rangle \geq 0$. Passing to the limit as $n' \to +\infty$, applying assumption (iii), we get $\forall u \in \mathcal{H},\ \langle Tu, u - v_\infty \rangle \geq 0$, that is, thanks to the maximality of T, $0 = Tv_\infty$.
3. $\{v_n\}$ HAS $u^\star$ AS UNIQUE WEAK LIMIT POINT.
3.1. Case where R is the subdifferential of a real valued strongly convex function g. So, $\forall u \in \mathcal{H},\ g(u) \geq g(v_n) + \langle z_n, u - v_n \rangle$, and, using (6), $g(u) \geq g(v_n) - \langle \frac{T_n u}{r_n}, u - v_n \rangle$. Therefore, thanks to the weak lower semi continuity of g and to assumption (iv), any weak limit point v_∞ of $\{v_n\}$ satisfies $\forall u \in S,\ g(v_\infty) \leq g(u)$, that is, $v_\infty = u^\star$.
3.2 Case where R is a bounded operator.
$z_n \in Rv_n$ being bounded, it has a weak limit point z_∞. Let us consider a subsequence $\{z_{n'}\}$ that converges weakly to z_∞, and such that $v_{n'}$ converges weakly to some $v_\infty \in S$. From (6) we have

$$\langle \frac{T_{n'} v_\infty}{r_{n'}}, v_\infty - v_{n'} \rangle + \langle z_{n'}, v_\infty \rangle \geq \langle z_{n'}, v_{n'} \rangle,$$

and so, thanks to assumption (iv),

$$\limsup_{n'\to+\infty} \langle z_{n'}, v_{n'}\rangle \leq \langle z_\infty, v_\infty\rangle.$$

Therefore ([5], proposition 2.5), $z_\infty \in Rv_\infty$ and $\langle z_{n'}, v_{n'}\rangle \to \langle z_\infty, v_\infty\rangle$. Passing to the limit in (6) for the considered subsequence, we get

$$\forall u \in S,\ \ \langle z_\infty, u - v_\infty\rangle \geq 0,$$

that is, $-z_\infty \in N_S(v_\infty)$, so $v_\infty = u^\star$.
4. STRONG CONVERGENCE.
From above, v_n converges weakly to $u^\star$. Strong convergence results from (7) and assumption (iv).

If S is empty, assuming that $\|v_n\| \not\to +\infty$, some subsequence $\{v_{n'}\}$ would be bounded. From step 2 above, $\{v_{n'}\}$ would have a weak limit point in S, a contradiction. □

Taking $P_n = P$ for all $n \in I\!N$ and noticing that in Theorem 2.1 the convergence property doesn't depend on the sequence $\{r_n\}$ tending to 0 lead immediately to the selection principle stated in the following corollary.

Corollary 2.1 *Let C, P, R, v_r (resp. w_r) defined as above and S be the fixed point set of P. Let us assume that R is the subdifferential of a continuous real valued strongly convex function on $\mathcal{H}$ or is an everywhere defined maximal and strongly monotone bounded operator on $\mathcal{H}$.*
If $S \neq \emptyset$, then $\lim_{r\to 0} v_r = \lim_{r\to 0} w_r = u^\star$ the unique solution to the variational inequality $0 \in (R + N_S)\, u$.
If $S = \emptyset$, then $\lim_{r\to 0} \|v_r\| = \lim_{r\to 0} \|w_r\| = +\infty$.

Remark 2.1 *(i) It results from Corollary 2.1 that $\{v_r\}$ is bounded iff $\{w_r\}$ is bounded iff S is nonempty. So, as $w_r \in C$, if C is bounded then S is nonempty, which is the Browder's theorem. Actually, the proof of Browder relies on the Tikhonov fixed point regularization ($R := I - \hat{u}$ with $\hat{u} \in C$). In ([6]) Browder proved only the weak convergence of a subsequence to some fixed point, in ([7, 8]) he proved the strong convergence to the fixed point closest to $\hat{u}$ that is, the selection principle in this particular case.*
(ii) When P is the projector onto a given nonempty closed convex subset S (so its fixed point set is S), the external regularization coincides with the partial Yosida approximation of the variational inequality (4): $0 \in (R + (N_S)_r)\, u$, where, for a maximal monotone operator A, A_r denotes its Yosida approximation $A_r := \dfrac{I - J_r^A}{r}$. As mentioned in ([21]), a direct consequence of ([5], Theorem 2.4) gives the estimate

$$\forall z \in (-Ru^\star) \cap N_S u^\star,\ \ \|v_r - u^\star\| \leq \|z\|\sqrt{r/\alpha}.$$

(iii) In ([24]) it is proposed, as a direct extension of Tikhonov fixed point regularization the following fixed point regularization scheme

$$v_r = \frac{1}{1+r}Pv_r + \frac{r}{1+r}\pi v_r$$

where π is a strongly contractive mapping from $\mathcal{H}$ into C, Tikhonov regularization corresponding to the particular case π constant equal to some $\hat{u}$ in C. Actually, a simple calculation shows that this scheme is nothing but our external fixed point regularization. Indeed, $v_r = J_r^R \circ Pv_r$ with $R := I - \pi$, which is an everywhere defined, bounded (because Lipschitzian), maximal and strongly monotone operator. Moreover, J_r^R maps C into C, so v_r is as well the unique fixed point of $J_r^R \circ P \circ \Pi_C$.
Therefore, applying Corollary 2.1 we get the strong convergence of v_r to the unique solution to the variational inequality $0 \in (I - \pi + N_S)\, u$.

Actually, the proof of Corollary 2.1 that is, the proof of Theorem 2.1 with $P_n = P$, relies essentially on the maximal monotonicity of $I - P \circ \Pi_C$ and the interpretation (5) of the external fixed point regularization. So, with a slight adaptation (for more details see [20] Theorem 1), this proof works to get, in the following proposition, the selection principle for the regularization of maximal monotone inclusions, so extending the Tikhonov one, a result strongly connected (but noncomparable) with Theorem 1 of [7].

Proposition 2.1 *Let T be a maximal monotone operator on $\mathcal{H}$ and u_r be the solution to the regularized inclusion problem*

$$0 \in (rR + T)\, u,$$

where R is the subdifferential of a real valued strongly convex function on $\mathcal{H}$ or is an everywhere defined maximal and strongly monotone bounded operator on $\mathcal{H}$. If $S := T^{-1}(0) \neq \emptyset$, then $\lim_{r \to 0} u_r = u^\star$ the unique solution to the variational inequality $0 \in (R + N_S)\, u$.
If S is empty then $\lim_{r \to 0} \|u_r\| = +\infty$.

3 Staircase Iteration

Let us consider some problem $\mathcal{P}$ with solution set S. The following is a standard situation in Numerical Analysis. There exists a sequence of approximate problems $\mathcal{P}_n$ of same type than $\mathcal{P}$ for which a notion of quasi solution $\tilde{u}_n$ controled by some positive ϵ_n such that, when ϵ_n tends to 0, then $\tilde{u}_n$ converges in a suitable sense to some solution $u^\star \in S$. Moreover, there exists an iterative method adapted to the considered type problem. A usual approach is to apply this iterative method successively to $\mathcal{P}_n$ for $n = 1, 2, \ldots$, but for each n stopping after a finite number k_n of iterates and restarting with a new approximate problem from the last iterate, giving so a "staircase iterative process":

$$\begin{cases} u_1^0 \text{ given}, \quad n = 1, 2, \dots \\ k = 1, \dots, k_n, \quad u_n^k = Q_n\, u_n^{k-1}; \\ u_{n+1}^0 := u_n^{k_n}; \end{cases}$$

where the iteration mapping Q_n is defined by the method.
Now, the main question is to deal with a "good" stopping rule that is, how defining k_n such that $u_n^{k_n}$ be an ϵ_n-solution of $\mathcal{P}_n$. Of course a possibility is to fix k_n in advance, which amounts to setting $k_n := 1$ for al n (considering k_n copies of $\mathcal{P}_n$), leading to what we can call a "diagonal iterative process". But, this is nonflexible and in general needs more restrictive assumptions than an adaptative choice to be a "good" stopping rule.

The purpose of this section is to show how fixed point regularization allows to design a simple adaptative stopping rule and get convergence results under mild assumptions. Let us first present this stopping rule in a general framework.
Let X be a real Banach space, $\{Q_n\}$ be a sequence of σ_n-strongly contractive self mappings on X, for all n, $\{\delta_n^k\}$ be a sequence of nonnegative reals such that $\lim_{k\to+\infty} \delta_n^k = 0$, $\{\epsilon_n\}$ be a sequence of positive reals. Let us consider the following staircase iterative process in which each iterate u_n^{k+1} is performed in an approximate way within the tolerance δ_n^k.

$$\begin{cases} u_1^0 \in X, \quad n = 1, 2, \dots \\ k = 1, \dots, k_n, \quad \|u_n^k - Q_n\, u_n^{k-1}\| \le \delta_n^k, \\ k_n := \text{first } k, \ \|u_n^k - u_n^{k-1}\| + \delta_n^k \le \epsilon_n; \\ u_{n+1}^0 := u_n^{k_n}; \end{cases}$$

It should be noticed that the stopping rule is effective because, thanks to the strong nonexpansiveness of Q_n and to the Hardy-Cesaro's theorem on convolution of real sequences, $\lim_{k\to+\infty} u_n^k = u_n$ the unique fixed point of Q_n, and so

$$\lim_{k\to+\infty} (\|u_n^k - u_n^{k-1}\| + \delta_n^k) = 0.$$

Proposition 3.1 *(Comparison result).*
Let $\{u_n^k\}$ be generated by the above staircase process. Then
(i) $\forall n,\ \|u_n^{k_n} - u_n\| \le \epsilon_n/(1-\sigma_n)$.
(ii) $\forall n,\ k = 0, \dots, k_n,\ \|u_n^k - u_n\| \le \dfrac{\epsilon_{n-1}}{1-\sigma_{n-1}} + \|u_{n-1} - u_n\| + \dfrac{\sup_k \delta_n^k}{1-\sigma_n}.$

Proof.
(i) $\|u_n^{k_n} - u_n\| \le \delta_n^{k_n} + \sigma_n \|u_n^{k_n-1} - u_n\| \le \delta_n^{k_n} + \|u_n^{k_n-1} - u_n^{k_n}\| + \sigma_n \|u_n^{k_n} - u_n\|$.
(ii) $\|u_n^k - u_n\| \le \|u_n^0 - u_n\| + \sup_k \delta_n^k \sum_{i=0}^k \sigma_n^i \le \|u_{n-1}^{k_{n-1}} - u_{n-1}\| + \|u_{n-1} - u_n\| + \sup_k \delta_n^k/(1-\sigma_n)$. □

Remark 3.1 *If $\forall n, k_n = 1$, case of diagonal iteration, an analogue result is given in ([18], Proposition 5.1) under more restrictive assumptions.*

In the following we specialize the basic iterative method (i.e. defined by Q_n) by mean of fixed point regularization. Let $\mathcal{H}$ be a real Hilbert space, C a nonempty closed convex subset of $\mathcal{H}$, $P, \{P_n\}$ be a sequence of nonexpansive self mappings on C, $\{r_n\}$ be a sequence of positive reals such that

$$\forall u \in C,\ P_n u \to Pu, \quad \forall u \in S,\ (P_n u - u)/r_n \to 0.$$

Let us define Q_n as follows:

$$\forall n, \quad Q_n := \begin{cases} J^R_{r_n} \circ P_n \circ \Pi_C \\ \text{or} \\ P_n \circ \Pi_C \circ J^R_{r_n} \end{cases}$$

where R satifies assumptions of Theorem 2.1. We have

$$u_n := v_n \ \text{ or } \ w_n, \quad \sigma_n = 1/(1 + \alpha\, r_n).$$

As the sequence $\{u_n\}$ is a mixing of the two sequences of (approximate) regularized fixed points, in the context of Theorem 2.1, it converges strongly to $u^\star$ the unique solution to the variational inequality (4). Therefore, the following proposition is a direct consequence of Theorem 2.1 and Proposition 3.1.

Proposition 3.2 *(i) $\epsilon_n/r_n \to 0 \ \Rightarrow\ u_n^{k_n} \to u^\star$.*
(ii) $(\epsilon_n + \sup_k \delta_n^k)/r_n \to 0 \ \Rightarrow\ u_n^k \to u^\star$.

Remark 3.2 *If $\forall n, k_n = 1$, case of diagonal iteration, $u_n^1 \to u^\star$ holds true under more restrictive assumptions: $\{r_n\} \notin \ell^1$, $|\frac{1}{r_n} - \frac{1}{r_{n-1}}| \to 0$ (with standard Tikhonov fixed point regularization see [19, 25], see also [4, 14] for a similar result). If moreover $\forall n, P_n = \Pi_S$, our diagonal iterative process coincides with the Passty's scheme based upon partial Yosida approximation of the variational inequality (4). In this case, the same result holds true under the assumption $\{r_n\} \in \ell^2 \backslash \ell^1$ ([26], Theorem 5).*

4 Degenerate Neumann Problem

In this section we are concerned with a typical example of variational problem which may have no solution or infinitely many ones and so is always TIP.
Let Ω be a bounded domain of $\mathbb{R}^p$ and $\mathcal{H}$ be the Sobolev space $H^1(\Omega)$. Let $f \in L^2(\Omega)$ and consider the degenerate continuous symmetric bilinear form a on $\mathcal{H}$:

$$\forall u, v \in V,\ \ a(u,v) := \sum_{i=1}^{p} \left(\frac{\partial u}{\partial x_i}, \frac{\partial v}{\partial x_i}\right),$$

$(.,.)$ denoting the inner product on $L^2(\Omega)$. This defines the real valued continuous quadratic convex function on $\mathcal{H}$:

$$\forall v \in V, \quad \phi(v) := \frac{1}{2}\, a(v,v) - (f,v).$$

It is known that $S :=$ Argmin$_{\mathcal{H}}$ $\phi \neq \emptyset$ iff $\int_\Omega f dx = 0$, in which case $S = \bar{u} + I\!R\, \mathbf{1}$ for any $\bar{u} \in S$. Such u solves in a weak sense the degenerate Neumann problem for the Laplacian:

$$\begin{cases} -\Delta u &= f \quad \text{in} \quad \Omega \\ \dfrac{\partial u}{\partial n} &= 0 \quad \text{on} \quad \partial\Omega \end{cases}$$

Let $Tu := \nabla\phi(u) = \Lambda_{\mathcal{H}}^{-1}(Au - f)$ where A denotes the linear operator associated with the bilinear form a and $\Lambda_{\mathcal{H}}$ the canonical isomorphism from $\mathcal{H}$ onto its dual. T is a maximal monotone (in fact semipositive definite continuous affine) operator on $\mathcal{H}$ and it is standard that $u \in S$ iff $u = J_\lambda^T\, u =$ prox$_{\lambda\phi} u$, $\lambda > 0$.

4.1 Tikhonov Regularization

In this example the standard Tikhonov regularization defines, for $r > 0$,

$$u_r := \ \text{argmin}_{\mathcal{H}}\ \{\phi + r\,\frac{|\,.\,|^2}{2}\} = \text{argmin}_{\mathcal{H}}\ \{\phi + \frac{r}{1-r}\,[\frac{\|\,.\,\|^2}{2} - (f,.)]\},$$

solution to the nondegenerate Neumann problem:

$$\begin{cases} -\Delta u_r + r\, u_r &= f \quad \text{in} \quad \Omega \\ \dfrac{\partial u_r}{\partial n} &= 0 \quad \text{on} \quad \partial\Omega \end{cases}$$

It is standard that, as r tends to 0, u_r converges strongly in $\mathcal{H}$ and in $L^2(\Omega)$ to $u^\star$ the projection (for both norms) of 0 onto S, that is, $\int_\Omega u^\star dx = 0$.

4.2 Discrete Approximation

We present here a simplified discussion of the discrete approximation, refering to ([27]) for more details on the finite elements method. First we note that for fixed $\lambda > 0$ and for all u in $\mathcal{H}$, Pu $(= J_\lambda^T u)$ is the unique minimizer of

$$\phi_{\lambda,u} := \frac{1}{2\lambda}\|. - u\|^2 + \phi$$

over $\mathcal{H}$. Let us consider, for fixed u, the first order finite element approximation of this variational problem. This defines a family $\{V_h\}_{h>0}$ of finite dimensional subspaces of $\mathcal{H}$, $P_h u := J_\lambda^{T_h} u$ being the unique minimizer of $\phi_{\lambda,u}$ over V_h, with $T_h := \nabla\phi + N_{V_h}$.

If we assume that Ω is polyhedral and $p \leq 3$, the following holds true ([27], Théorème 5.1-4). $\forall u \in \mathcal{H}$, $\lim_{h\to 0} P_h u = Pu$, and if $S \subset H^2(\Omega)$ (regularity condition), then $\forall u \in S$, $\|P_h u - u\| \leq c(\lambda, u)\, h$ for a suitable constant $c(\lambda, u)$.

4.3 Prox + Tikhonov + Approximation

Let us assume that S is nonempty and that the regularity condition holds true. Let us consider a sequence of approximation parameters $\{h_n\}$ and a sequence of regularization parameters $\{r_n\}$ such that $r_n \to 0$ and $h_n/r_n \to 0$. We are in the following situation. $T, T_n (:= T_{h_n}), n = 1, \ldots$ are maximal monotone operators on the real Hilbert space $\mathcal{H}$, $\lambda > 0$, $0 < r_n \to 0$,

$$\forall u \in \mathcal{H},\ J_\lambda^{T_n} u \to J_\lambda^T u, \qquad \forall u \in S,\ \|J_\lambda^{T_n} u - u\|/r_n \to 0.$$

We can consider the following particular staircase iterative process:

$$\begin{cases} u_1^0 \in \mathcal{H}, \quad n = 1, 2, \ldots \\ k = 1, \ldots, k_n, \quad \|u_n^k - J_\lambda^{T_n}\ u_n^{k-1}/(1 + r_n)\| \leq \delta_n^k, \\ k_n := \text{first } k, \ \|u_n^k - u_n^{k-1}\| + \delta_n^k \leq \epsilon_n; \\ u_{n+1}^0 := u_n^{k_n}; \end{cases}$$

where, for each n, $J_\lambda^{T_n}\ u_n^k/(1 + r_n)$ must be understood as $(J_\lambda^{T_n}\ u_n^k)/(1 + r_n)$ (external regularization) or as $J_\lambda^{T_n}\ (u_n^k/(1 + r_n))$ (internal regularization). Applying Proposition 3.2, we get

$$\epsilon_n/r_n \to 0 \Rightarrow u_n^{k_n} \to \text{proj}_S 0, \quad (\epsilon_n + \sup_k \delta_n^k)/r_n \to 0 \Rightarrow u_n^k \to \text{proj}_S 0.$$

Remark 4.1 *(i) The above particular staircase process compares with the multistep proximal method of Kaplan-Tichatschke ([13]) in which no Tikhonov regularization is involved.*
(ii) If $\forall n, k_n = 1$, case of diagonal iteration, the same convergence result holds true for u_n^1 ([15, 18]) under more restrictive assumptions:
$\sum_{n=1}^{+\infty} r_n = +\infty$, $|\frac{1}{r_n} - \frac{1}{r_{n-1}}| \to 0$.

5 Periodic Solutions

In this section we consider an example of fixed point problem for a nonexpansive mapping which is not the resolvent of a maximal monotone operator.
Let $\mathcal{H}$ be a real Hilbert space. Let be given a function $f : I\!R^+ \to \mathcal{H}$ and, for all $t \geq 0$, a closed proper convex function φ_t on $\mathcal{H}$. Let us assume that f and $t \to \varphi_t$ are τ-periodic for some positive τ, and consider the first order differential inclusion with initial value:

$$\frac{du}{dt} + \partial\varphi_t(u(t)) \ni f(t), \quad t > 0, \quad u(0) := x. \tag{9}$$

It is known ([3], Théorème 1A) that under a suitable growth condition on $t \to \varphi_t$, if $f \in L^2(0, \tau; \mathcal{H})$ and $x \in C := \overline{D(\varphi_0)}$, then there exists $u(x, .)$ unique strong solution to (9) on $[0, \tau]$.

$\forall x \in C$, let $Px := u(x,\tau)$. P maps C into $D(\varphi_\tau)$ $(= D(\varphi_0))$ and, thanks to the monotonicity of the subdifferential, is nonexpansive. Moreover the fixed point set S of P is the set of all initial values of a τ-periodic solution to (9). So, (9) has periodic solutions if, for instance C is bounded ([6]), or if, on each interval of periodicity, $\varphi_t = \varphi$, a strongly coercive closed proper convex function on $\mathcal{H}$ ([5], Corollaire 3.4).

5.1 Discrete Approximation

We present an approximation scheme defining a family $\{P_h\}_{h>0}$ of approximate fixed point mappings by discretizing the interval $[0,\tau]$.
Let $N \in I\!N\backslash\{0\}$, $h := \tau/N$, $t_i := ih$, $i = 0, ..., N$, and consider the implicite Euler discretization scheme of (9):

$$\frac{u_i - u_{i-1}}{h} + \partial\varphi_{t_i}(u_i) \ni f(t_i),\ i = 1, ..., N, \quad u_0 := x,$$

which is equivalent to:

$$u_0 := x,\ \ u_i = J_h^{\partial\tilde{\varphi}_{t_i}}\, u_{i-1},\ i = 1, ..., N, \ \ \text{with}\ \ \tilde{\varphi}_t := \varphi_t - \langle f(t), .\rangle.$$

Therefore, P_h is defined as follows:

$$P_h := J_h^{\tilde{\varphi}_\tau} \circ ... \circ J_h^{\tilde{\varphi}_{t_1}}.$$

Clairly, P_h maps C into C and, as a finite composition of nonexpansive mappings, is nonexpansive.
The following proposition states an approximation result which will be used in the following subsection for applying Proposition 3.2.

Proposition 5.1 *Let $u(x,.)$ be the solution to (9). Let us assume (regularity condition) that for all $\forall x \in D(\varphi_0)$, $\frac{d}{dt}u(x,.)$ is absolutely continuous on $[0,\tau]$. Then*

$$\forall x \in C,\ \lim_{h\to 0} P_h x = Px, \qquad \forall x \in S,\ \|P_h x - x\| \le c(x)\, h,$$

for a suitable constant $c(x)$.

Proof. The result follows from the estimate:

$$\forall x \in D(\varphi_0),\ \|Px - P_h x\| \le c(x)\, h, \tag{10}$$

and the facts that C is the closure of $D(\varphi_0)$ and that S is included in $D(\varphi_0)$. So, let us prove (10).
Thanks to the monotonicity of $\partial\tilde{\varphi}_t$, we have

$$\langle u_{i-1} - u_i + h\frac{du(x,.)}{dt}(t_i), u_i - u(x,t_i)\rangle \ge 0,$$

from which we get easily the estimate

$$\|u(x,t_i)-u_i\| \le \|u(x,t_{i-1})-u_{i-1}\| + \|u(x,t_i)-u(x,t_{i-1})-h\frac{du(x,.)}{dt}(t_i)\|.$$

As $u(x,.)$ is C^1,

$$\|u(x,t_i)-u(x,t_{i-1})-h\frac{du(x,.)}{dt}(t_i)\| \le h\|\frac{du(x,.)}{dt}(t_{i-1}+\theta_i t_i)-\frac{du(x,.)}{dt}(t_i)\|,$$

with $0<\theta_i<1$. Finally,

$$\|u(x,\tau)-u_N\| \le h\|\frac{d^2u(x,.)}{dt^2}\|_{L^1(0,\tau;\mathcal{H})},$$

which is precisely (10). □

Remark 5.1 *Generally ([3], Théorème 1A),* $\frac{d}{dt}u(x,.) \in L^2(0,\tau;\mathcal{H})$. *The regularity condition holds true, for instance, if f is absolutely continuous on $[0,\tau]$ and if, on each interval of periodicity, $\varphi_t=\varphi$ with $\varphi \in C^2(\mathcal{H})$.*

5.2 Iteration + Tikhonov + Approximation

Let $N_n \in I\!N\backslash\{0\}$, $h_n := \tau/N_n$, $\epsilon_n > 0$, $0 < r_n \to 0$,

$$\frac{h_n+\epsilon_n}{r_n} \to 0, \ \forall n \in I\!N\backslash\{0\},\ i=1,...,N_n,\ 0 \le \delta_n^{i,k} \xrightarrow{k} 0.$$

The general staircase process of section 3 in the framework of Proposition 3.2, using standard Tikhonov fixed point regularization and the previous approximation scheme specializes as follows.

$$\begin{cases} \hat{x},\ u_1^0 \in D(\varphi_0), \quad n=1,2,... \\ k=1,...,k_n, \quad x(0):=u_n^{k-1}\ \ [x(0):=\frac{1}{1+r_n}u_n^{k-1}+\frac{r_n}{1+r_n}\hat{x}], \\ i=1,...,N_n,\ \|x(i)-J_{h_n}^{\tilde{\varphi}_{t_i}}\ x(i-1)\| \le \delta_n^{i,k}; \\ u_n^k := \frac{1}{1+r_n}x(N_n)+\frac{r_n}{(1+r_n)}\hat{x} \quad [u_n^k := x(N_n)], \\ k_n := \text{first } k,\ \|u_n^k-u_n^{k-1}\|+\sum_{i=1}^{N_n}\delta_n^{i,k} \le \epsilon_n; \\ u_{n+1}^0 := u_n^{k_n}; \end{cases}$$

where, for each n, the terms in [ ] correspond to internal regularization of P_{h_n}, while otherwise external regularization is involved. Moreover, thanks to composition of nonexpansive mappings, the tolerance δ_n^k in the general scheme is given here by $\sum_{i=1}^{N_n}\delta_n^{i,k}$ explaining this sum in the stopping rule. When $\mathcal{H}$ is finite dimensional and φ_t is everywhere finite, the computation of $x(i)$ that is, the approximate computation of $J_{h_n}^{\tilde{\varphi}_{t_i}}\ x(i-1)$ within the tolerance $\delta_n^{i,k}$ can

be performed by cutting plane techniques like in bundle methods for convex minimization ([10]).
If we assume the regularity condition in (9), the convergence of this scheme results from Propositions 3.2 and 5.1: $u_n^{k_n} \to \text{proj}_S\, \hat{x}$.

Acknowledgements. Many thanks to A. Moudafi and to the referee for having indicated useful references.

References

[1] Ya. I. Alber and S. Guerre-Delabrière, "Problems of fixed point theory in Hilbert and Banach spaces" *Funct. Diff. Equations, Proceedings of the Israel Seminar* 2 (1994) 5-10.

[2] H. Attouch, "Viscosity solutions of minimization problems" *SIAM J. Optimization* 6, 3 (1996) 769-806.

[3] H. Attouch and A. Damlamian, "Problèmes d'évolution dans les Hilberts et applications" *J. Math. Pures et Appl.* 54 (1975) 53-74.

[4] A.B. Bakushinskii, "Iterative methods for solving nonlinear operator equations in the absence of regularity. A new approach" *Russian Acad. Sci. Dokl. Math.* 47,3 (1993) 451-454.

[5] H. Brezis, *Opérateurs maximaux monotones et semi-groupes de contractions dans les espaces de Hilbert*, (Cours polycopié, Université Paris VI et CNRS-ERA 215, 1971) (North Holland, 1973).

[6] F.E. Browder, "Fixed point theorems for nonlinear semicontractive mappings in Banach spaces" *Arch. for Rational Mech. and Anal.* 21 (1966) 259-269.

[7] F.E. Browder, "Existence and approximation of solutions of nonlinear variational inequalities" *Proc. Nat. Acad. Sci. U.S.A.* 56 (1966) 1080-1086.

[8] F.E. Browder, "Convergence of approximants to fixed points of nonexpansive nonlinear mappings in Banach spaces" *Arch. for Rational Mech. and Anal.* 24 (1967) 82-90.

[9] R. Cominetti, "Coupling the proximal point algorithm with approximation methods" to appear in JOTA.

[10] R. Correa and C. Lemaréchal, "Convergence of some algorithms for convex minimization" *Math. Programming* 62 (1993) 261-276.

[11] A. Dontchev and T. Zolezzi, *Well-posed Optimization Problems*, (Lecture Notes in Mathematics 1543, Springer, 1993).

[12] K. Goebel and Reich, "Uniform convexity, hyperbolic geometry and non-expansive mappings" *Pure and applied Math.* 83 (1984).

[13] A. Kaplan and R. Tichatschke, *Stable methods for ill-posed problems*, (Akademie Verlag, Berlin, 1994).

[14] B. Halpern, "Fixed points of nonexpanding maps" *Bull. AMS* 73 (1967) 957-961.

[15] N. Lehdili and A. Moudafi, "Combining the proximal algorithm and Tikhonov regularization" *Optimization* 37, 3 (1996) 239-252.

[16] B. Lemaire, "Régularisation et pénalisation en optimisation convexe" *Séminaire d'Analyse Convexe de Montpellier* 1 (1971) exposé 17.

[17] B. Lemaire, "On the convergence of some iterative methods for convex minimization" *Recent Developments in Optimization, Lecture Notes in Economics and Mathematical Systems, Springer* 429 (1995) 252-268.

[18] B. Lemaire, "Which fixed-point does the iteration method select?" *Recent Advances in Optimization, Lecture Notes in Economics and Mathematical Systems* 452 (1997) 154-167.

[19] B. Lemaire, "Well-posedness, conditioning and regularization of minimization, inclusion and fixed point problems" *to appear in Pliska Studia Mathematica Bulgarian Journal.*

[20] B. Lemaire, "From Fixed Point Regularization to Constraints Decomposition in Variational Inequalities" *to appear in Proc. of the 9th Belgian-French-German Conference on Optimization, Namur 1998.*

[21] P. Mahey and Pham Dinh Tao, "Partial regularization of the sum of two maximal monotone operators" *Math. Model. and Num. Anal.* (M^2AN) 03 (1993) 375-392.

[22] V.A. Morosov, *Regularization Methods for Ill posed-problems*, (CRC Press, Florida, 1993).

[23] A. Moudafi, "Coupling proximal algorithm and Tikhonov method" *Nonlinear Times and Digest* 1 (1994) 203-210.

[24] A. Moudafi, "Personal communication".

[25] C. Ould Ahmed Salem, *Approximation de points fixes d'une contraction*, (Thèse de Doctorat, Université Montpellier 2, 1998).

[26] G.B. Passty, "Ergodic convergence to a zero of the sum of monotone operators in Hilbert spaces" *J. of Math. Anal. and Appl.* 72 (1979) 383-390.

[27] P.A. Raviart and J.M. Thomas, *Introduction à l'analyse numérique des équations aux dérivées partielles*, (Masson, 1983).

[28] R.T. Rockafellar, *Convex Analysis*, (Princeton University Press, 1970).

[29] P. Tossings, "The perturbed Tikhonov's algorithm and some of its applications" *Math. Model. and Num. Anal. (M^2AN)* 28 (1994) 189-221.

[30] A. Tikhonov and V. Arsenine, *Méthodes de résolution de problèmes mal posés*, (MIR, 1974).

[31] V.V. Vasin, "Ill-posed problems and iterative approximation of fixed points of pseudo-contractive mappings" *in: Proc. Intern. Conf. on Ill-posed Problems, August 1991* (ed. A.N. Tikhonov), Moscow (1992).

$\mathcal{VU}$-Decomposition Derivatives for Convex Max-Functions

Robert Mifflin * Claudia Sagastizábal **

Abstract

For minimizing a convex max-function f we consider, at a minimizer, a space decomposition. That is, we distinguish a subspace $\mathcal{V}$, where f's nonsmoothness is concentrated, from its orthogonal complement, $\mathcal{U}$. We characterize smooth trajectories, tangent to $\mathcal{U}$, along which f has a second order expansion. We give conditions (weaker than typical strong second order sufficient conditions for optimality) guaranteeing the existence of a Hessian of a related $\mathcal{U}$-Lagrangian. We also prove, under weak assumptions and for a general convex function, superlinear convergence of a conceptual algorithm for minimizing f using $\mathcal{VU}$-decomposition derivatives.

Keywords: Max-functions, second-order derivatives, $\mathcal{U}$-Lagrangian.

AMS subject classification: Primary: 49K35, 49M27. Secondary: 65K10, 90C25.

1 Introduction

1.1 Motivation

Many convex minimization problems have an objective function that is not smooth but is of the max type:

$$\min_{x\in\mathbb{R}^N} f(x)\,, \quad \text{where } f(x) = \max\{f_i(x) : i = 0, \cdots, m\} \tag{1}$$

is convex and each f_i is C^2 on $\mathbb{R}^N$. One of the main difficulties with such problems is that f is not differentiable at those points where more than one underlying function f_i defining the max is active (typically, such is the case at a minimizer $\bar{x}$ in (1)). Furthermore, there is no well-defined Hessian, or second-order object, and a straightforward application of a Newton-type method is not possible.

*Mathematics, Washington State University, Pullman, WA 99164-3113. Research supported by the National Science Foundation under Grant No. DMS-9703952 and by FAPERJ (Brazil) under Grant No. E-26/171.393/97.

**DEE/PUC-Rio. Rua Marquês de São Vicente, 225. 22453-000 Rio de Janeiro RJ, Brazil Research supported by FAPERJ (Brazil) under Grant No.E26/150.205/98.

Since the existence of well-defined Hessians is crucial for developing fast algorithms, this paper focuses on identifying second-order information **contained** in f, in spite of its nonsmoothness. In our effort we follow the lead of [4], where $\mathbb{R}^n$ is decomposed into two orthogonal subspaces $\mathcal{V}$ and $\mathcal{U}$ in such a way that, near $\bar{x}$, all of f's nonsmoothness is concentrated in $\mathcal{V}$. More precisely, for an arbitrary $g \in \partial f(\bar{x})$, the subdifferential of f at $\bar{x}$, define

$$\mathcal{V} := \text{lin}(\partial f(\bar{x}) - g) \qquad \text{and} \qquad \mathcal{U} := \mathcal{V}^{\perp}. \tag{2}$$

Note that the linear space $\mathcal{V}$, depending on $\bar{x}$, is independent of the particular choice of $g \in \partial f(\bar{x})$. Because f often appears to be smooth on the $\mathcal{U}$-subspace, it may have some kind of Hessian, which we call a "$\mathcal{U}$-Hessian" and denote by $\text{H}_{\mathcal{U}} f$, see § 1.2 below. In this paper, using the $\mathcal{VU}$-theory following from (2), we give rather weak sufficient conditions for the existence of a $\mathcal{U}$-Hessian matrix at $\bar{x}$ and certain smooth trajectories, tangent to $\mathcal{U}$, yielding a second order expansion for a max-function f as in (1).

We say a minimax problem (1) is *well-posed* if f has a $\mathcal{U}$-Hessian at a minimizer $\bar{x}$ with respect to $0 \in \partial f(\bar{x})$. Our conditions for implying that (1) is well-posed are weaker than assuming that the gradients of the active f_i's at $\bar{x}$ are affinely independent with corresponding multipliers that are strictly positive.

Next we describe the basics of the $\mathcal{VU}$-space decomposition, in particular the important concept of the $\mathcal{U}$-Lagrangian and its gradient and Hessian.

1.2 $\mathcal{VU}$-Theory

Relative to a general convex function f with subdifferential ∂f, we recall some definitions and results from [4], starting with the $\mathcal{VU}$-space decomposition of §2 therein. We equip the space $\mathbb{R}^N$ with a scalar product $\langle \cdot, \cdot \rangle$ and $\|\cdot\|$ is the associated norm; in a subspace $\mathcal{S}$, we will write $\langle \cdot, \cdot \rangle_{\mathcal{S}}$, $\|\cdot\|_{\mathcal{S}}$ and $B_{\mathcal{S}}$, respectively, for the induced scalar product, norm and unit ball. For a set S, we let conv$S$ and lin$S$ denote, respectively, the convex and linear hull of S.

Given $\bar{x} \in \mathbb{R}^N$, not necessarily a minimizer of f, the orthogonal subspaces of (2) are such that $\mathbb{R}^N = \mathcal{U} \oplus \mathcal{V}$. Hence, every $x \in \mathbb{R}^N$ can be decomposed into components $x_{\mathcal{U}}$ and $x_{\mathcal{V}}$, where $x_{\mathcal{U}}(x_{\mathcal{V}})$ stands for the projection of x onto $\mathcal{U}(\mathcal{V})$. We will often use the notation $x_{\mathcal{U}} \oplus x_{\mathcal{V}}$ for the vector in $\mathbb{R}^N$ with components $x_{\mathcal{U}}$ and $x_{\mathcal{V}}$. With this convention, $\mathcal{U}$ and $\mathcal{V}$ are considered themselves as vector spaces, with induced scalar products satisfying $\langle g, x \rangle = \langle g_{\mathcal{U}} \oplus g_{\mathcal{V}}, x_{\mathcal{U}} \oplus x_{\mathcal{V}} \rangle = \langle g_{\mathcal{U}}, x_{\mathcal{U}} \rangle_{\mathcal{U}} + \langle g_{\mathcal{V}}, x_{\mathcal{V}} \rangle_{\mathcal{V}}$. Throughout, we assume $\mathcal{U} \neq \{0\}$ so that $\mathcal{V} \neq \mathbb{R}^N$.

Relative to a fixed $\bar{x} \in \mathbb{R}^N$ and a subgradient $\bar{g} \in \partial f(\bar{x})$, the *$\mathcal{U}$-Lagrangian* of f is defined as follows:

$$\mathcal{U} \ni u \mapsto L_{\mathcal{U}}(u) := \inf_{v \in \mathcal{V}} \{ f(\bar{x} + u \oplus v) - \langle \bar{g}_{\mathcal{V}}, v \rangle_{\mathcal{V}} \}. \tag{3}$$

Its associated *W*inning set of minimizers (possibly empty) is given by

$$W(u) := \text{Argmin}_{v \in \mathcal{V}} \{ f(\bar{x} + u \oplus v) - \langle \bar{g}_{\mathcal{V}}, v \rangle_{\mathcal{V}} \}. \tag{4}$$

We call the graph of the point-to-set mapping $u \hookrightarrow \bar{x} + u \oplus W(u)$ a *Cornucopia* and denote it by $\mathcal{C}$, see Fig. 1 for the Cornucopia corresponding to Example 17. Elements in this set will provide us with the smooth trajectories mentioned above. It is shown in [4, Theorem 3.2] that when

$$\bar{g} \in \mathrm{ri}\partial f(\bar{x}) \tag{5}$$

the set $W(u)$ is nonempty. However, we show below via Example 15 (Case 2), that (5) is only a sufficient condition. A different sufficient condition for a function f defined by (1) is given in § 2 below.

The convex function $L_{\mathcal{U}}$ is such that $L_{\mathcal{U}}(0) = f(\bar{x})$ and $0 \in W(0)$. It also has interesting smoothness properties. First, $L_{\mathcal{U}}$ has a gradient at 0:$\nabla L_{\mathcal{U}}(0) = \bar{g}_{\mathcal{U}}$, which, because $\bar{g}_{\mathcal{U}} = g_{\mathcal{U}}$ for all $g \in \partial f(\bar{x})$, is called the *$\mathcal{U}$-gradient* of f at $\bar{x}$. When $L_{\mathcal{U}}$ has a Hessian at 0 we say that $\nabla^2 L_{\mathcal{U}}(0)$ is the *$\mathcal{U}$-Hessian* of f at $\bar{x}$ relative to $\bar{g}$. Whenever $\nabla^2 L_{\mathcal{U}}(0)$ exists, those $x(u) := \bar{x} + u \oplus w$ with $w \in W(u)$ yield the following second order expansion of f:

$$\begin{aligned} f(x(u)) &= f(\bar{x}) + \langle \nabla L_{\mathcal{U}}(0), u\rangle_{\mathcal{U}} + \langle \bar{g}_{\mathcal{V}}, w\rangle_{\mathcal{V}} + \tfrac{1}{2}\left\langle \nabla^2 L_{\mathcal{U}}(0)u, u\right\rangle_{\mathcal{U}} + o(\|u\|_{\mathcal{U}}^2) \\ &= f(\bar{x}) + \langle \bar{g}, u \oplus w\rangle + \tfrac{1}{2}\langle \mathrm{H}_{\mathcal{U}} f(\bar{x})u, u\rangle_{\mathcal{U}} + o(\|u \oplus w\|^2)\,. \end{aligned} \tag{6}$$

When the function f is strongly convex with constant $c > 0$, so is $L_{\mathcal{U}}$:

$$L_{\mathcal{U}}(u) \geq L_{\mathcal{U}}(0) + \langle \nabla L_{\mathcal{U}}(0), u\rangle_{\mathcal{U}} + \frac{c}{2}\|u\|_{\mathcal{U}}^2\,. \tag{7}$$

When there exists a constant $c > 0$ such that (7) holds, we say that f is *$\mathcal{U}$-strongly convex* at $\bar{x}$ with respect to $\bar{g} \in \partial f(\bar{x})$. With this property, when the $\mathcal{U}$-Hessian exists, it is also positive definite.

Our paper is organized as follows. We first consider a max-function as given in (1). In Section 2 we give conditions for the existence of a $\mathcal{U}$-Hessian along with corresponding explicit expressions for the $\mathcal{U}$-Lagrangian and its derivatives. In particular, we establish in Theorem 11 the relation between the $\mathcal{U}$-Hessian and the Hessian of the ordinary Lagrangian. Then, for a general convex problem, in §3 we describe a conceptual algorithm from [4, § 4.3] and we provide a rate of convergence result under assumptions weaker than those stated in the cited paper. Finally, in § 4 we give some examples to illustrate our approach and the generality of our assumptions.

2 $\mathcal{VU}$-Space Decomposition Derivatives

Suppose f is the convex max-function given in (1). Let $I(x) :=\{i \in \{0, \cdots, m\} : f_i(x) = f(x)\}$ be the set of active indices at x; given $\bar{x} \in \mathbb{R}^N$, we use the notation $\bar{I}$ for the set $I(\bar{x})$. Without loss of generality we may assume that index $0 \in \bar{I}$. We use the notation $g_i(x)^T := \nabla f_i(x)$ and $\bar{g}_i^T = \nabla f_i(\bar{x})$ for all i. We consider gradient vectors to be row vectors, so each g_i is a column vector. Note that we do not assume convexity of each f_i on $\mathbb{R}^N$. Our assumption that each f_i is C^2 on $\mathbb{R}^N$ implies that the Clarke subdifferential [1] at a point is the convex hull of the gradients of the f_i's that are active at the point. Our additional assumption that f is convex implies that the Convex Analysis subdifferential equals the Clarke subdifferential. With the above notation, this means that $\partial f(\bar{x}) = \mathrm{conv}\{\bar{g}_i^T\}_{i \in \bar{I}}$.

2.1 Spanning the $\mathcal{V}$-Subspace

Our first result states that $\mathcal{U}$ is the tangent space to the set of active gradient differences. Its proof is straightforward from (2).

Proposition 1 *In terms of the above notation, the subspaces $\mathcal{V}$ and $\mathcal{U}$ from* (2) *satisfy* $\mathcal{V} = \mathrm{lin}\{\bar{g}_i - \bar{g}_0\}_{i \in \bar{I}}$, $\mathcal{U} = \{d \in \mathbb{R}^N : \langle d, \bar{g}_i - \bar{g}_0 \rangle = 0 \text{ for all } i \in \bar{I}\}$. □

To give formulas for the $\mathcal{U}$-Lagrangian and its derivatives, we need to identify elements in $W(u)$. We achieve this in two steps. First, given a particular subset of active indices and corresponding f_i's, we define a system of nonlinear equations whose solution generates a candidate for being an element of $W(u)$. Then, in § 2.4 we give conditions for such a set, denoted by K, to produce a successful candidate.

Definition 2 A set $\emptyset \neq K \subseteq \bar{I}$ is called a *basic index set* (relative to $\bar{x}$) if
(i) the vectors $\{\bar{g}_i\}_{i \in K}$ are affinely independent.
K is called a *feasible basic index set* relative to $\bar{g}^T \in \partial f(\bar{x})$ if, in addition,
(ii) the linear system with nonnegative variables

$$\sum_{i \in K} \alpha_i \bar{g}_i = \bar{g}\,, \sum_{i \in K} \alpha_i = 1 \text{ and } \alpha_i \geq 0 \text{ for } i \in K$$

has a (unique) solution $\{\alpha_i = \bar{\alpha}_i\}_{i \in K}$. □

When $\bar{I}$ is a basic index set observe that (i), together with Proposition 1, implies that the dimension of $\mathcal{V}$ is exactly $|\bar{I}| - 1$, i.e., one less than the number of active functions. Furthermore, in this case $\bar{I}$ is a feasible basic index set for all $\bar{g}^T \in \partial f(\bar{x})$. Next we show the existence of such index sets.

Lemma 3 *Relative to each $\bar{g}^T \in \partial f(\bar{x})$ there exists at least one feasible basic index set $K \subseteq \bar{I}$ satisfying Definition 2.*

Proof. Apply Corollary 17.1.1 in [10]: $\bar{g}^T$ is an element of the convex hull $\partial f(\bar{x}) = \mathrm{conv}\{\bar{g}_i^T\}_{i \in \bar{I}}$, so it can be expressed as a convex combination of affinely independent generators, say $\{\bar{g}_i^T\}_{i \in K}$ (with $K \subseteq \bar{I}$). □
We have already stressed the dependency of our $\mathcal{U}$-objects on the choice of $\bar{g}^T \in \partial f(\bar{x})$. From now on, we will also fix $K \subseteq \bar{I}$ and assume, without loss of generality, that index $0 \in K$.

Matrix notation 4 We adopt the following convention: given a set of column vectors $\{v_1, \ldots, v_\ell\}$, we denote the corresponding matrix by $[\{v_1, \ldots, v_\ell\}]$. Given a basic index set K, let $\bar{V} := [\{\bar{g}_i - \bar{g}_0\}_{0 \neq i \in K}]$. Let $\mathcal{V}_K$ denote the subspace with basis matrix $\bar{V}$: $\mathcal{V}_K := \text{lin}\{\bar{g}_i - \bar{g}_0\}_{0 \neq i \in K}$. If $\mathcal{V}_K$ is the whole subspace $\mathcal{V}$, let $\bar{Z} := [\emptyset]$, a vacuous matrix. Otherwise, let $\bar{Z}$ be a basis matrix for $\mathcal{V} \backslash \mathcal{V}_K$. Then, in both cases, we have augmented $\{\bar{g}_i - \bar{g}_0\}_{0 \neq i \in K}$ to form a set of vectors spanning $\mathcal{V}$. Then $[\bar{U}|\bar{V}|\bar{Z}]$ is a basis matrix for $\mathbb{R}^n$, where $\bar{V}$ and $\bar{Z}$ depend on the particular basic index set K under consideration and $\bar{U}$ is a basis matrix for $\mathcal{U}$. □

2.2 Generating $\mathcal{C}$ornucopia Candidates

In order to construct a candidate for being an element of $W(u)$ we change notation slightly and employ Matrix notation 4 corresponding to a basic index set K as follows:

- we replace $(u, v) \in \mathcal{U} \times \mathcal{V}$ by a column vector $\begin{pmatrix} u \\ v \end{pmatrix} \in \mathbb{R}^{\dim \mathcal{U}} \times \mathbb{R}^{\dim \mathcal{V}}$;
- we let $u \oplus v := u \oplus (v_K, z) = \bar{U}u + \bar{V}v_K + \bar{Z}z$, where the column vector $v = \begin{pmatrix} v_K \\ z \end{pmatrix} \in \mathbb{R}^{\dim \mathcal{V}_K} \times \mathbb{R}^{\dim \mathcal{V} - \dim \mathcal{V}_K}$.

For example, with this notation, (3) becomes

$$L_{\mathcal{U}}(u) = \min_{(v_K, z)^T \in \mathbb{R}^{\dim \mathcal{V}}} \{f(\bar{x} + \bar{U}u + \bar{V}v_K + \bar{Z}z) - \bar{g}^T(\bar{V}v_K + \bar{Z}z)\} \tag{8}$$

for $u \in \mathbb{R}^{\dim \mathcal{U}}$. So vectors in $W(u) \subset \mathcal{V}$ are of the form $\bar{V}v_K + \bar{Z}z$ with $(v_K, z)^T$ a minimizer in (8). The idea is to characterize one such minimizer via an implicit function $v_K(u)$ and it will turn out that this particular minimizer has $z = 0$. Accordingly, for u near $0 \in \mathbb{R}^{\dim \mathcal{U}}$, we associate a trajectory $x(u)$ with a basic index set K by letting

$$x(u) := \bar{x} + \bar{U}u + \bar{V}v_K(u), \tag{9}$$

where the function $v_K : \mathbb{R}^{\dim \mathcal{U}} \mapsto \mathbb{R}^{\dim \mathcal{V}_K}$ is defined in the next theorem.

Theorem 5 *Let K be a nonsingleton basic index set as described in Definition 2(i). For each u small enough,*
(i) the nonlinear system with variables (u, v)

$$f_i(\bar{x} + \bar{U}u + \bar{V}v) - f_0(\bar{x} + \bar{U}u + \bar{V}v) = 0 \text{ for all } 0 \neq i \in K \tag{10}$$

has a unique solution $v = v_K(u)$ with $\bar{V}v_K(u)$ in the subspace $\mathcal{V}_K \subseteq \mathcal{V}$.
(ii) $v_K(\cdot)$ has a continuous Jacobian $\mathrm{J}v_K(u) = -(V(u)^T\bar{V})^{-1}V(u)^T\bar{U}$, where

$$V(u) := [\{g_i(x(u))^T - g_0(x(u))^T\}_{0 \neq i \in K}];$$

(iii) the trajectory $x(\cdot)$ *has a continuous Jacobian* $\mathrm{J}x(u) = \bar{U} + \bar{V}\mathrm{J}v_K(u)$.
(iv) In particular $v_K(0) = 0$, $x(0) = \bar{x}$, $V(0) = \bar{V}$, $\mathrm{J}v_K(0) = 0$ *and* $\mathrm{J}x(0) = \bar{U}$.

Proof. Differentiating (10) with respect to v gives the the rows of the Jacobian of the left hand side as

$$g_i(\bar{x} + \bar{U}u + \bar{V}v)^T\bar{V} - g_0(\bar{x} + \bar{U}u + \bar{V}v)^T\bar{V} \quad \text{for all } 0 \neq i \in K.$$

This Jacobian at $(u, v) = (0, 0)$ is $\bar{V}^T\bar{V}$, which is nonsingular. There is also a Jacobian with respect to u, so, by the Implicit Function Theorem, there is a C^1 function $v_K(u)$ defined on a neighbourhood of $u = 0$ such that $v_K(0) = 0$ and

$$f_i(\bar{x}+\bar{U}u+\bar{V}v_K(u))-f_0(\bar{x}+\bar{U}u+\bar{V}v_K(u)) = 0 \text{ for all } 0 \neq i \in K. \tag{11}$$

Since v_K is C^1, the Jacobians $\mathrm{J}v_K(u)$ and $\mathrm{J}x(u)$ exist and are continuous. Differentiating (11) with respect to u gives

$$\left(g_i(x(u))^T - g_0(x(u))^T\right)\mathrm{J}x(u) = 0 \quad \text{for all } 0 \neq i \in K,$$

or, in matrix notation, $V(u)^T\mathrm{J}x(u) = 0$. Using $\mathrm{J}x(u) = \bar{U} + \bar{V}\mathrm{J}v_K(u)$ we have that $V(u)^T(\bar{U} + \bar{V}\mathrm{J}v_K(u)) = 0$. By continuity, $V(u)^T\bar{V}$ is nonsingular, hence $\mathrm{J}v_K(u) = -(V(u)^T\bar{V})^{-1}V(u)^T\bar{U}$. Since $\mathcal{V}_K \perp \mathcal{U}$, $\bar{V}^T\bar{U} = 0$, so $\mathrm{J}v_K(0) = 0$ and $\mathrm{J}x(0) = \bar{U}$. □
Note that when K is a singleton then $\mathcal{V}_K = \{0\}$ and $\bar{V}$ is a vacuous matrix; in this case the trajectory and its Jacobian are simply defined by

$$x(u) = \bar{x} + \bar{U}u \tag{12}$$

and $\mathrm{J}x(u) \equiv \bar{U}$. In view of (9) it is useful to define $\bar{V}v_K(u) := 0$ for this case. In order to proceed further on our way to showing $\bar{V}v_K(u) \in W(u)$ for some feasible basic index set K and for having a second order expansion for f on the associated $x(u)$ as in (6) we need some multipliers.

2.3 Multiplier Functions

We now define multiplier functions $\alpha_i(u)$ that will be useful for expressing the $\mathcal{U}$-gradient and $\mathcal{U}$-Hessian as combinations of the gradients and Hessians of the functions f_i. The next theorem shows how to compute these multipliers that are themselves smooth functions depending on $x(u)$ and $\bar{V}^T\bar{g}$.

Theorem 6 *Let* K *be a feasible basic index set relative to* $\bar{g}^T \in \partial f(\bar{x})$. *For each* u *small enough, the following hold:*
(i) the linear system $\left\{\begin{array}{llll} \bar{V}^T & (\sum_{i\in K} & \alpha_i g_i(x(u)) - \bar{g}) & = 0 \\ & \sum_{i\in K} & \alpha_i & = 1 \end{array}\right\}$ *has a unique solution* $\{\alpha_i = \alpha_i(u)\}_{i\in K}$, *given by*

$$\begin{array}{rcl} \{\alpha_i(u)\}_{0\neq i\in K} & = & -(\bar{V}^T V(u))^{-1}\bar{V}^T(g_0(x(u)) - \bar{g}), \\ \alpha_0(u) & = & 1 - \sum_{0\neq i\in K}\alpha_i(u). \end{array}$$

In particular, for all $i \in K$ we have $\alpha_i(0) = \bar{\alpha}_i$, introduced in Definition 2(ii) above.

(ii) Let $\alpha(u)$ be the $|K|$-dimensional column vector formed from $\{\alpha_i(u)\}_{i\in K}$ with index 0 coming first, e^T be a $(|K|-1)$-dimensional row vector of ones and $I_{|K|-1}$ be a $(|K|-1) \times (|K|-1)$ identity matrix. Define the $N \times |K|$ matrix function $G(u) := [0|V(u)] + g_0(x(u))[1|e^T] = [\{g_i(x(u))\}_{i\in K}]$. Then

$$G(u)\mathrm{J}\alpha(u) = [0|V(u)]\mathrm{J}\alpha(u)\,, \tag{13}$$

where the Jacobian of α is given by

$$\mathrm{J}\alpha(u) = \begin{bmatrix} e^T \\ -I_{|K|-1} \end{bmatrix} [\bar{V}^T V(u)]^{-1} \bar{V}^T \sum_{i\in K} \alpha_i(u) \nabla^2 f_i(x(u)) \mathrm{J}x(u) \tag{14}$$

with $\mathrm{J}\alpha(u) = 0$ if K is a singleton.

Proof. If K is a singleton, then $\bar{V}$ is vacuous and $\alpha_0(u) = 1 = \bar{\alpha}_0$ for all u gives the desired results. Suppose K is not a singleton. Then the linear system defining $\alpha(u)$ can be written as

$$\begin{array}{rcl} \bar{V}^T G(u) \quad \alpha(u) & = & \bar{V}^T \bar{g} \\ [1|e^T] \quad \alpha(u) & = & 1\,. \end{array} \tag{15}$$

Computing $\alpha_0(u) = 1 - \sum_{0\neq i\in K} \alpha_i(u)$ from the last equation and rearranging terms in the first gives

$$\bar{V}^T \textstyle\sum_{0\neq i\in K} \alpha_i(u) g_i(x(u)) = -\bar{V}^T \left((1 - \sum_{0\neq i\in K} \alpha_i(u)) g_0(x(u)) - \bar{g}\right) .$$

Subtracting $\bar{V}^T \sum_{0\neq i\in K} \alpha_i(u) g_0(x(u))$ from both sides yields

$$(\bar{V}^T V(u))\{\alpha_i(u)\}_{0\neq i\in K} = -\bar{V}^T \left(g_0(x(u)) - \bar{g}\right) ,$$

and (i) follows. In particular, for $u = 0$, recall $x(0) = \bar{x}$ and Definition 2(ii): $\alpha_i = \bar{\alpha}_i$ for $i \in K$ therein satisfies linear system (i) above; therefore $\alpha_i(0) = \bar{\alpha}_i$. Let us now prove (ii). Differentiate(15) with respect to u and use the definition of $G(u)$:

$$\begin{array}{lllll} \bar{V}^T G(u) & \mathrm{J}\alpha(u) & +\bar{V}^T \sum_{i\in K} \alpha_i(u) \nabla^2 f_i(x(u)) \mathrm{J}x(u) & = 0 & \text{(a)} \\ [1|e^T] & \mathrm{J}\alpha(u) & & = 0 & \text{(b)}\,. \end{array} \tag{16}$$

The definition of $G(u)$ and (16)(b) imply (13). Furthermore, (16)(b) and (13) combined with (16)(a) imply

$$\begin{bmatrix} 1 & | & e^T \\ 0 & | & \bar{V}^T V(u) \end{bmatrix} \mathrm{J}\alpha(u) = \begin{bmatrix} 0 \\ -\bar{V}^T \sum_{i\in K} \alpha_i(u) \nabla^2 f_i(x(u)) \mathrm{J}x(u) \end{bmatrix} .$$

The final desired result (14) is obtained by multiplying this expression on the left hand side by the inverse matrix $\begin{bmatrix} 1 & | & -e^T [\bar{V}^T V(u)]^{-1} \\ 0 & | & [\bar{V}^T V(u)]^{-1} \end{bmatrix}$. □

2.4 $\mathcal{V}$-Optimal Index Sets and $\mathcal{U}$-Lagrangian Derivatives

When $\bar{I}$ is a basic index set and, hence, $\mathcal{V}_{\bar{I}} = \mathcal{V}$, we say f satisfies *strong transversality* at $\bar{x}$. If, in addition, corresponding to $\bar{g}^T \in \partial f(\bar{x})$ $\bar{\alpha}_i > 0$ for all $i \in \bar{I}$ in Definition 2 with $K = \bar{I}$, we say $\bar{I}$ satisfies *strong feasibility* with respect to $\bar{g}^T$. Similar strong assumptions were used to show results similar to the ones we give in Theorems 9 and 11 below for an exact penalty function in [4, Sec.4.1] and for the maximum eigenvalue of a symmetric matrix whose elements are affine functions in [8]. Next we consider less restrictive conditions that allow us to give expressions for the $\mathcal{U}$-Lagrangian and its derivatives. Our conditions aim at excluding unfavourable situations for a K-trajectory $x(u)$ when some of the functions f_j that are active at $\bar{x}$ are larger than $f_0(x(u))$, some of the corresponding multipliers are negative, and/or $\mathcal{V}_K$ is not equal to $\mathcal{V}$.

Definition 7 We say that a feasible basic index set $\bar{K}$ is $\mathcal{V}$-*optimal* relative to $\bar{g}^T \in \partial f(\bar{x})$ if for all u small enough, the associated vector functions $x(u)$ from (9) or (12) and $\alpha(u)$ from Theorem 6 with $K = \bar{K}$ satisfy
(i) $f_j(x(u)) \le f_0(x(u))$ for all $j \in \bar{I}\backslash\bar{K}$,
(ii) $\alpha_i(u) \ge 0$ for all $i \in \bar{K}$,
(iii) $\bar{Z}^T \left(\sum_{i\in\bar{K}} \alpha_i(u) g_i(x(u)) - \bar{g}\right) = 0$. □

Remark 8 If the multipliers $\bar{\alpha}_i$ from Definition 2(ii) are strictly positive then, by continuity, the multipliers $\alpha_i(u)$ from Theorem 6 are also strictly positive for all u near 0. As a result, condition (ii) of Definition 7 only restricts the multipliers with indexes i in $\bar{K}$ such that $\bar{\alpha}_i = 0$.
Note that Definition 7(i) only depends on the trajectory $x(u)$ from (9) or (12) and not on the multipliers $\alpha_i(u)$. To state conditions (ii) and (iii) with such dependence, consider the following combined system:

$$\begin{array}{rlcll}
[\bar{V}|\bar{Z}]^T & (\sum_{i\in\bar{K}} \alpha_i\, g_i(x(u)) - \bar{g}) & = & 0 & \\
 & \sum_{i\in\bar{K}} \alpha_i & = & 1 & \\
 & \alpha_i & \ge & 0 & \text{for all } i \in \bar{K} \\
 & \alpha_i & > & 0 & \text{for all } i \text{ such that } \bar{\alpha}_i > 0.
\end{array}$$

By Tucker's theorem of the alternative, see for instance [5, p. 29], applied to this expanded system, conditions (ii) and (iii) are equivalent to the following:
For each u near 0 there is no $d \in \mathcal{V}$ such that $\langle g_i(x(u)) - \bar{g}, d\rangle \ge 0$ for all $i \in \bar{K}$ and $\langle g_i(x(u)) - \bar{g}, d\rangle > 0$ for some i such that $\bar{\alpha}_i > 0$.
See Remark 16(Case 2) below for an illustration of such an alternative system having a solution. The associated example also illustrates the need for allowing $\bar{K}$ to correspond to a proper subspace $\mathcal{V}_{\bar{K}} \subsetneq \mathcal{V}$ and, hence, to require condition (iii) above in such a case. □

We now characterize an element in the Winning set $W(u)$.

Theorem 9 *Suppose $\bar{K}$ is a $\mathcal{V}$-optimal index set relative to $\bar{g}^T \in \partial f(\bar{x})$. Let $x(u) = \bar{x} + \bar{U}u + \bar{V}v_{\bar{K}}(u)$, where $v_{\bar{K}}(u)$ is as defined in Theorem 5 if $K = \bar{K}$ is not a singleton or $\bar{V}v_{\bar{K}}(u) := 0$ otherwise. Then for u small enough,*
(i) $\bar{V}v_{\bar{K}}(u)$ is an element of $W(u)$ defined in (4).
(ii) The $\mathcal{U}$-Lagrangian can be expressed as

$$L_{\mathcal{U}}(u) = f_i(x(u)) - \langle \bar{g}, \bar{V}v_{\bar{K}}(u)\rangle \quad \text{for each } i \in \bar{K}\,.$$

Proof. With $\alpha_i(u)$ from Theorem 6, set $g(u) := \sum_{i\in\bar{K}} \alpha_i(u) g_i(x(u))$. Because of (10) with $K = \bar{K}$ and Definition 7(i),

$$f(x(u)) = f_i(x(u)) \text{ for each } i \in \bar{K}\,. \tag{17}$$

So, for each $i \in \bar{K}$, $g_i(x(u))^T$ lies in $\partial f(x(u))$. Then Definition 7(ii) implies $g(u)^T \in \partial f(x(u))$. Furthermore, Theorem 6 and Definition 7 (iii) imply that $[\bar{V}|\bar{Z}]^T(g(u) - \bar{g}) = 0$. Altogether, $(v, z) = (v_{\bar{K}}(u), 0) \in \mathbb{R}^{\dim \mathcal{V}}$ and $g = g(u)$ satisfy $\exists (v, z) \in \mathbb{R}^{\dim\mathcal{V}}$ and $g^T \in \partial f(\bar{x} + u \oplus (v, z))$ such that $g - \bar{g} \perp \mathcal{V}$, which is the optimality condition for the minimization problem in (8). Therefore (i) is proved and, hence, by (8) $L_{\mathcal{U}}(u) = f(\bar{x} + \bar{U}u + \bar{V}v_{\bar{K}}(u)) - \bar{g}^T\bar{V}v_{\bar{K}}(u) = f(x(u)) - \bar{g}^T\bar{V}v_{\bar{K}}(u)$ and (ii) follows from (17). □
When strong transversality and feasibility hold we specialize our result:

Theorem 10 *If f satisfies strong transversality at $\bar{x}$ and $\bar{I}$ satisfies strong feasibility with respect to $\bar{g}^T \in \partial f(\bar{x})$ then $\bar{I}$ is the only feasible basic index set relative to $\bar{g}$ and it is a $\mathcal{V}$-optimal index set. Moreover, for all u small enough $W(u)$ is the singleton $\{\bar{V}v_{\bar{I}}(u)\}$.*

Proof. From Definition 2, if there was a feasible basic index set $K \subsetneq \bar{I}$, then zero multipliers could be associated with $\bar{g}_i$ for $i \in \bar{I}\backslash K$, contradicting the strong feasibility assumption on $\bar{I}$. As for $\mathcal{V}$-optimality of $\bar{I}$, it is straightforward from Definition 7 written with $\bar{K} = \bar{I}$, remembering that $\mathcal{V}_{\bar{I}} = \mathcal{V}$ and $\bar{Z}$ is vacuous.
Let us prove that $W(u)$ is a singleton. Apply Theorem 9 (i) with $\bar{K} = \bar{I}$ to obtain $\bar{V}v_{\bar{I}}(u) \in W(u)$. Since $\bar{\alpha}_i > 0$ for all $i \in \bar{I}$, the subgradient $\bar{g}^T$ lies in the relative interior of $\partial f(\bar{x})$. From Theorem 3.2(iv) in [4], $W(0) = \{0\}$; together with Corollary 3.4 therein, this implies that $w(u) = o(||u||)$ for all $w(u) \in W(u)$. In particular, any such $w(u)$ tends to 0 together with u.
Take u near 0, and assume for contradiction purposes that $w(u) \neq \bar{V}v_{\bar{I}}(u)$. Since $\bar{Z}$ is vacuous, $w(u)$ can be written as $\bar{V}\bar{w}(u)$. Then the uniqueness of $v_{\bar{I}}(u)$ in Theorem 5 (i) written with $K = \bar{I}$ implies that either

$$f(\bar{x} + \bar{U}u + \bar{V}\bar{w}(u)) = f_j(\bar{x} + \bar{U}u + \bar{V}\bar{w}(u)) \text{ for some } j \notin \bar{I}. \tag{18}$$

or

$$f(\bar{x} + \bar{U}u + \bar{V}\bar{w}(u)) \neq f_i(\bar{x} + \bar{U}u + \bar{V}\bar{w}(u)) \text{ for some } i \in \bar{I}. \tag{19}$$

If (18) holds for infinitely many $u = u_n \to 0$, then, since $w(u_n) \to 0$, a continuity argument gives the desired contradiction: $f_j(\bar{x}) = f(\bar{x})$ for some $j \notin \bar{I}$.

Therefore, $f(\bar{x}+\bar{U}u_n+\bar{V}\bar{w}(u_n)) > f_j(\bar{x}+\bar{U}u_n+\bar{V}\bar{w}(u_n))$ for all $j \notin \bar{I}$ and, thus, each optimal multiplier (associated with $w(u_n)$) on $g_j(\bar{x}+\bar{U}u_n+\bar{V}\bar{w}(u_n))$ for $j \notin \bar{I}$ is zero.
Alternatively, if (19) holds for infinitely many $u = u_n \to 0$, then the optimal multiplier (associated with $w(u_n)$) on $g_i(\bar{x}+\bar{U}u_n+\bar{V}\bar{w}(u_n))$ is zero. In this case, another continuity argument argument gives the following contradiction of strong feasibility of $\bar{I}$: $\bar{\alpha}_i = 0$ for some $i \in \bar{I}$. □
See Example 15 (Case 2) for a situation where $\bar{\alpha}_i = 0$ for some $i \in \bar{K}$ and $W(u)$ is a singleton.
We see in the general situation of Theorem 9 that the cornucopia set $\mathcal{C}$ *contains* the trajectory defined in (9) or (12) associated with $\bar{K}$. Now we give expressions for the $\mathcal{U}$-gradient and the $\mathcal{U}$-Hessian of f.

Theorem 11 *If the assumptions of Theorem 9 hold then, for u small enough,*
(i) the gradient of $L_{\mathcal{U}}$ *is given by* $\nabla L_{\mathcal{U}}(u) = \sum_{i\in\bar{K}} \alpha_i(u) g_i(x(u))^T \bar{U}$. *In particular,* $\nabla L_{\mathcal{U}}(0) = \bar{g}^T\bar{U} = g^T\bar{U}$ *for all* $g \in \partial f(\bar{x})$.
(ii) The Hessian of $L_{\mathcal{U}}$ *is given by*

$$\nabla^2 L_{\mathcal{U}}(u) = \mathrm{J}x(u)^T \left(\sum_{i\in\bar{K}} \alpha_i(u)\nabla^2 f_i(x(u))\right) \mathrm{J}x(u)\,;$$

as a result, f has a $\mathcal{U}$*-Hessian at* $\bar{x}$*, corresponding to the* $\mathcal{UU}$*-block of the ordinary Lagrangian:*

$$\mathrm{H}_{\mathcal{U}} f(\bar{x}) = \nabla^2 L_{\mathcal{U}}(0) = \bar{U}^T \left(\sum_{i\in\bar{K}} \bar{\alpha}_i \nabla^2 f_i(\bar{x})\right) \bar{U}\,.$$

Proof. First suppose that $\bar{K}$ is not a singleton. To prove (i), differentiate the expression for the $\mathcal{U}$-Lagrangian from Theorem 9(ii) with respect to u to obtain

$$\nabla L_{\mathcal{U}}(u) = g_i(x(u))^T \mathrm{J}x(u) - \bar{g}^T \bar{V} \mathrm{J}v_{\bar{K}}(u) \quad \text{for each } i \in \bar{K}\,.$$

Using the expression for $\mathrm{J}x(u)$ given in Theorem 5(iii), we obtain

$$\nabla L_{\mathcal{U}}(u) = g_i(x(u))^T \bar{U} + (g_i(x(u)) - \bar{g})^T \bar{V} \mathrm{J}v_{\bar{K}}(u) \quad \text{for } i \in \bar{K}\,. \tag{20}$$

Multiplying each equation above by $\alpha_i(u)$, summing and using Theorem 6 (i) gives

$$\begin{aligned}\nabla L_{\mathcal{U}}(u) &= \textstyle\sum_{i\in\bar{K}} \alpha_i(u) g_i(x(u))^T \bar{U} + (\sum_{i\in\bar{K}} \alpha_i(u) g_i(x(u)) - \bar{g})^T \bar{V} \mathrm{J}v_{\bar{K}}(u) \\ &= \textstyle\sum_{i\in\bar{K}} \alpha_i(u) g_i(x(u))^T \bar{U}\,.\end{aligned}$$

Now we prove (ii). Since $x(u)$ is C^1, $V(u)$ is C^1 and $G(u)$ from (i) is C^1. Differentiating $\nabla L_{\mathcal{U}}(u)^T = \bar{U}^T G(u)\alpha(u)$ gives

$$\nabla^2 L_{\mathcal{U}}(u) = \bar{U}^T \left(G(u)\mathrm{J}\alpha(u) + \sum_{i\in\bar{K}} \alpha_i(u)\nabla^2 f_i(x(u))\mathrm{J}x(u)\right)\,. \tag{21}$$

Combining (13) and (14) with $K = \bar{K}$ yields

$$G(u)\ J\alpha(u) = -V(u)[\bar{V}^T V(u)]^{-1}\bar{V}^T \sum_{i\in\bar{K}} \alpha_i(u)\nabla^2 f_i(x(u))\mathrm{J}x(u)\,. \qquad (22)$$

Also, parts (*ii*) and (*iii*) of Theorem 5 give $\mathrm{J}x(u) = \bar{U} - \bar{V}[V(u)^T\bar{V}]^{-1}V(u)^T\bar{U}$. The transpose of this expression, together with (21) and (22), gives the desired expression for $\nabla^2 L_{\mathcal{U}}(u)$.
Finally, suppose that $\bar{K}$ is a singleton. Then the proof follows as above, with $\mathrm{J}\alpha(u) = 0$, $\mathrm{J}x(u) = \bar{U}$ for all u and all terms involving $\bar{V}$ replaced by zero. □

3 A Fast Conceptual Minimization Algorithm

After some introductory remarks, we consider here Algorithm 4.5 of [4] which will be applied to the examples in the following section. It is not the intent of our paper to give a practical algorithm. Instead its purpose is to establish the necessary theoretical basis for further problem extensions and future algorithmic work. Along these lines the conceptual Algorithm 12 given below contains the elements essential for obtaining superlinear convergence.
In this section we return to the $(u, v) \in \mathcal{U} \times \mathcal{V}$ notation of Section 1.2, where we stated that existence of a $\mathcal{U}$-Hessian implies the expansion (6) of f for $u \in \mathcal{U}$ and all $w \in W(u)$. This existence also implies that there are some particular subgradients which can be expanded up to first order (see [2] and [4]):

$$\{g_{\mathcal{U}} : g_{\mathcal{U}} \oplus \bar{g}_{\mathcal{V}} \in \partial f(\bar{x} + u \oplus w)\,, w \in W(u)\} \subset \bar{g}_{\mathcal{U}} + \mathrm{H}_{\mathcal{U}} f(\bar{x})u + o(\|u\|_{\mathcal{U}})B_{\mathcal{U}}\,. \qquad (23)$$

It is proved in [4, Theorem 5.3] than when

$$\|w\|_{\mathcal{V}} = o(\|u\|_{\mathcal{U}}) \qquad (24)$$

holds for at least one $w \in W(u)$, $\mathrm{H}_{\mathcal{U}} f(\bar{x})$ can be expressed in terms of the Hessian of the Moreau-Yosida regularization of f. Another important consequence of (24) holding and f being $\mathcal{U}$-strongly convex at a point with respect to a zero subgradient is the superlinear convergence of the conceptual algorithm stated below (see also Theorem 4.7 in [4]). A sufficient condition for (24) to hold is that (5) holds. However it may happen that at a minimizer $\bar{x}$ of (1), 0 is not in the relative interior of $\partial f(\bar{x})$ and yet the algorithm given below converges superlinearly (such is the case for Example 15 with $\bar{g} = 0$). Here we only assume the following condition that is weaker than (24) in the sense of being set-valued and having a $\mathcal{U}$-rate that is linear rather than superlinear:

$$\exists\, C \geq 0\,, \delta > 0 \text{ such that } \|u\|_{\mathcal{U}} < \delta \;\Rightarrow\; W(u) \subset W(0) + C\|u\|_{\mathcal{U}} B_{\mathcal{V}}\,. \qquad (25)$$

Let $\mathcal{S}$ be the set of minimizers of a general convex function f defined on $\mathbb{R}^N$. Suppose that $\bar{x} \in \mathcal{S}$ is such that f has a $\mathcal{U}$-Hessian and is $\mathcal{U}$-strongly convex at

$\bar{x}$ with respect to $\bar{g} = 0$. Then, since $\mathrm{H}_{\mathcal{U}} f(\bar{x})$ is positive definite, it is easy to see from (6) and (8) that

$$\mathcal{S} = \bar{x} + 0 \oplus W(0) = \bar{x}_{\mathcal{U}} \oplus (\bar{x}_{\mathcal{V}} + W(0)), \tag{26}$$

where $W(0)$ depends on $\bar{x}$ and $\bar{g} = 0$.

Algorithm 12 Suppose that $\mathcal{V}$, $\mathcal{U}$ and $\mathrm{H}_{\mathcal{U}} f(\bar{x})$ are known. Let $x_0 \in \mathbb{R}^N$ be an initial point close enough to $\mathcal{S}$. Set $k = 0$.
$\mathcal{V}$-Step. Solve the $\mathcal{V}$-space subproblem and find a $\mathcal{C}$ornucopia point by computing a solution $v = v_k \in \mathcal{V}$ of

$$\min\{f(x_k + 0 \oplus v) : v \in \mathcal{V}\} \tag{27}$$

and then setting $x_k^{\mathcal{C}} := x_k + 0 \oplus v_k$. Let $(g^{\mathcal{C}})^T \in \partial f(x_k^{\mathcal{C}})$ satisfy the associated optimality condition $g_{\mathcal{V}}^{\mathcal{C}} = 0$, i.e., $g^{\mathcal{C}} = g_{\mathcal{U}}^{\mathcal{C}} \oplus 0$. If $g_{\mathcal{U}}^{\mathcal{C}} = 0$, stop with $x_k^{\mathcal{C}}$ optimal.
$\mathcal{U}$-Step. Solve the $\mathcal{U}$-space subproblem and make a $\mathcal{U}$-Newton move by computing the solution $u = u_k \in \mathcal{U}$ of

$$\mathrm{H}_{\mathcal{U}} f(\bar{x}) u = -g_{\mathcal{U}}^{\mathcal{C}} \tag{28}$$

and then adding $u_k \oplus 0$ to $x_k^{\mathcal{C}}$.
Update. Set $x_{k+1} := x_k^{\mathcal{C}} + u_k \oplus 0 = x_k + u_k \oplus v_k$; $k := k + 1$ and go to *$\mathcal{V}$-Step.*
□

Figure 1 shows finite convergence of this iterative process for Example 17. There are two $\mathcal{V}$-steps and one intermediate $\mathcal{U}$-step.

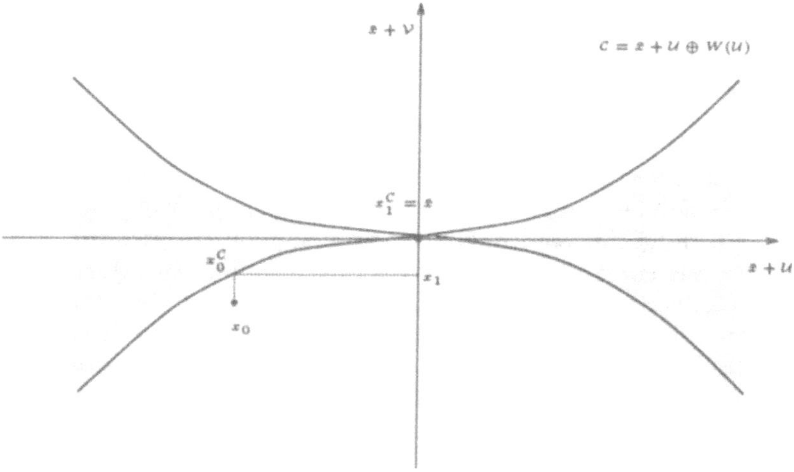

Figure 1: $\mathcal{VU}$-decomposition, $\mathcal{C}$ornucopia and iterates of Algorithm 12 for Example 17.

Now we prove superlinear convergence of $\{x_k^{\mathcal{C}}\}$ to $\mathcal{S}$, assuming only (25). We denote by $d(\cdot, \mathcal{S})$ the distance function to the set of minimizers.

Theorem 13 *Assume that* $\mathrm{H}_{\mathcal{U}} f(\bar{x})$ *is positive definite and* (25) *holds. Then Algorithm 12 generates a sequence* $\{x_k^{\mathcal{C}}\}$ *such that* $d(x_{k+1}^{\mathcal{C}}, \mathcal{S}) = o(d(x_k^{\mathcal{C}}, \mathcal{S}))$.

Proof. Use (26) to decompose $d(\cdot, \mathcal{S})$ as follows:

$$\begin{aligned} d(\cdot, \mathcal{S})^2 &= d_{\mathcal{U}}(\cdot, \mathcal{S})^2 + d_{\mathcal{V}}(\cdot, \mathcal{S})^2 = \|(\cdot - \bar{x})_{\mathcal{U}}\|_{\mathcal{U}}^2 + d_{\mathcal{V}}(\cdot, \mathcal{S})^2 \\ &= \|(\cdot - \bar{x})_{\mathcal{U}}\|_{\mathcal{U}}^2 + \|(\cdot - \bar{x})_{\mathcal{V}} - w_{\cdot}(0)\|_{\mathcal{V}}^2 , \end{aligned}$$

for some $w_{\cdot}(0) \in W(0)$ depending on the argument in $d(\cdot, \mathcal{S})$.
Now, consider the error $e := x_k^{\mathcal{C}} - \bar{x} = e_{\mathcal{U}} \oplus e_{\mathcal{V}}$ and note that $e_{\mathcal{U}} = (x_k^{\mathcal{C}} - x^*)_{\mathcal{U}}$ for all $x^* \in \mathcal{S}$. To prove the desired result we show that

$$d(x_{k+1}^{\mathcal{C}}, \mathcal{S}) = o(\|e_{\mathcal{U}}\|_{\mathcal{U}}) . \tag{29}$$

First, consider the $\mathcal{U}$-step defined by solving (28) at iteration k: because of (23) with $\bar{g} = 0$, we have that $-\mathrm{H}_{\mathcal{U}} f(\bar{x}) u_k = g_{\mathcal{U}}^{\mathcal{C}} \in \mathrm{H}_{\mathcal{U}} f(\bar{x}) e_{\mathcal{U}} + o(\|e_{\mathcal{U}}\|_{\mathcal{U}}) B_{\mathcal{U}}$. The positive-definiteness of $\mathrm{H}_{\mathcal{U}} f(\bar{x})$ gives

$$\|e_{\mathcal{U}} + u_k\|_{\mathcal{U}} = o(\|e_{\mathcal{U}}\|_{\mathcal{U}}) . \tag{30}$$

Now apply the update formula relating cornucopia elements

$$x_{k+1}^{\mathcal{C}} = x_k^{\mathcal{C}} + u_k \oplus v_{k+1} \tag{31}$$

to write $(x_{k+1}^{\mathcal{C}} - \bar{x})_{\mathcal{U}} = (x_k^{\mathcal{C}} - \bar{x})_{\mathcal{U}} + u_k = e_{\mathcal{U}} + u_k$. With (30), this yields $d_{\mathcal{U}}(x_{k+1}^{\mathcal{C}}, \mathcal{S}) = o(\|e_{\mathcal{U}}\|_{\mathcal{U}})$.
Second, consider the $\mathcal{V}$-step from x_{k+1}. Observe that $x_{k+1} + 0 \oplus v = \bar{x} + (e_{\mathcal{U}} + u_k) \oplus (e_{\mathcal{V}} + v)$ give in (27)

$$e_{\mathcal{V}} + v_{k+1} \in \operatorname*{Argmin}_{v \in \mathcal{V}} \{f(\bar{x} + (e_{\mathcal{U}} + u_k) \oplus (e_{\mathcal{V}} + v))\} = W(e_{\mathcal{U}} + u_k) .$$

This, using (25) written with $u = e_{\mathcal{U}} + u_k$, yields $e_{\mathcal{V}} + v_{k+1} = w_0 + w_1$, for some $w_0 \in W(0)$ and $w_1 \in \mathcal{V}$ with $\|w_1\|_{\mathcal{V}} \leq C\|e_{\mathcal{U}} + u_k\|_{\mathcal{U}} = o(\|e_{\mathcal{U}}\|_{\mathcal{U}})$, where the last equality follows from (30) . Applying (31) again leads to $(x_{k+1}^{\mathcal{C}} - \bar{x})_{\mathcal{V}} = w_0 + o(\|e_{\mathcal{U}}\|_{\mathcal{U}})$, so that $d_{\mathcal{V}}(x_{k+1}^{\mathcal{C}}, \mathcal{S}) = o(\|e_{\mathcal{U}}\|_{\mathcal{U}})$. Altogether, (29) is proved. □
Our scheme above is highly conceptual, because a practical algorithm needs to generate convergent estimates of $\mathcal{V}$, $\mathcal{U}$, a subset of $\mathcal{C}$ and a positive definite $\mathcal{U}$-Hessian corresponding to an optimal solution and a zero subgradient. Relevant ideas for doing this are contained in [3], [6] and [7].

4 Examples

The following example functions are all defined on $\mathbb{R}^2$ and considered at the point $\bar{x} = (0, 0)^T$ which is also a minimizing point for each function f. The variables are denoted by $(x, y) \in \mathbb{R}^2$, so x in the pair (x, y) is a scalar throughout this section and $f(x, y)$ means $f(\begin{pmatrix} x \\ y \end{pmatrix})$, and the same for $f_i(x, y)$ and $g_i(x, y)$. The first example has a singleton $\partial f(\bar{x}) = \{\bar{g}^T\}$, but no $\mathcal{V}$-optimal index set $\bar{K}$ and no $\mathcal{U}$-Hessian at $\bar{x}$. However, since it is strongly convex on $\mathbb{R}^2$, it is $\mathcal{U}$-strongly convex at $\bar{x}$ with respect to $\bar{g}^T \in \partial f(\bar{x})$ in the sense defined in (7). This example appears in [9, p. 103].

Example 14 Consider the function

$$f(x, y) = \max\{f_0(x, y), f_1(x, y)\} := \max\left\{\frac{1}{2}x^2 + y^2,\ x^2 + \frac{1}{2}y^2\right\}.$$

Then $\partial f(\bar{x}) = \{\bar{g}^T = (0, 0)\}$, $\mathcal{V} = \{(0, 0)^T\}$, $\mathcal{U} = \mathbb{R}^2$, $L_\mathcal{U} \equiv f$ and (7) is satisfied with $c = 1$.
At $\bar{x}$ the active index set $\bar{I}$ is $\{0, 1\}$. Any set K satisfying Definition 2(i) is a singleton, so (12) gives the trajectory $x(u) = u \in \mathbb{R}^2$. Furthermore, $\bar{I} \backslash K$ is also a singleton, so Definition 7(i) cannot be satisfied for all u close to $(0, 0)^T$.
Also, this example function does not have a $\mathcal{U}$-Hessian at $\bar{x} = (0, 0)^T$, because a second order expansion of $f(x, y)$ at $(0, 0)^T$ requires a Hessian of the form

$$\begin{bmatrix} 2 & 0 \\ 0 & 1 \end{bmatrix} \quad \text{if } |y| \leq |x| \quad \text{and} \quad \begin{bmatrix} 1 & 0 \\ 0 & 2 \end{bmatrix} \quad \text{if } |y| \geq |x|.$$

□

The next two examples illustrate that for different values of $\bar{g}^T \in \partial f(\bar{x})$ there are different $\mathcal{V}$-optimal sets $\bar{K}$ satisfying Definition 7 and different expressions for the respective $\mathcal{U}$-derivatives. They also illustrate several other important features, including finiteness of the conceptual minimization Algorithm 12. Neither function is strongly convex on $\mathbb{R}^2$, but both are $\mathcal{U}$-strongly convex at $\bar{x}$ with respect to subgradients in subsets of $\partial f(\bar{x})$ containing the zero vector. In both examples all of the underlying C^2-functions are active at $\bar{x} = (0, 0)^T$, and we have

$$\mathcal{V} = \text{lin}\{(0, 1)^T\} \quad \text{and} \quad \mathcal{U} = \text{lin}\{(1, 0)^T\}. \tag{32}$$

Concerning the Matrix Notation 4, we use the matrices $\bar{U} = [(1, 0)^T]$ and either $\bar{V} = [(0, 1)^T]$ or $\bar{Z} = [(0, 1)^T]$, depending on the set $\bar{K}$ under consideration.

Example 15 Define $f(x, y) := \max\{\frac{1}{2}(x^2 + y^2), y\}$ in (1). Then the relevant C^2-functions and gradients are

$$\begin{array}{lll} f_0(x, y) := \frac{1}{2}x^2 + \frac{1}{2}y^2 & \text{and} & f_1(x, y) := y\,; \\ g_0(x, y) = (x, y)^T & \text{and} & g_1(x, y) = (0, 1)^T. \end{array}$$

Note that at $\bar{x} = (0, 0)^T$ the active index set is $\bar{I} = \{0, 1\}$, so

$$\partial f(\bar{x}) = \text{conv}\,\{(0, 0), (0, 1)\} = \{(0, \gamma) : \gamma \in [0, 1]\}$$

and $\mathcal{V}$ and $\mathcal{U}$ are given by (32).
We separate our analysis of this example into two cases, and first consider $\gamma \neq 0$.

Case 1: $\gamma \in (0,1]$ For this case we consider $u \in S_\gamma := \{u : |u| < \gamma\}$. Performing the minimization in (8) with $\bar{g} = (0,\gamma)^T$ and $\bar{x} + u \oplus v = (0,0)^T + [(1,0)^T]u + [(0,1)^T]v = (u,v)^T$ yields

$$W(u) = \begin{cases} \{1 - \sqrt{1-u^2}\} & \text{if } \gamma \in (0,1) \\ [1 - \sqrt{1-u^2}, 1 + \sqrt{1-u^2}] & \text{if } \gamma = 1, \end{cases}$$

and $$\mathcal{C} = \begin{cases} \{(u, 1 - \sqrt{1-u^2}) : u \in S_\gamma\} & \text{if } \gamma \in (0,1) \\ \{\{u\} \times [1 - \sqrt{1-u^2}, 1 + \sqrt{1-u^2}] : u \in S_\gamma\} & \text{if } \gamma = 1. \end{cases}$$

Thus, if $\gamma \in (0,1)$, (24) holds. If $\gamma = 1$, $W(u) \subseteq W(0) = [0,2]$, so (25) holds with $C = 0$ and $\delta = \gamma$ and the $\mathcal{U}$-rate is superlinear as in (24). However, this is not a situation where Algorithm 12 applies, because $\bar{g} = (0,\gamma)^T \neq (0,0)^T$. Now we apply the theory of § 2 to track a smooth trajectory in a $\mathcal{C}$ornucopia corresponding to $\bar{g} = (0,\gamma)^T$. For $\gamma \in (0,1]$ the following analysis shows that $\bar{I}$ is a $\mathcal{V}$-optimal index set, but it does not satisfy strong feasibility for $\gamma = 1$.

1. Space Decomposition Take $\bar{K} = \bar{I}$ so that $\bar{V} = [\{\bar{g}_1 - \bar{g}_0\}]$ and we have the following space decomposition quantities:

$\gamma \in$	$\bar{K}$	$\mathcal{V}_{\bar{K}}$	$\bar{V}$	$\bar{Z}$	$\bar{U}$
$(0,1]$	$\{0,1\}$	$\mathcal{V}$	$[(0,1)^T]$	vacuous	$[(1,0)^T]$

2. Function $v_K(u)$ and Trajectory $x(u)$ The function $v = v_{\bar{K}}(u)$ defined by (10) of Theorem 5(i) solves $\frac{1}{2}u^2 + \frac{1}{2}v^2 = v$. Hence, for all $u \in [-1,1] \supseteq S_\gamma$

$$v_{\bar{K}}(u) = 1 - \sqrt{1-u^2}, \quad \text{and} \quad \mathrm{J}v_{\bar{K}}(u) = [\frac{u}{\sqrt{1-u^2}}].$$

The trajectory from (9) is

$$\begin{aligned} x(u) = \bar{x} + u \oplus v_{\bar{K}}(u) &= (0,0)^T + [(1,0)^T]u + [(0,1)^T]v_{\bar{K}}(u) \\ &= (u, 1 - \sqrt{1-u^2})^T, \end{aligned}$$

and $\mathrm{J}x(u) = U(0) + \bar{V}\mathrm{J}v_{\bar{K}}(u) = [(1,0)^T] + [(0,1)^T]\mathrm{J}v_{\bar{K}}(u) = [(1, u(1-u^2)^{-\frac{1}{2}})^T]$. Note that these quantities do not depend on the particular value of $\gamma \in (0,1]$, whereas the following quantities do.

3. Multipliers from Theorem 6 Multipliers α_i solve the system

$$\begin{aligned} [(0,1)](\ \alpha_0(u, 1 - \sqrt{1-u^2})^T &+ \alpha_1(0,1)^T - (0,\gamma)^T) &= 0 \\ \alpha_0 &+ \alpha_1 &= 1. \end{aligned} \tag{33}$$

The solutions are given by $$\alpha_0(u) = \frac{1-\gamma}{\sqrt{1-u^2}}, \alpha_1(u) = \frac{\sqrt{1-u^2} - 1 + \gamma}{\sqrt{1-u^2}}, \tag{34}$$

which are $\mathcal{V}$-optimal for all $u \in (-1,1) \supseteq S_\gamma$. In particular, the linear systems of Definition 2(ii) have the solutions $\alpha_0(0) = \bar{\alpha}_0 = 1 - \gamma \geq 0$,$\alpha_1(0) = \bar{\alpha}_1 = \gamma > 0$.

4. $\mathcal{V}$-Optimality Since $\bar{K} = \bar{I}$, Definition 7(i) is satisfied for all $u \in [-1,1] \supset S_\gamma$. From (34), we see that (ii) is satisfied for all $u \in (-\gamma, \gamma) = S_\gamma$, since $\gamma \in (0,1]$. As for part (iii), it is trivially satisfied, because $\bar{Z}$ is vacuous. Note that for $\gamma = 1$ (34) gives $\alpha_0(0) = \bar{\alpha}_0 = 0$. Hence, $\bar{I}$ does not satisfy the strong feasibility hypothesis of Theorem 10 for $\gamma = 1$ even though $\bar{I}$ is $\mathcal{V}$-optimal .

5. $\mathcal{U}$-Lagrangians and Derivatives Since $\bar{K} = \bar{I}$ is $\mathcal{V}$-optimal for $u \in S_\gamma$, Theorem 9 gives

$$v_{\bar{K}}(u) = 1 - \sqrt{1-u^2} \in W(u)\,, \quad \{(u, 1-\sqrt{1-u^2}) : u \in S_\gamma\} \subseteq \mathcal{C}\,,$$

and $L_{\mathcal{U}}(u) = (1-\gamma)(1-\sqrt{1-u^2})$.
Furthermore, Theorem 11 gives us the following expressions for the $\mathcal{U}$-gradients and the $\mathcal{U}$-Hessians:

$$\nabla L_{\mathcal{U}}(u)^T = [(1,0)]\Big(\alpha_0(u)(u, 1-\sqrt{1-u^2})^T + \alpha_1(u)(0,1)^T\Big) = \alpha_0(u)u = \frac{(1-\gamma)u}{\sqrt{1-u^2}}\,,$$

$$\begin{aligned}\nabla^2 L_{\mathcal{U}}(u) &= [(1, u\sqrt{1-u^2})]\Big(\alpha_0(u)\begin{bmatrix}1 & 0\\ 0 & 1\end{bmatrix} + \alpha_1(u)\begin{bmatrix}0 & 0\\ 0 & 0\end{bmatrix}\Big)\begin{bmatrix}1\\ u\sqrt{1-u^2}\end{bmatrix}\\ &= [\frac{\alpha_0(u)}{1-u^2}]\,,\end{aligned}$$

that is,

$$\nabla^2 L_{\mathcal{U}}(u) = [\frac{1-\gamma}{(1-u^2)^{\frac{3}{2}}}]\,. \tag{35}$$

Finally, observe that each $\mathcal{U}$-Hessian corresponding to $\gamma < 1$ is positive definite. Thus, this function f is $\mathcal{U}$-strongly convex at $(0,0)^T$, with respect to $\bar{g}^T = (0,\gamma) \in \partial f(\bar{x})$ for all $\gamma \in (0,1)$. The constant c in (7) is $(1-\gamma)$.
These results clearly illustrate the dependence of the $\mathcal{U}$-derivatives on $\bar{g}$, via γ.

Remark 16 The reason that $\bar{K} = \bar{I} = \{0,1\}$ is not a $\mathcal{V}$-optimal index set for $\gamma = 0$ is that in (34) $\alpha_1(u) < 0$ for all nonzero $u \in (-1,1)$. An alternative way to state this is as follows:
When $\gamma = 0$ and u is near 0, solving (33) with the additional constraints $\alpha_0 \geq 0$ and $\alpha_1 \geq 0$ is equivalent (via normalization) to solving

$$\begin{array}{lcrcl}\alpha_0 & (1-\sqrt{1-u^2}) & + \;\alpha_1 & = & 0\\ \alpha_0 & & & > & 0\,,\\ & & \alpha_1 & \geq & 0\,.\end{array} \tag{36}$$

By Tucker's theorem of the alternative, (36) has no solution if the following system has a solution $d \in \mathbb{R}$:

$$\begin{cases}(1-\sqrt{1-u^2})\,d & > \;0\\ \phantom{(1-\sqrt{1-u^2})}\,d & \geq \;0\,.\end{cases}$$

For all nonzero $u \in (-1,1)$, this latter system does have a solution, so (36) has no solution in this case. Hence, for $\gamma = 0$ we need to find a *singleton* set $\bar{K}$ with a corresponding nonvacuous matrix $\bar{Z}$ such that Definition 7(iii) is nontrivially satisfied. □

Case 2: $\gamma = 0$ For this case, we consider all $u \in \mathbb{R}$. Performing the minimization in (8) with $\bar{g} = (0,0)^T$ and $\bar{x} + u \oplus v = (0,0)^T + [(1,0)^T]u + [(0,1)^T]v = (u,v)^T$ yields

$$W(u) = \{0\} \quad \text{and} \quad \mathcal{C} = \{(u,0)^T : u \in \mathbb{R}\}.$$

Thus, (24) is satisfied in this case. As discussed in Remark 16, we need a singleton for $\bar{K}$ in order to apply the theory of Section 2 for this case where $\bar{g} = (0,0)^T$. Recall that for a singleton $\bar{K}$, $\bar{V}v_{\bar{K}}(u) = 0$.

1. Space Decomposition Take $\bar{K} = \{0\}$ so that we have the following space decomposition quantities:

γ	$\bar{K}$	$\mathcal{V}_{\bar{K}}$	$\bar{V}$	$\bar{Z}$	$\bar{U}$
0	$\{0\}$	$\{(0,0)^T\}$	vacuous	$[(0,1)^T]$	$[(1,0)^T]$

2. Trajectory The trajectory from (12) is

$$x(u) = (0,0)^T + [(1,0)^T]u = (u,0)^T,$$

and $Jx(u) \equiv \bar{U} = [(1,0)^T]$ for all u.

3. Multipliers from Theorem 6 The linear system is solved by the multiplier $\alpha_0(u) = 1$ for all u.

4. $\mathcal{V}$-Optimality Since $f_0(x(u)) = \frac{1}{2}u^2 \geq 0 = f_1(x(u))$, Definition 7(i) is satisfied for all u. As for *(ii)*, it is trivially satisfied. Finally, because $\bar{Z} = [(0,1)^T]$ and

$$\alpha_0(u)g_0(u,0) - \bar{g} = 1(u,0)^T - (0,0)^T = (u,0)^T,$$

(iii) is also satisfied for all u.

5. $\mathcal{U}$-Lagrangian and Derivatives Because Theorem 9 applies for any u, we have that $\bar{V}v_{\bar{K}}(u) = 0 \in W(u)(= \{0\})$ and $\{(u,0)^T : u \in \mathbb{R}\} \subseteq \mathcal{C}(= \mathbb{R} \times \{0\})$, where the equalities in parentheses follow from the minimization in (8). Together with Theorem 11 we obtain, for all $u \in \mathbb{R}$,

$$L_{\mathcal{U}}(u) = f_0(x(u)) = \frac{1}{2}u^2, \quad \nabla L_{\mathcal{U}}(u)^T = u \quad \text{and} \quad \nabla^2 L_{\mathcal{U}}(u) = [1].$$

Observe that f is clearly $\mathcal{U}$-strongly convex at $(0,0)^T$ with respect to $\bar{g}^T = (0,0)$. Note that in either case $\nabla L_{\mathcal{U}}(0) = 0$ and for all $\gamma \in [0,1]$ $\mathrm{H}_{\mathcal{U}} f(\bar{x}) = \nabla^2 L_{\mathcal{U}}(0) = [1-\gamma]$. Only the gradient at 0 ($= 0 \in \mathcal{U}$) is independent of γ. Furthermore, this is an example where the $\mathcal{U}$-Hessian corresponding to one extreme subgradient, namely $(0,0)^T$, is (trivially) positive definite and corresponding to another, namely $(0,1)^T$, is singular: in (35), $\nabla^2 L_{\mathcal{U}}(0) = [0]$ when $\gamma = 1$.

6. **Algorithm 12** When $\bar{g} = (0,0)^T$, $\gamma = 0$, $\mathrm{H}_{\mathcal{U}} f(\bar{x}) = \nabla^2 L_{\mathcal{U}}(0) = [1]$, $W(u) = \{0\}$ and (25) holds, because (24) holds. Thus, Algorithm 12 generates a superlinearly convergent $\mathcal{C}$ornucopia sequence $\{x_k^{\mathcal{C}}\}$. Indeed, the algorithm minimizes this example function f in a *finite* number of steps as follows: Starting from any point $x_0 = (x,y)^T$, minimization along $\mathcal{V}$ in (27) gives

$$v_0 = \operatorname{argmin}\{f\left((x,y)^T + [(0,1)^T]v\right) : v \in \mathbb{R}\} = -y$$

and leads to the point $x_0^{\mathcal{C}} = (x,0)^T$, which is optimal if $x = 0$. If $x \neq 0$ then

$$f(x,0) = f_0(x,0) = \frac{1}{2}x^2 > 0 = f_1(x,0) \;\Rightarrow\; \partial f(x,0) = \{g_0(x,0)^T\} = \{(x,0)\}\,.$$

We then have $g^{\mathcal{C}} = (x,0)^T$, with $\mathcal{U}$-component $g_{\mathcal{U}}^{\mathcal{C}} = x$.
Solving (28) and recalling $\mathrm{H}_{\mathcal{U}} f(\bar{x}) = \nabla L_{\mathcal{U}}(0) = [1]$ gives

$$u_0 = -[\mathrm{H}_{\mathcal{U}} f(\bar{x})]^{-1} g_{\mathcal{U}}^{\mathcal{C}} = -[1]^{-1}x = -x\,,$$

and the $\mathcal{U}$-Newton step leads to

$$x_1 = x_0^{\mathcal{C}} + u_0 \oplus 0 = (x,0)^T + [(1,0)^T](-x) = (0,0)^T\,,$$

which is the optimal solution.
This is an example where Algorithm 12 converges rapidly and (5) does not hold, since here $\bar{g}^T = (0,0)$ is not in the relative interior of $\partial f(\bar{x})$. □

The final example does not satisfy strong transversality and it demonstrates that more than one feasible basic index set $\bar{K}$ can be $\mathcal{V}$-optimal for a particular $\bar{g}^T \in \partial f(\bar{x})$. As shown in Figure 1, it also illustrates the existence of a "thick" $\mathcal{C}$ornucopia $\mathcal{C}$, i.e., the graph of $u \hookrightarrow \bar{x} + u \oplus W(u) = \operatorname{conv}\{u \oplus -\frac{1}{2}u^2, u \oplus \frac{1}{2}u^2\}$, with every corresponding $w \in W(u)$ satisfying (24).

Example 17 Let the function of (1) be defined as follows:

$$f(x,y) := \max\{\frac{1}{2}x^2, |y|\}\,.$$

Then the relevant C^2-functions and gradients are

$$\begin{array}{lll} f_0(x,y) = \frac{1}{2}x^2\,, & f_1(x,y) = y & \text{and} \quad f_2(x,y) = -y\,; \\ g_0(x,y) = (x,0)^T\,, & g_1(x,y) = (0,1)^T & \text{and} \quad g_2(x,y) = (0,-1)^T\,. \end{array}$$

Note that at $\bar{x} = (0,0)^T$, the active set is $\bar{I} = \{0,1,2\}$,

$$\partial f(\bar{x}) = \text{conv}\,\{(0,0),(0,1),(0,-1)\} = \{(0,\gamma) : \gamma \in [-1,1]\}\ ,$$

and $\mathcal{V}$ and $\mathcal{U}$ are given by (32).
Clearly, this f at $(0,0)$ does not satisfy the strong transversality assumption of Theorem 10 and, hence, for $\bar{K}$ to be a basic index set it must satisfy $\bar{K} \subsetneq \bar{I}$.
We now describe the various features of this example with $\bar{g}^T = (0,\gamma)$, $|\gamma| \le 1$.

1. Space Decomposition Corresponding to different sets of γ-values are the following $\mathcal{V}$-optimal sets $\bar{K}$ and space-decomposition objects:

$\gamma \in$	$\bar{K}$	$\mathcal{V}_{\bar{K}}$	$\bar{V}$	$\bar{Z}$	$\bar{U}$
[0,1]	$\{0,1\}$	$\mathbb{R}$	$[(0,1)^T]$	vacuous	$[(1,0)^T]$
[-1,0]	$\{0,2\}$	$\mathbb{R}$	$[(0,1)^T]$	vacuous	$[(1,0)^T]$
$\{0\}$	$\{0\}$	$\{(0,0)^T\}$	vacuous	$[(0,1)^T]$	$[(1,0)^T]$

2. Trajectories For $\gamma \in [-1,1]$ the trajectories from (9) and (12) are

$$x(u) = (u, \frac{1}{2} sign(\gamma) u^2)^T\ ,$$

with Jacobians $\mathrm{J}x(u) = [(1, sign(\gamma)u)^T]$ for $u \in \mathbb{R}$.

3. Multipliers from Theorem 6 For $\gamma \in [-1,1]$, the linear systems defining the multipliers α_i are solved by

$$\alpha_0(u) = 1 - |\gamma|\,, \quad \alpha_1(u) = \max\{\gamma, 0\}\,, \quad \alpha_2(u) = \max\{-\gamma, 0\}\,, \text{ for } u \in \mathbb{R}\,.$$

4. $\mathcal{V}$-Optimality Using analysis as for Example 15, it can be seen that these $\bar{K}$-sets are indeed $\mathcal{V}$-optimal . Note that for $\gamma = 0$ there are three $\mathcal{V}$-optimal sets and for each of them $\alpha_0(u) = 1$ for all $u \in \mathbb{R}$.

5. $\mathcal{U}$-Lagrangians and Derivatives From Theorems 9 and 11 we find $L_{\mathcal{U}}(u) = \frac{1}{2}(1-|\gamma|)u^2$, $\nabla L_{\mathcal{U}}(u) = (1-|\gamma|)u$ and $\nabla^2 L_{\mathcal{U}}(u) = [1-|\gamma|]$, for $u \in \mathbb{R}$. Note that only the $\mathcal{U}$-Hessians at $\bar{x} = (0,0)^T$ corresponding to $\gamma = \pm 1$ are singular. Moreover, only the ones corresponding to $|\gamma| < 1$ are positive definite. Hence, f is $\mathcal{U}$-strongly convex at $(0,0)^T$ with respect to $\bar{g}^T = (0,\gamma) \in \partial f(\bar{x})$ if and only if $|\gamma| < 1$. In particular, (7) holds with equality for $c = 1 - |\gamma|$.

6. Algorithm 12 When $\bar{g} = (0,0)^T$, $\gamma = 0$, $\mathrm{H}_{\mathcal{U}} f(\bar{x}) = \nabla L_{\mathcal{U}}(0) = [1]$ and $\alpha_0(u) = 1$ for all $u \in \mathbb{R}$. Furthermore, minimization in (8) with $\bar{g} = (0,0)^T$ and $\bar{x} + u \oplus v = (0,0)^T + \bar{U}u + [\bar{V}|\bar{Z}]v$ yields $W(u) = \mathrm{conv}\{-\frac{1}{2}u^2, \frac{1}{2}u^2\}$ and $\bar{U}^T g = u$ for all $g^T \in \partial f(\bar{x} + u \oplus w)$ and $w \in W(u)$. Note that $W(u)$ is a singleton if and only if $u = 0$. Due to these facts, Algorithm 12 minimizes this function in at most three steps as follows:
From any starting point $x_0 = (x,y)^T$, the $\mathcal{V}$-*Step* leads to $x_0^{\mathcal{C}} = (x,w)^T$ for some $w \in W(x)$ where $g_{\mathcal{U}}^{\mathcal{C}} = x$ for all subgradients $(g^{\mathcal{C}})^T \in \partial f(x_0^{\mathcal{C}})$ satisfying the optimality condition of (27). Then the $\mathcal{U}$-Newton step $u_0 = -H_{\mathcal{U}} f(0,0)^{-1} g_{\mathcal{U}}^{\mathcal{C}} = -[1]^{-1}x$ gives $x_1 = (0,w)^T$. Since $W(0) = \{0\}$, the next minimization along $\mathcal{V}$ leads to the optimal solution, $(0,0)^T$. A particular realization of this process is shown in Figure 1. □

Acknowledgments. We want to thank Claude Lemaréchal for his assistance and helpful contributions, including the "corne d'abondance".

References

[1] F.H. Clarke. *Optimization and Nonsmooth Analysis*. Wiley, 1983. Reprinted by SIAM, 1990.

[2] J.-B. Hiriart-Urruty. The approximate first-order and second-order directional derivatives for a convex function. In J.-P. Cecconi and T. Zolezzi, editors, *Mathematical Theories of Optimization*, number 979 in Lecture Notes in Mathematics, pages 154–166. Springer-Verlag, 1983.

[3] C. Lemaréchal and R. Mifflin. Global and superlinear convergence of an algorithm for one-dimensional minimization of convex functions. *Mathematical Programming*, 24:241–256, 1982.

[4] C. Lemaréchal, F. Oustry, and C. Sagastizábal. The $\mathcal{U}$-Lagrangian of a convex function. *Transactions of the AMS*, 1997. Accepted for publication.

[5] O.L. Mangasarian. *Nonlinear Programming*. McGraw - Hill, 1969.

[6] R. Mifflin. A quasi-second-order proximal bundle algorithm. *Mathematical Programming*, 73(1):51–72, 1996.

[7] R. Mifflin, D.F. Sun, and L.Q. Qi. Quasi-Newton bundle-type methods for nondifferentiable convex optimization. *SIAM Journal on Optimization*, 8(2):583–603, 1998.

[8] F. Oustry. The $\mathcal{U}$-Lagrangian of the maximum eigenvalue function. *SIAM Journal on Optimization*, 1998. To appear.

[9] M.L. Overton and X.J. Ye. Towards second-order methods for structured nonsmooth optimization. In S. Gomez and J.P. Hennart, editors, *Advances in Optimization and Numerical Analysis*, pages 97–109. Kluwer, 1994.

[10] R.T. Rockafellar. *Convex Analysis*. Number 28 in Princeton Mathematics Ser. Princeton University Press, 1970.

Proximal and Dynamical Approaches to Equilibrium Problems

Abdellatif Moudafi *   Michel Théra **

1 Introduction

The theory of equilibrium problems has emerged as an interesting branch of applied mathematics, permitting the general and unified study of a large number of problems arising in mathematical economics, optimization and operations research. Inspired by numerical methods developed for variational inequalities and motivated by recent advances in this field, we propose several ways (including an auxiliary problem principle, a selection method, as well as a dynamical procedure) to solve the following equilibrium problem:

$$(GEP) \qquad \text{Find } \bar{x} \in C \quad \text{such that} \quad F(\bar{x}, x) + \langle G(\bar{x}), x - \bar{x} \rangle \geq 0 \quad \forall x \in C,$$

where C is a nonempty convex closed subset of a real Hilbert space X, $F : C \times C \to \mathbb{R}$ is a given bivariate function with $F(x, x) = 0$ for all $x \in C$ and $G : C \to \mathbb{R}$ is a continuous mapping. This problem has useful applications in nonlinear analysis, including as special cases optimization problems, variational inequalities, fixed-point problems and problems of Nash equilibria.
Throughtout the paper, X is a real Hilbert space, $\langle \cdot, \cdot \rangle$ denotes the associated inner product and $|\cdot|$ stands for the corresponding norm. From now on, we assume that the solution set, S, of problem (GEP) is nonempty. This corresponds to some important situations such as linear programming and semi-coercive minimization problems.

2 A Proximal Method

Let λ be a positive number and consider a given iterate x_k. The auxiliary equilibrium problem consists of determining $x_{k+1} \in C$ which satisfies

$$(AEP) \quad F(x_{k+1}, x) + \langle G(x_k) + \lambda^{-1}(h'(x_{k+1}) - h'(x_k)), x - x_{k+1} \rangle \geq 0 \quad \forall x \subset C,$$

where h' is the derivative of a given strictly convex function on C. We will use a co-coercivity condition on G to analyze the convergence behavior of the

*Université des Antilles et de la Guyane, 97159 Pointe-à-Pitre, Guadeloupe, France
**Université de Limoges, LACO, 123 avenue Albert Thomas, F-87060 Limoges Cedex, France

proposed iterative method based on an algorithmic approach introduced by G. Cohen as "the auxiliary problem principle" for optimization problems in [13], [14], and variational inequalities in [15].

Theorem 1 *Let F be a upper hemicontinuous*[1] *monotone*[2] *bivariate function over a closed convex C such that $F(x,\cdot)$ is convex and lower semicontinuous for all $x \in C$. Let h be strongly convex over C (with modulus α) and its derivative, h', be Lipschitz continuous (with constant β). We suppose that G is co-coercive over C (with modulus γ), namely, there exists $\gamma > 0$ such that*

$$\langle G(x) - G(y), x - y\rangle \geq \gamma |G(x) - G(y)|^2 \quad \forall x, y \in C.$$

Then, there exists a unique solution x_{k+1} to the auxiliary problem (AEP). Furthermore, if $0 < \lambda < 2\alpha\gamma$, then the sequence $\{x_k\}_{k\in\mathbb{N}}$ weakly converges to a solution of the considered problem.

Proof The existence and uniqueness of x_{k+1} follow by invoking ([8], Theorem 1). Let x^* denote a fixed solution of (GEP) and consider, for $x \in C$, the function

$$\Lambda(x) = h(x^*) - h(x) - \langle h'(x), x^* - x\rangle \geq \frac{\alpha}{2}|x - x^*|^2. \tag{1}$$

We have

$$\begin{aligned} \Lambda(x_k) - \Lambda(x_{k+1}) &= h(x_{k+1}) - h(x_k) - \langle h'(x_k), x_{k+1} - x_k\rangle \\ &\quad + \langle h'(x_{k+1}) - h'(x_k), \overline{x} - x_{k+1}\rangle \\ &\geq \frac{\alpha}{2}|x_{k+1} - x_k|^2 - \lambda F(x_{k+1}, \overline{x}) - \lambda\langle G(x_k), \overline{x} - x_{k+1}\rangle. \end{aligned}$$

By setting $x = x_{k+1}$ in (GEP) and taking into account the monotonicity of G, we get

$$-F(x_{k+1}, \overline{x}) + \langle G(\overline{x}), x_{k+1} - \overline{x}\rangle \geq 0. \tag{2}$$

Combining the last inequalities, we obtain

$$\Lambda(x_k) - \Lambda(x_{k+1}) \geq \frac{\alpha}{2}|x_{k+1} - x_k|^2 + \lambda\langle G(x_k) - G(\overline{x}), x_{k+1} - \overline{x}\rangle. \tag{3}$$

Now, by setting $E = \lambda\langle G(x_k) - G(\overline{x}), x_{k+1} - \overline{x}\rangle$,

$$E = \lambda\langle G(x_k) - G(\overline{x}), x_k - \overline{x}\rangle + \lambda\langle G(x_k) - G(\overline{x}), x_{k+1} - x_k\rangle.$$

As G is co-coercive, we get

$$\begin{aligned} E &\geq \lambda\left(\gamma|G(x_k) - G(\overline{x})|^2 + \langle G(x_k) - G(\overline{x}), x_{k+1} - x_k\rangle\right) \\ &\geq -\frac{\lambda}{4\gamma}|x_{k+1} - x_k|^2. \end{aligned}$$

[1] $\limsup_{t\to 0} F(tz + (1-t)x, y) \leq F(x,y) \quad \forall x, y, z \in C$
[2] $F(x,y) + F(y,x) \leq 0 \quad \forall x, y \in C$

Therefore

$$\Lambda(x_k) - \Lambda(x_{k+1}) \geq \frac{1}{2}\left(\alpha - \frac{\lambda}{2\gamma}\right)|x_{k+1} - x_k|^2. \tag{4}$$

The assumption $\lambda < 2\alpha\gamma$ implies that the sequence $\{\Lambda(x_k)\}_{k\in\mathbb{N}}$ decreases and likewise $l(x^*) := \lim_{k\to+\infty}\Lambda(x_k)$ exists and is finite, and in light of (4), we must have $\lim_{k\to+\infty}|x_{k+1} - x_k| = 0$. Furthermore, since $|x_k - x^*|^2 \leq \frac{2}{\alpha}\Lambda(x_k)$, we can assert that the sequence $\{x_k\}_{k\in\mathbb{N}}$ is bounded.
Now, let $\tilde{x}$ be any weak cluster point of $\{x_k\}_{k\in\mathbb{N}}$. Using (AEP), monotonicity of F and Lipschitz continuity of h' and G, we can write

$$\begin{aligned}(|G(x_k)| + \beta|x - x_{k+1}|)\,|x_k - x_{k+1}| &\geq \langle h'(x_{k+1}) - h'(x_k), x - x_{k+1}\rangle \\ &\geq \lambda\left(F(x, x_{k+1}) + \langle G(\tilde{x}), x_k - x\rangle\right),\end{aligned}$$

Passing to the limit in last inequality and taking into account the lower semicontinuity of $F(x,.)$, we obtain

$$F(x,\tilde{x}) + \langle G(\tilde{x}), \tilde{x} - x\rangle \leq 0 \quad \forall x \in C. \tag{5}$$

Now, let $x_t = tx + (1-t)\tilde{x}, 0 < t \leq 1$, from the properties of F follows then for all t

$$\begin{aligned}0 = F(x_t, x_t) &\leq tF(x_t, x) + (1-t)F(x_t, \tilde{x}) \\ &\leq tF(x_t, x) + t(1-t)\langle G(\tilde{x}), x - \tilde{x}\rangle.\end{aligned}$$

Dividing by t and letting $t \downarrow 0$, we get $x_t \to \tilde{x}$, which together with the upper hemicontinuity of F yields

$$F(\tilde{x}, x) + \langle G(\tilde{x}), x - \tilde{x}\rangle \geq 0 \quad \forall\, x \in C, \tag{6}$$

that is $\tilde{x}$ is a solution of the problem (GEP). To prove that there is no more than one weak cluster is standard (see, Opial [28]). Thus the entire sequence weakly converges to a solution of (GEP) which completes the proof. ◇

Remark 1 (i) The previous convergence theorem is an extension of a result by Zhu and Marcotte [34] and it is worth mentioning that if T is a single-valued operator, by taking $F(x,y) = \langle Tx, y - x\rangle$, we recover as a special case the auxiliary problem principle of G. Cohen [15].
(ii) It should be noticed that inexact versions are essential for producing implementable methods since, except for very special situations, the auxiliary equilibrium problem may not be solved exactly. So it is essential to replace the iteration scheme (AEP) by a relation which takes into account an approximate calculation of the iterates. More precisely, one can consider the following method: at stage k, starting from x_k, compute a solution of

$$F(x_{k+1}, x) + \langle G(x_k) + \lambda^{-1}(h'(x_{k+1}) - h'(x_k)), x - x_{k+1}\rangle \geq -\varepsilon_k \quad \forall x \in C,$$

where $\{\varepsilon_k\}_{k\in\mathbb{N}}$ is a sequence of positive numbers which goes to zero.
The convergence result above still holds true provided that $\sum_{k=1}^{+\infty} \varepsilon_k < +\infty$.
(iii) If we set $F(x, y) = \varphi(y) - \varphi(x)$, where φ is a real-valued function, then the relation above reduces to Lemaire's scheme [24], namely

$$x_{k+1} \in (h' + \lambda\partial_{\varepsilon_k}\tilde{\varphi})^{-1}(h'(x_k) - \lambda G(x_k)), \tag{7}$$

where, $\tilde{\varphi} = \varphi + i_C$, i_C denoting the indicator function of C. If T is a maximal monotone operator, taking $C = X$ and $F(x, y) = \sup_{z\in T(x)}\langle z, y - x\rangle$, it leads to

$$x_{k+1} \in (h' + \lambda T^{\varepsilon_k})^{-1}(h'(x_k) - \lambda G(x_k)), \tag{8}$$

where T^{ε} stands for the ε-enlargement of the operator T (see for instance [12]), namely

$$T^{\varepsilon}(x) := \{y : \langle y - v, x - u\rangle \geq -\varepsilon \quad \forall v \in T(u)\}. \tag{9}$$

The next theorem improves a convergence result by G. Cohen [15].

Theorem 2 *In addition to hypothesis on F, C and h, we assume that G is strongly monotone over C (with modulus γ), namely, there exists a positive constant γ such that*

$$\langle G(x) - G(y), x - y\rangle \geq \gamma|x - y|^2 \quad \forall x, y \in C,$$

and Lipschitz continuous (with constant β), that is, there exists a positive constant β such that

$$|G(x) - G(y)| \leq \beta|x - y| \quad \forall x, y \in C.$$

If $0 < \lambda < \frac{2\alpha\gamma}{\beta^2}$, then the sequence $\{x_k\}_{k\in\mathbb{N}}$ strongly converges to the unique solution of (GEP).

Proof Relation (3) remains valid. From properties of F, we have

$$E \geq \varepsilon\left(\gamma\,|x_k - \overline{x}|^2 + |x_k - \overline{x}|\,|x_{k+1} - x_k|\right) \geq -\frac{\varepsilon\beta^2}{4\gamma}\,|x_{k+1} - x_k|^2. \tag{10}$$

Therefore

$$\Lambda(x_k) - \Lambda(x_{k+1}) \geq \frac{1}{2}\left(\alpha - \frac{\varepsilon}{2\gamma}\right)|x_{k+1} - x_k|^2. \tag{11}$$

From which we infer, by using the same argument as in the proof of theorem 2.1, that the sequence $\{x_k\}_{k\in\mathbb{N}}$ weakly converges to a solution of (GEP). On the other hand, according to the strong convexity of h, we can write

$$h(\overline{x}) \geq h(x_k) + \langle h'(\overline{x}), x_k - \overline{x}\rangle + \frac{\alpha}{2}\,|x_k - \overline{x}|^2, \tag{12}$$

thus

$$h(\overline{x}) \geq h(x_k) + \langle h'(\overline{x}), x_k - \overline{x} \rangle . \tag{13}$$

Passing to the limit in the last inequality, we successively obtain

$$\begin{aligned} h(\overline{x}) &\geq \limsup_{k \to +\infty} h(x_k) + \lim_{k \to +\infty} \langle h'(\overline{x}), x_k - \overline{x} \rangle \\ &\geq \liminf_{k \to +\infty} h(x_k) + \lim_{k \to +\infty} \langle h'(\overline{x}), x_k - \overline{x} \rangle , \end{aligned}$$

from which we deduce

$$\lim_{k \to +\infty} h(x_k) = h(\overline{x}), \tag{14}$$

since h is lower semicontinuous and $\{x_k\}_{k \in \mathbb{N}}$ weakly converges to $\overline{x}$. Returning to (13), by letting $k \to +\infty$ and taking into account relation (14), we obtain the desired result. ◇

3 Splitting Methods

In this section, we would like to mention that for solving equilibrium problems involving the sum of two bivariate functions, namely,

$$(EP) \qquad \text{find} \quad \bar{x} \in C \quad \text{such that} \quad F_1(\bar{x}, y) + F_2(\bar{x}, y) \geq 0 \quad \forall y \in C,$$

one can propose some splitting methods based upon partial or complete regularization of (EP). Here, we just indicate briefly the main idea. To begin with, let us give an extension of the notion of Yosida approximate. To this end, let λ be a positive real number. For a given bivariate function F we define the associated *Yosida approximate*, F_λ, over C as follow

$$F_\lambda(x, y) := \langle A_\lambda^F(x), y - x \rangle \quad \text{in which} \quad A_\lambda^F(x) := \lambda^{-1}(x - J_\lambda^F(x))$$

where $J_\lambda^F(x) \in C$ (the *resolvent* of F with parameter λ) is the unique solution of

$$\lambda F(J_\lambda^F(x), y) + \langle J_\lambda^F(x) - x, y - J_\lambda^F(x) \rangle \geq 0 \quad \forall y \in C. \tag{15}$$

It is worth noting that when T is a maximal monotone operator, $C = X$ and $F(x, y) = \sup_{z \in T(x)} \langle z, y - x \rangle$, it directly yields

$$J_\lambda^F(x) = (I + \lambda T)^{-1}(x) \quad \text{and} \quad A_\lambda^F(x) = (\lambda I + T^{-1})^{-1}(x).$$

Hence, we recover the classical concepts. Furthermore, a simple calculation involving (15) and monotonicity of F yields

$$\forall \lambda > 0, \forall x, y \in X \qquad |J_\lambda^F(x) - J_\lambda^F(y)| \leq |x - y|, \tag{16}$$

and

$$\forall x \in X, \forall \lambda, \mu > 0 \qquad J_\lambda^F(x) = J_\mu^F\left(\frac{\mu}{\lambda}x + (1-\frac{\mu}{\lambda})J_\lambda^F(x)\right). \tag{17}$$

If we want to regularize (EP) one can replace, for example, one of bivariate functions by its Yosida approximate. An analogue of the Passty method [30] is nothing but the corresponding iteration method with variable parameter. More precisely at stage k, knowing x_k compute the solution of

$$\lambda_k F_2(x_{k+1}, y) + \langle x_{k+1} - J_{\lambda_k}^{F_1}(x_k), y - x_{k+1}\rangle \geq 0 \quad \forall y \in C,$$

in which $J_{\lambda_k}^{F_1}(x_k)$ solves

$$\lambda_k F_1(J_{\lambda_k}^{F_1}(x_k), y) + \langle J_{\lambda_k}^{F_1}(x_k) - x_k, y - J_{\lambda_k}^{F_1}(x_k)\rangle \geq 0 \quad \forall y \in C,$$

where $\{\lambda_k\}_{k\in\mathbb{N}}$ is a sequence of positive numberswith limit zero.
In the case where F_1 and F_2 are defined by two maximal monotone operators A and B with maximal monotone sum, (EP) is nothing but the problem of finding a zero of $A+B$ and it is easy to check that the proposed method reduces to the following iterative scheme introduced by Passty [30]

$$x_{k+1} = (I + \lambda_k A)^{-1} \circ (I + \lambda_k B)^{-1}(x_k).$$

The regularization of both bifunctions leads to a splitting scheme that extends the barycentric proximal method proposed by Lehdili-Lemaire in [23].
On the other hand, it is easy to see that finding a solution of (GEP) with $G \equiv 0$ is equivalent to finding a zero of A_γ^F, that is, to solving the following unconstrained regularized problem

$$(REP) \qquad \text{find} \quad \bar{x} \in X \quad \text{such that} \quad A_\gamma^F(\bar{x}) = 0.$$

Then, instead of using an algorithm to compute a solution of (GEP), we can use the basic algorithm with $F \equiv 0$ and $G = A_\gamma^F$ to solve the regularized problem (REP) since A_γ^F is co-coercive. This results in the following scheme

$$(RAEP) \qquad \min_{x\in X}(h(x) + \langle \lambda A_\gamma^F(x_k) - h(x_k), x\rangle).$$

Moreover, from Theorem 2.1, we derive the immediate corollary hereafter:

Corollary 1 *If $0 < \lambda < 2\beta\gamma$, then the sequence $\{x_k\}_{k\in\mathbb{N}}$ converges weakly to a solution of (GEP).*

For the sake of simplicity, in the sequel we set $G \equiv 0$.

4 A Selection Method

The Tikhonov regularization method for ill-posed problems is well known for minimization, monotone inclusion and fixed-point problems. The aim of this section is to extend this method to the problem of finding

$$(EP) \qquad \bar{x} \in C \quad \text{such that} \quad F(\bar{x}, x) \geq 0 \quad \forall x \in C.$$

Let us introduce the basic algorithm. Let $\{\lambda_k\}_{k\in\mathbb{N}}$ be a sequence of real numbers tending to $+\infty$. The selection method consists of finding the point x_k solution of the problem

$$(TEP) \qquad F(x_k, x) + \lambda_k^{-1}\langle x_k, x - x_k\rangle \geq 0 \quad \forall x \in C, k \in \mathbb{N}.$$

The next result gives a convergence property of (TEP):

Theorem 3 *Assume that F satisfies assumptions of Theorem 2.1. Then, there exists a unique solution x_k to the auxiliary problem (TEP) and the sequence $\{x_k\}_{k\in\mathbb{N}}$ strongly converges to $\tilde{x}$ the element of minimal norm of the solution set, S, of problem (EP), i.e., $|\tilde{x}| = \min_{x\in S}|x|$.*

Proof The existence and uniqueness of x_k follow by invoking ([8], theorem 1). Under the hypothesis made in the above theorem, it is easy to check that the solution set S of (EP) is a nonempty closed convex subset of X. So, there is a unique $\overline{x} \in S$ such that $|\overline{x}| = \min_{x\in S}|x|$.
By setting $x = \overline{x}$ in (TEP), $x = x_k$ in (EP) and adding the resulting inequalities, we obtain

$$F(\overline{x}, x_k) + F(x_k, \overline{x}) + \langle \gamma_k^{-1} x_k, \overline{x} - x_k\rangle \geq 0 \quad \forall k \in \mathbb{N}. \tag{18}$$

From which we deduce, thanks to the monotonicity of F,

$$|x_k| \leq |\overline{x}| \quad \forall\, k \in \mathbb{N}. \tag{19}$$

This implies that the sequence $\{x_k\}_{k\in\mathbb{N}}$ is bounded.
Let $\widetilde{x}$ be any weak cluster point of $\{x_k\}_{k\in\mathbb{N}}$. According to (TEP) and to the fact that F is monotone, we can write

$$-F(x, x_k) + \langle \gamma_k^{-1} x_k, x - x_k\rangle \geq 0 \;\; \forall x \in K. \tag{20}$$

Passing to the limit, on a subsequence, in the last inequality and using to the weak lower semicontinuity of F, we obtain

$$F(x, \widetilde{x}) \leq 0 \;\; \forall x \in K. \tag{21}$$

Using the same device as did to obtain (6) from (5), we get

$$F(\tilde{x}, x) \geq 0 \quad \forall x \in C, \tag{22}$$

that is $\tilde{x}$ is a solution of the problem (EP).
It remains to prove that there is no more than one weak cluster point. To this end, passing to the limit, on a subsequence, in (19) and taking into account the weak lower semicontinuity of the norm, we derive

$$|\widetilde{x}| \leq \lim\inf_{\nu\to+\infty} |x_{k_\nu}| \leq \lim\sup_{\nu\to+\infty} |x_{k_\nu}| \leq |\overline{x}|, \tag{23}$$

which together with (4) yields $\widetilde{x} = \overline{x}$.

Now working as above, but on the whole sequence $\{x_k\}_{k\in\mathbb{N}}$, we obtain

$$|\overline{x}| \leq \lim_{k\to+\infty}\inf |x_k| \leq \lim_{k\to+\infty}\sup |x_k| \leq |\overline{x}|. \tag{24}$$

Therefore, we have

$$\lim_{k\to+\infty} |x_k| = |\overline{x}|.$$

This property, combined with the weak convergence of the sequence $\{x_k\}_{k\in\mathbb{N}}$ to $\overline{x}$, allows us to conclude. ◇

Remark 2 Let us notice that by taking $F(x, y) = \sup_{z\in T(x)}\langle z, y - x\rangle$ and $C = X$, where T a maximal monotone operator, (TEP) reduces to the classical Tikhonov method for monotone inclusion, namely $x_k = (I + \lambda_k T)^{-1}(0)$ and by setting $F(x, y) = \langle x - Px, y - x\rangle$, (TEP) reduces to the method of continuous regularization

$$x_k + \lambda_k^{-1} x_k = P(x_k)$$

given in [20] (see also [29]), and we recover its convergence results as a special case. Let us note that in this case F is monotone if, and only if, $\langle Px - Py, x - y\rangle \leq |x - y|$ for all $x, y \in C$, hence in particular if P is nonexpansive.
We would like to mentionthe possibility of considering perturbed versions of the algorithms introduced above by replacing the function F by a sequence of functions F_k which approximate F in the following senses (for some $\lambda > 0$)

$$\lim_{k\to+\infty} J_\lambda^{F_k}(x) = J_\lambda^F(x) \quad \forall x \in X \tag{25}$$

or

$$\lim_{k\to+\infty} \sup_{|x|\leq r} |J_\lambda^{F_k}(x) - J_\lambda^F(x)| = 0 \quad \forall r \geq 0. \tag{26}$$

Under either condition (25) or 26) the above convergence result still holds true. Furthermore, we observe that when $F_k(x, y) = \sup_{z\in T_k(x)}\langle z, y-x\rangle$ and $F(x, y) = \sup_{z\in T(x)}\langle z, y-x\rangle$, T_k, T being maximal monotone operators, the first condition is equivalent to the graph convergence of the sequence $\{T_k\}_{k\in\mathbb{N}}$ to T and the second one reduces to the convergence in the Attouch-Wets variational distance sense (see for example [4]). In this case we recover results obtained among others in [32] and [33] as a particular case.

5 A Continuous Process

In order to make a connection with a method studied by Flam [19], let us notice that (EP) can be rewritten in the following equivalent form:

$$\text{find} \quad \bar{x} \in C \quad \text{such that} \quad \bar{x} \in \operatorname{argmin}\{F(\bar{x}, x) : x \in C\}.$$

For each $\varepsilon > 0$, we consider the approximate equilibrium problem:

$$(EP)_\varepsilon \quad x(\varepsilon) \in \operatorname{argmin}\{F(x(\varepsilon), x, \varepsilon) : x \in C\},$$

where $F(\cdot,\cdot,\varepsilon) : C \times C \times \mathbb{R}^+ \to \mathbb{R}$ is an approximation of F, and we suppose the existence of a path $x(\varepsilon)$ which is an optimal solution of $(EP)_\varepsilon$, which is absolutely continuous on every interval of the form $[\varepsilon_1, \varepsilon_0]$ with $0 < \varepsilon_1 < \varepsilon_0$, and converges as $\varepsilon \to 0$ towards an equilibrium point $x^* \in S$.
Our purpose is to construct a general dynamical method which allows us to compute the particular solution x^*. This solution obtained by approximation usually enjoys nice geometrical or variational properties, see for example [17].
When considering problem (EP), the proximal method may be interpreted as a first order implicit discretization of the differential inclusion

$$(DM) \quad \frac{du(t)}{dt} \in P_{Tx}\left(-\partial F(u(t), \cdot)u(t)\right),$$

where $Tx = \operatorname{cl}\mathbb{R}_+(C - x)$ is the tangent cone of C at $x \in C$ and the operator P_K stands for the orthogonal projection onto a closed convex set K.
Flam [19] showed that the trajectory defined by this "idealized gradient method" weakly converges to some $\bar{x} \in S$ as t goes to $+\infty$. The idea to be developed here relies on the approach by Attouch and Cominetti. We will take advantage of the regularizing properties of the approximations $(EP)_\varepsilon$, and we will consider the following approximate dynamical method: solve for $t \geq 0$ the differential inclusion where $u(0) \in C$ is arbitrary

$$(ADM) \qquad \frac{du(t)}{dt} \in P_{Tx}\left(-\partial F\left(u(t), \cdot, \varepsilon(t)\right) u(t)\right),$$

with the assumption that $\varepsilon : [0, +\infty[\to \mathbb{R}_+$ is some strictly positive fonction decreasing to zero as $t \to \infty$ and $\partial F(u(0), \cdot)u(0) \neq \emptyset$.
We are mainly concerned with the asymptotical behavior of the solutions of (ADM) as t goes to $+\infty$ and we are specially interested in conditions ensuring the strong convergence of $u(t)$ towards the particular element x^*.
Hereafter, we also assume that (ADM) has a (unique) strong global solution, that is, there exists an absolutely continuous function $u(\cdot) : [0, +\infty[\to C$ satisfying the inclusion (ADM) almost everywhere. It follows from viability theory that $u(\cdot)$ so generated remains in C for all t (see Aubin [5]).

6 The Slow Convergence Results

A basic property of the approximate problem which ensures the uniqueness of the solution $x(\varepsilon)$ of $(EP)_\varepsilon$ is the strong convexity of the function F with respect to the second variable. This is equivalent to the strong monotonicity of its partial subdifferential, that is, for every $\varepsilon > 0$ there exists $\beta(\varepsilon) > 0$ such that

$$\langle \partial F(x, \cdot, \varepsilon)x - \partial F(y, \cdot, \varepsilon)y, x - y\rangle \geq \beta(\varepsilon)|x - y|^2 \quad \forall x, y \in C. \tag{27}$$

A weaker condition is the strong monotonicity over bounded sets, namely, for every $\varepsilon > 0, K \in \mathbb{N}^*$ there exists $\beta_K(\varepsilon) > 0$ such that

$$\langle \partial F(x, \cdot, \varepsilon)x - \partial F(y, \cdot, \varepsilon)y, x - y \rangle \geq \beta_K(\varepsilon)|x - y|^2 \quad \forall x, y \in B(0, K). \tag{28}$$

Since nonuniqueness in (EP) is an important issue, the function $\beta(\varepsilon)$ is allowed to go to zero as ε tends to zero. We will see that the rate of convergence of $\beta(\varepsilon)$ to 0 plays a crucial role in the next results which are mainly inspired from Attouch-Cominetti [3].

Let us first consider the case of a short optimal trajectory:

Theorem 4 *In addition to assumptions of Theorem 2.1 on F, let us suppose that*

1. $\forall \varepsilon_0 > 0 \quad \int_0^{\varepsilon_0} |\frac{dx}{d\varepsilon}| d\varepsilon < +\infty.$
2. $\forall K > 0$ *the local strong monotonicity condition (28) holds true, $\beta_K(\cdot)$ is measurable and $\varepsilon(t)$ is such that* $\int_0^{+\infty} \beta_K(\varepsilon(s)) ds = +\infty.$

Then $u(t)$ solution of (ADM) norm converges to x^* as $t \to +\infty$.

Proof rm Since $x(\varepsilon(t))$ converges to x^*, it suffices to show that the Liapunov function $\varphi(t) := \frac{1}{2}|u(t) - x(\varepsilon(t))|^2 \to 0$ as $t \to 0$. For almost all $t \in [0, +\infty[$ we have

$$\dot{\varphi}(t) = \langle \dot{u}(t) - \dot{\varepsilon}(t)\frac{dx}{d\varepsilon}(\varepsilon(t)), u(t) - x(\varepsilon(t)) \rangle$$

which implies

$$\dot{\varphi}(t) \leq \langle \dot{u}(t), u(t) - x(\varepsilon(t)) \rangle - \dot{\varepsilon}(t)|\frac{dx}{d\varepsilon}(\varepsilon(t))||u(t) - x(\varepsilon(t))|.$$

Since for almost every $t > 0$

$$-\dot{u}(t) \in \partial F(u(t), \cdot, \varepsilon(t))u(t) + N_C(u(t)), \tag{29}$$

and by a well-known optimality condition

$$0 \in \partial F(x(\varepsilon(t)), \cdot, \varepsilon(t))x(\varepsilon(t)) + N_C(x(\varepsilon(t)), \tag{30}$$

the monotonicity of both the partial subdifferential of F and the normal cone to C implies

$$\dot{\varphi}(t) \leq -\dot{\varepsilon}(t)|\frac{dx}{d\varepsilon}(\varepsilon(t))||u(t) - \varepsilon(t)| = -\dot{\varepsilon}(t)|\frac{dx}{d\varepsilon}(\varepsilon(t))|\sqrt{2\varphi(t)}.$$

Using ([3], Lemma 3.1), we obtain

$$\sqrt{\varphi(t)} \leq \sqrt{\varphi(0)} + \frac{1}{\sqrt{2}}\int_0^{\varepsilon(0)} |\frac{dx}{d\varepsilon}| d\varepsilon.$$

Thus $u(t)$ remains bounded. Let $K > 0$ be such that $|u(t)| \leq K$ and $|x(\varepsilon(t)| \leq K$ for all $t > 0$. Invoking the strong monotonicity of $\partial F + N_C$ we get

$$\dot{\varphi}(t) \leq -\beta_K(\varepsilon(t))|u(t) - x(\varepsilon(t))|^2 - \dot{\varepsilon}(t)|\frac{dx}{d\varepsilon}(\varepsilon(t))||u(t) - x(\varepsilon(t))|,$$

that is to say

$$\dot{\varphi}(t) + 2\beta_K(\varepsilon(t))\varphi(t) \leq -\sqrt{2}|\dot{\varepsilon}(t)||\frac{dx}{d\varepsilon}(\varepsilon(t))|\sqrt{\varphi(t)}. \tag{31}$$

Now we use an argument used is in Attouch-Cominetti [3] and which is detailed here for sake of completeness. Denoting $E(t) = \int_0^t \beta_K(\varepsilon(s))ds$ and multiplying the above inequality by $e^{2E(t)}$ we get

$$\frac{d}{dt}\left(e^{2E(t)}\varphi(t)\right) \leq -\sqrt{2}e^{E(t)}|\frac{dx}{d\varepsilon}(\varepsilon(t))|\dot{\varepsilon}(t)\sqrt{e^{2E(t)}\varphi(t)}.$$

Applying ([3], Lemma 3.1) with $\theta(t) = 2e^{2E(t)}\varphi(t)$, we deduce that for each xt such that $0 \leq t_0 < t < +\infty$

$$e^{E(t)}\sqrt{\varphi(t)} \leq e^{E(t_0)}\sqrt{\varphi(t_0)} - \frac{1}{\sqrt{2}}\int_{t_0}^{t} e^{E(s)}|\frac{dx}{d\varepsilon}(\varepsilon(s))|\dot{\varepsilon}(s)ds.$$

Multiplying the last inequality by $e^{-E(t)}$ and taking into account that $E(\cdot)$ is increasing wederive

$$\sqrt{\varphi(t)} \leq e^{-(E(t)-E(t_0))}\sqrt{\varphi(t_0)} + \frac{1}{\sqrt{2}}\int_{\varepsilon(t)}^{\varepsilon(t_0)} |\frac{dx}{d\varepsilon}|d\varepsilon.$$

Using the fact that $E(t) \to +\infty$ and then letting $t_0 \to +\infty$ and using the finite lenght assumption we obtain

$$\limsup_{t\to+\infty} \sqrt{\varphi(t)} = 0. \tag{32}$$

◇

Let us consider now the case where the trajectory $\varepsilon \to x(\varepsilon)$ may have an infinite length:

Theorem 5 *In addition to assumptions of Theorem 2.1 on F, let us assume that $x(\varepsilon)$ satisfies*

$$|\frac{dx}{d\varepsilon}(\varepsilon)| \leq \frac{1}{\gamma(\varepsilon)} \quad a.e. \text{ on }]0, +\infty[, \tag{33}$$

for some function γ possibly tending to 0 as $\varepsilon \to 0$. Suppose (27) is satisfied and assume $\varepsilon(t)$ satisfies:

$$\int_0^{+\infty} \beta(\varepsilon(s))ds = +\infty \quad and \quad \lim_{t\to+\infty} \frac{\dot{\varepsilon}(t)}{\beta(\varepsilon(t))\gamma(\varepsilon(t))} = 0. \tag{34}$$

Then, the (ADM) trajectory $u(t)$ norm converges to x^* as $t \to +\infty$.

Proof Proceeding as in the proof of Theorem 4.1 and replacing β_K by β we obtain

$$\sqrt{\varphi(t)} \leq e^{-(E(t)-E(t_0))}\sqrt{\varphi(t_0)} - \frac{1}{\sqrt{2}}e^{-E(t)}\int_{t_0}^{t} e^{-E(s)}|\frac{dx}{d\varepsilon}(\varepsilon(s))|\dot{\varepsilon}(s)ds$$

where $E(t) = \int_0^t \beta(\varepsilon(s))ds$.
Define $h(t_0) = \sup_{s\geq t_0} \frac{|\dot{\varepsilon}(s)|}{\beta(\varepsilon(s))\gamma(\varepsilon(s))}$. Since $E(s) = \beta(\varepsilon(s))$ it follows that

$$\sqrt{\varphi(t)} \leq e^{-(E(t)-E(t_0))}\sqrt{\varphi(t_0)} + \frac{1}{\sqrt{2}}h(t_0)$$

Thanks to (34) we have

$$\limsup_{t\to+\infty} \sqrt{\varphi(t)} \leq \frac{1}{\sqrt{2}}h(t_0)$$

which in turn implies

$$\limsup_{t\to+\infty} \sqrt{\varphi(t)} \leq 0$$

since $h(t_0) \to 0$ as $t_0 \to +\infty$.

7 Special Cases

7.1 Convex Minimization

Let $\varphi : X \to \mathbb{R}$ be a convex and lower semicontinous function, then if we take $F(x,y) = \varphi(y) - \varphi(x)$ and $F(x,y,\varepsilon) = \varphi(y,\varepsilon) - \varphi(x,\varepsilon)$, (EP) becomes in this context

$$\bar{x} \in \operatorname{argmin}\{\varphi(x);\ x \in C\} = \operatorname{argmin}\{\tilde{\varphi}(x);\ x \in X\},$$

where $\tilde{\varphi} = \varphi + i_C$, i_C denoting the indicator function of C. (ADM) reduces to

$$-\frac{du(t)}{dt} \in \partial\tilde{\varphi}(\cdot,\varepsilon)u(t). \tag{35}$$

So we recover results of Attouch and Cominetti, that is, $u(t)$ norm converges to x^* a minimizer of $\tilde{\varphi}$ provided that $\varepsilon(t)$ converges slowly to zero.

7.2 Fixed-Points Problems

Let $P : C \to C$ be a given mapping, if we set $F(x,y) = \langle x - Px, y - x\rangle$ and $F(x,y,\varepsilon) = \langle x - P(\cdot,\varepsilon)x, y - x\rangle$, in this setting (EP) is nothing but the problem of finding fixed points of P and (ADM) takes the following form

$$-\frac{du(t)}{dt} - u(t) + P(\cdot,\varepsilon)u(t) \in N_C(u(t)). \tag{36}$$

We recall that monotonicity of F is equivalent to saying

$$\langle P(x) - P(y), x - y\rangle \le |x - y|^2,$$

which is clearly satisfied if P is nonexpansive. On the other hand, strong monotonicity of the operator $\partial_y F(\cdot,\cdot,\varepsilon) = I - P(\cdot,\epsilon)$ is ensured by the condition

$$\langle P(x,\varepsilon) - P(y,\varepsilon), x - y\rangle \le (1 - \beta(\varepsilon))\,|x - y|^2,$$

which is in particular verified if

$$|P(x,\varepsilon) - P(y,\varepsilon)| \le (1 - \beta(\varepsilon))\,|x - y|.$$

In the present case our main results ensure the norm convergence of $u(t)$ to a fixed point of P provided that $\varepsilon(t)$ tends not too fast to zero.

8 Tikhonov Regularization

We consider the case where $F(x,y,\varepsilon) = F(x,y) + \varepsilon\langle x, y - x\rangle$ so that $x(\varepsilon)$ is uniquely determined by

$$\partial F(x(\varepsilon),\cdot)x(\varepsilon) + \varepsilon x(\varepsilon) \ni 0.$$

$x(\varepsilon)$ is bounded if and only if $S \neq \emptyset$ and in that case (see Theorem 4.1.)

$$\lim_{\varepsilon\to 0} x(\varepsilon) = proj_S 0 = x^*.$$

Now we will study the differentiability properties of the mapping $\varepsilon \to x(\varepsilon)$. The proof relies on the resolvent equation (17). In view of relation (15) we have $x(\varepsilon) = J^F_{\frac{1}{\varepsilon}}(0)$ and from (16) and (17) we infer

$$|J^F_\lambda(x) - J^F_\mu(x)| \le |\mu - \lambda|\left|\frac{x - J^F_\lambda}{\lambda}\right|.$$

Therefore,

$$|x(\varepsilon) - x(\varepsilon')| = |J^F_{\frac{1}{\varepsilon}}(0) - J^F_{\frac{1}{\varepsilon'}}(0)| \le \left|\frac{1}{\varepsilon} - \frac{1}{\varepsilon'}\right|\varepsilon|x(\varepsilon)|.$$

As a result, there exists a constant $C > 0$ such that

$$|x(\varepsilon) - x(\varepsilon')| \le \frac{C}{\varepsilon'}|\varepsilon - \varepsilon'|.$$

Hence $x(\varepsilon)$ is locally Lipschitz on $]0,+\infty[$ and

$$\left|\frac{dx}{d\varepsilon}(\varepsilon)\right| \le \frac{C}{\varepsilon} \quad \text{a.e.}$$

Returning to the corresponding condition (34) for the parametrization $\varepsilon(t)$ to have $\lim_{t\to+\infty} u(t) = x^*$. We can take $\beta(\varepsilon) = \varepsilon$ so that conditions in theorem 5.2 become:

$$\int_0^{+\infty} \varepsilon(s)ds = +\infty \quad \text{and} \quad \lim_{t\to+\infty} \frac{\dot\varepsilon(t)}{\varepsilon^2(t)} = 0.$$

If we take for example $\varepsilon(t) = \frac{1}{t^\alpha}$ the condition holds true as far as $0 < \alpha < 1$,

Proposition 1 *For every $u_0 \in C$ and $0 < \alpha < 1$, the solution of*

$$\begin{cases} \dot{u}(t) + \partial F(u(t), \cdot)u(t) + \frac{1}{t^\alpha}u(t) = 0 \\ u(0) = u_0 \end{cases} \tag{37}$$

tends to x^* element of the minimal norm of the solution set S as $t \to +\infty$.

References

[1] Alvarez, F. (1998), On the minimizing property of a second order dissipative system in Hilbert space. Prépublications de l'Université Montpelier II.

[2] Antipin, A. S. and Flam, S. D. (1997), Equilibrium programming using proximal-like algorithms. Math.- Programming, **78**, (1), 29–41.

[3] Attouch H. and Cominetti R. (1996), A dynamical approach to convex minimization coupling approximation with the steepest method. J.l of Differential Equations, **128**, 2, 519–540.

[4] Attouch H., Moudafi A. and Riahi H. (1993) Quantitative stability analysis for maximal monotone operators and semigroups of contractions. Nonlinear Analysis: Theory, Methods and Appl.,**21**, 9, 697–723.

[5] Aubin J.- P. (1991) Viability theory, Birkhauser, Basel.

[6] Auslender A. (1987) Numerical methods for non differentiable convex optimization. Math. Prog. Study **30**, 102–126.

[7] Bianchi M. and Schaible S. (1996) Generalized monotone bifunction and equilibrium problems. J. Optim. Theory Appli., **90**, 1, 31–43.

[8] Blum E. and Oettli W. (1994) From optimization and variational inequalities to equilibrium problems. The Math. Students, **63**, 123–145.

[9] Brézis H. (1971) Monotonicity to nonlinear partial differential equations. Contribution to nonlinear analysis, Academic Press, New York, 101–156.

[10] Browder F. E. (1965) Existence of periodic solutions for nonlinear equations of evolution, Proc. N. A. S., **53**, 1100–1103.

[11] Bruck R. E. (1975) Asymptotic convergence of nonlinear contraction semigroups in Hilbert space. J. Funct. Anal., **18**, 15–26.

[12] Burachik R. S., Iusem A. N. and Svaiter B. F. (1997) Enlargement of monotone operators with applications to variational inequalities. Set-Valued Analysis, **5**, 159–180.

[13] Cohen G. (1978) Optimization by decomposition and coordination: a unified approach. IEEE Transactions on Automatic Control AC-23, 222–232.

[14] Cohen G. (1980) Auxiliary problem principle and decomposition of optimization problems. J. Optim. Theory Appl., **32**, 277–305.

[15] Cohen G. (1988) Auxiliary problem principle extended to variational inequalities. J. of Optimization Theory and Applications, **59**, 325–333.

[16] Cominetti R. and San Martin J. (1994) Asymptotical analysis of the exponential penalty trajectory in linear programming. Mathematical Programming, **67**, 169–187.

[17] Cominetti R. (1995) Asymptotic convergence of the steepest descent method for exponential penalty in linear programming. J. Convex Anal. 2, **112**, 145–152.

[18] Flam S. D. and Greco G. (1991) Noncooperative games; methods for subgradient projection and proximal point. In:W. Oettli and D. Pallaschke, (Eds) Advances in Optimization, Lambrecht, Lecture Notes in Econom. and Math. Systems, **382**, 406–419, Springer Verlag, Berlin.

[19] Flam S. D. (1997) Gradient approches to equilibrium. Lecture Notes In Economics and Mathematical Systems, **452**, Springer-Verlag, Berlin, 49–60.

[20] Halpern B. (1967) Fixed points of nonexpansive maps. Bull. Amer. Math. Soc., **73**, 957–961.

[21] Kaplan A. and Tichatschke R. (1994) Stable methods for ill-posed problems. Akademie Verlag, Berlin.

[22] Lassonde M. (1983) On the use of KKM multifunctions in fixed point theory and related topics. J. Math. Anal. and Appli., **97**, 1, 151–201.

[23] Lehdili N. and Lemaire B. (1998) The barycentric proximal method. To appear in Communications on Applied Nonlinear Analysis.

[24] Lemaire B. (1991) About the convergence of the proximal method. Advances in Optimization, Lecture Notes in Economics and Mathematical Systems **382**, Springer-Verlag, 39–51.

[25] Lemaire B. (1997) On the convergence of some iterative methods for convex minimization. Recent Developements in Optimization, Lecture Notes in Economics and Mathematical Systems, **452**, 154–167.

[26] Mouallif K., Nguyen V. H. and Strodiot J.-J. (1991) A perturbed parallel decomposition method for a class of nonsmooth convex minimization problems. Siam J. Control Opt., **29**, 822–847.

[27] Moudafi A. and Théra M. (1997) Finding a zero of the sum of two maximal monotone operators. J. Optim. Theory Appl., **94**, 425–448.

[28] Opial G. B. (1967) Weak convergence of sequence of successive approximations for nonexpansive mapping. Bull. Amer. Math. Soc., **77**, 591–597.

[29] Ould Ahmed Salem C. (1998) Approximation de points fixes d'une contraction. Thèse de Doctorat, Université Montpellier II.

[30] Passty G. B. (1979) Ergodic convergence to a zero of the sum of monotone operators. J. of Math. Anal. and App., **72**, 383–390.

[31] Rockafellar R. T. (1976) Monotone operators and the proximal algorithm. Siam J. Control. Opt., **14**, (5), 877–898.

[32] Tikhonov A.N. and Arsenine V. Ya (1974) Methods for solving ill-posed problems. Nauka, Moscow.

[33] Tossings P. (1994) The perturbed Tikhonov's algorithm and some of its applications. M^2AN, **28**, 2, 189–221.

[34] Zhu D. L. and Marcotte P. (1996) Co-coercivity and the role in the convergence of iterative schemes fo solving variational inequalities. Siam Journal on Optimization, **6**, 3, 714–726.

Some Links Between Approximation, Nonsmooth Analysis, Penalization and Regularization

Jean-Paul Penot *

Abstract

We point out some connections between the different subjects of the title. We also present a simple approach to the viscosity selection principle of H. Attouch which avoids the use of epi-convergence.

Keywords: approximation, nonsmooth analysis, optimization, penalization, regularization, well-posedness, Yosida approximation.
AMS subject classification: 65J20, 65J15, 49R20, 90C99.

1 Introduction

We endeavour to delineate some connections between the different themes of the title. Some of these connections are well-known; others are unexpected. We start with a reformulation of the bright version given by Attouch of the regularization technique ([2], [3]). We adopt a simple, direct approach, avoiding the use of epi-convergence. This variational convergence is a powerful tool, especially when used in conjunction with duality ([42]). However, the fact that it can be bypassed allows one to concentrate on the assumptions which have to be imposed. In particular, we show how the Tikhonov regularization technique enters this framework and we sketch elements which may be useful for the study of convergence rates.
We consider the question of regularizing functions using either the Moreau-Yosida scheme or other schemes which are more adapted to nonconvex functions. We mention the related question of the Yosida type regularization for operators of monotone or non monotone type between a Banach space X and its dual which is treated in several monographs and in the recent paper [36] under weak assumptions.
In the last section we evoke the use of penalization and decoupling techniques in nonsmooth analysis.

*Mathématiques Appliquées, UPRES CNRS 5033, Av. de l'Université 64000 PAU, France

For the links between various techniques for studying conditioning and growth properties and the question of well-behavior and well-posedness we refer to [34] and its references.

2 A Viscosity Selection Principle

It is the purpose of this section to present a simple approach to the viscosity selection principle. This principle has been put in light in the important articles [2], [3] by H. Attouch by the means of epi-convergence techniques. Here we intend to give a direct, elementary approach which avoids such techniques.
Let us address the minimization problem

$$(\mathcal{P}) \qquad \text{Minimize} \quad f(x) \qquad x \in X,$$

where X is a topological space (or a convergence space) and f is a real-valued function on X. Throughout we suppose $\inf f(X)$ is finite. The regularization technique consists in introducing a family $(\mathcal{P}_t)$ of minimization problems parametrized by a subset T of a topological space P of parameters and in looking at possible limits of (approximate) solutions (x_t) of $(\mathcal{P}_t)$ as $t \to 0$, a given base point of P in the closure of T, for a well chosen topology on X. When X is a normed vector space, unless otherwise mentioned, we use the weak topology on X. For what concerns the parameter space, the following two cases are the most usual ones:
(a) $P := \mathbb{R}_+$, $T := \mathbb{P} :=]0, \infty[$ and the base point is 0;
(b) $P := \mathbb{N} \cup \{\infty\}$, $T := \mathbb{N}$, and the base point is ∞.
Of course, our framework is versatile enough to encompass other cases such as the case in which two real parameters are present. We suppose we dispose of a penalization function $k : P \times X \to \mathbb{R}_+$ and we take as regularized problem the problem

$$(\mathcal{P}_t) \qquad \text{Minimize} \quad f(x) + k_t(x) \qquad x \in X,$$

where $k_t(x) := k(t, x)$. In the classical case, X is a Hilbert space, $T := \mathbb{P}$ and $k(t, x) = \gamma(t)g(x)$ where $\gamma : T \to \mathbb{P}$, $g : X \to \mathbb{R}_+$, the usual case being $\gamma(t) = t$, $g(x) = \|x\|^2$, so that $k(t, x) = t\|x\|^2$. We will see in the next section that it is useful to consider cases in which k is not in such a simple decomposed form. In order to have some flexibility, we also suppose that we are given a weight $\omega : T \to \mathbb{R}_+$ which will be related to a convergence rate of the values of the objective functions.
The essence of the method is contained in the following result.

Theorem 1 *Let $\varepsilon : T \to \mathbb{R}_+$ be such that $\varepsilon(t) \to 0$ as $t \to 0$ in T and let x_t be an $\varepsilon(t)$-approximate solution to $(\mathcal{P}_t)$. Suppose $k(t, x) \to 0$ as $t \to 0$ for each $x \in X$. Let $\widehat{x}$ be a cluster point of (x_t) as $t \to 0$. Then $\widehat{x}$ is in the set $\overline{S}$ of minimizers of the lower semicontinuous hull $\overline{f}$ of f, $f(x_t)$ converges to $m := \inf f(X)$ and for each $t \in T$ one has $k_t(x_t) \leq \inf_{x \in S} k_t(x) + \varepsilon(t)$.*

In particular, if f is lower semicontinuous at $\widehat{x}$, then $\widehat{x}$ belongs to the set S of solutions to $(\mathcal{P})$.
If S is nonempty and if there exists $\omega : T \to \mathbb{R}_+$ such that
(a) $\omega(t)\varepsilon(t) \to 0$ as $t \to 0$,
(b) for each $x \in X$, $\omega(t)k(t,x)$ has a finite limit $g(x)$ as $t \to 0$,
(c) for each $x \in \overline{S}$ one has $g(x) \leq \liminf_{(t,v)\to(0,x)} \omega(t)k(t,v)$,
then $\widehat{x}$ satisfies the viscosity selection principle:

$$g(\widehat{x}) \leq g(x) \quad \text{for each } x \in S \tag{1}$$

and furthermore, when $\widehat{x} \in S$,

$$\omega(t)k(t,x_t) \to g(\widehat{x}) \quad \text{as } t \to 0,\ (x_t) \to \widehat{x}, \tag{2}$$

$$\omega(t)\left(f(x_t) - f(\widehat{x})\right) \to 0 \quad \text{as } t \to 0,\ (x_t) \to \widehat{x}. \tag{3}$$

In particular, if f is lower semicontinuous at $\widehat{x}$, then $\widehat{x}$ is a minimizer of f on X and a minimizer of g on S, and (2), (3) hold.

Assertion (2) means that for any net $(t_i)_{i\in I} \to 0$ such that $(x_{t_i})_{i\in I} \to \widehat{x}$ one has $\omega(t_i)k(t_i,x_{t_i}) \to g(\widehat{x})$; the interpretation of (3) is similar. Note that (2) cannot be changed into $\omega(t)k(t,x_t) \to g(\widehat{x})$ as $t \to 0$, as the following example shows.
Example 1. Let $X = \mathbb{R}_+$, $f(x) = xe^{-x} + \sin^2 \pi x$, $g(x) = e^{-x}\cos^2 \pi x$ for $x \in X$, $P = \mathbb{R}_+$, $T = \{1/n : n \in \mathbb{N},\ n \neq 0\}$, $x_t = t$ when t is an even integer, $x_t = 1/t$ when t is an odd integer. Setting $k(t,x) = tg(x)$ for $(t,x) \in T \times X$, $\omega(t) = 1/t$ as in the classical case, we see that x_t is an $\varepsilon(t)$-approximate solution for $\varepsilon(t) := f(x_t) + tg(x_t) \to 0$, $(x_{2p}) \to \widehat{x} = 0$ as $p \to \infty$, but $g(x_t) = e^{-t} \to 0 \neq g(0) = 1$ when $t = 2p$ with $p \to \infty$.
Also, note that $\overline{S}$ may be be different from the closure of S and that the conclusions (2), (3) may not hold when f is not lower semicontinuous at $\widehat{x}$.
Example 2. Let $X = [0,1]$, $f(x) = x$ for $x \in]0,1[$, $f(0) = 1$, $f(1) = 0$, let $k(t,x) = tg(x)$ for $t \in T :=]0,+\infty[$ with $g(x) = 1$ for $x \in]0,1]$, $g(0) = 0$. Then $x_t := t$ is an $\varepsilon(t)$-approximate solution to $(\mathcal{P}_t)$ with $\varepsilon(t) = t$ and $(x_t) \to \widehat{x} = 0 \notin S = \{1\}$. Here, for $\omega(t) = t^{-1}$, we have $\omega(t)k(t,x_t) = g(x_t) = 1 \neq g(\widehat{x})$ and $\omega(t)\left(f(x_t) - f(\widehat{x})\right) \to -\infty$.
Note that assumption (a) bounds the accurateness one has to impose on the approximate solution to $(\mathcal{P}_t)$. Observe that this assumption is trivially satisfied when $\varepsilon(t) = 0$ for each t, i.e. when x_t is an exact solution of $(\mathcal{P}_t)$, as in [2]. Assumption (b) is also obviously satisfied when k has the decomposed form $k(t,x) = \gamma(t)g(x)$ with $\gamma(t) > 0$ for each $t \in T$ and $\omega(t) = 1/\gamma(t)$, as in the classical case. If moreover g is l.s.c., then assumption (c) is also satisfied. Note that here we do not make any compactness assumption as in [2] (but of course, we do assume that (x_t) has at least one cluster point) and we deal with approximate solutions. For practical purposes, such an extension is important, as numerical algorithms seldom provide exact solutions. Moreover,

we do not make the assumption that g is continuous, as in Theorem 2.2 of [2]. Such an assumption is difficult to satisfy when X is a Banach space endowed with its weak topology. It has been observed by an anonymous referee that assumptions (b), (c) imply that g is lower semicontinuous at each point of $\overline{S}$; such a consequence is not too restrictive.

Proof. The assumption that x_t is an $\varepsilon(t)$-approximate solution to $(\mathcal{P}_t)$ can be written

$$\forall\, v \in X \qquad f(x_t) + k(t, x_t) \leq f(v) + k(t, v) + \varepsilon(t). \tag{4}$$

Since k takes nonnegative values, passing to the limit inferior over some net $(x_{t_i})_{i\in I}$ with limit $\widehat{x}$, where $(t_i)_{i\in I} \to 0$, for each $v \in X$ we get

$$\begin{aligned}\overline{f}(\widehat{x}) &\leq \liminf_{i\in I} f(x_{t_i}) \leq \liminf_{i\in I} (f(x_{t_i}) + k(t_i, x_{t_i})) \\ &\leq \limsup_{i\in I} (f(v) + k(t_i, v) + \varepsilon(t_i)) = f(v).\end{aligned}$$

Given $x \in X$, taking the limit inferior as $v \to x$, we obtain

$$\overline{f}(\widehat{x}) \leq \liminf_{v\to x} f(v) = \overline{f}(x),$$

so that $\widehat{x} \in \overline{S}$. A similar argument shows that

$$m = \overline{f}(\widehat{x}) \leq \liminf_{t\to 0} f(x_t) \leq \limsup_{t\to 0} f(x_t) \leq \limsup_{t\to 0} (f(x) + k(t, v) + \varepsilon(t)) = f(v),$$

and, as v is arbitrary, we get $(f(x_t)) \to m$ as $t \to 0$. When f is lower semicontinuous at $\widehat{x}$, we get $f(\widehat{x}) = \overline{f}(\widehat{x}) \leq \overline{f}(x) \leq f(x)$ for each $x \in X : \widehat{x} \in S$. Taking $v = x \in S$ in (4), and using the inequality $f(x) \leq f(x_t)$ we get

$$\forall x \in S \qquad k_t(x_t) \leq k_t(x) + \varepsilon(t), \tag{5}$$

hence $k_t(x_t) \leq \inf_{x\in S} k_t(x) + \varepsilon(t)$. Note that when S is empty the preceding inequality is trivial. In the remaining part of the proof we assume that S is nonempty.

Suppose now that ω satisfies our assumptions. Let $(t_i)_{i\in I}$ be an arbitrary net with limit 0 such that $(x_{t_i})_{i\in I} \to \widehat{x}$. Then, by (5), for each $x \in S$, one has

$$\omega(t_i)k(t_i, x_{t_i}) \leq \omega(t_i)k(t_i, x) + \omega(t_i)\varepsilon(t_i).$$

Passing to the limits, and using assumptions (a), (b), (c) and relation (5) we get

$$\begin{aligned}g(\widehat{x}) &\leq \liminf_{(t,v)\to(0,\widehat{x})} \omega(t)k(t, v) \\ &\leq \liminf_{i\in I} \omega(t_i)k(t_i, x_{t_i}) \leq \limsup_{t\to 0} \omega(t)k(t, x_t) \\ &\leq \lim_{t\to 0} (\omega(t)k(t, x) + \omega(t)\varepsilon(t)) = g(x).\end{aligned}$$

When $\widehat{x} \in S$, taking $x = \widehat{x}$ in these inequalities we get (2). Taking $v = \widehat{x}$ in (4), so that

$$\omega(t)\left(f(x_t) - f(\widehat{x}) + k(t, x_t)\right) \leq \omega(t)\left(k(t, \widehat{x}) + \varepsilon(t)\right), \tag{6}$$

and using (c), (a), (b) we obtain

$$\begin{aligned}\limsup_{i \in I} \omega(t_i)\left(f(x_{t_i}) - f(\widehat{x})\right) + g(\widehat{x}) &\leq \limsup_{i \in I} \omega(t_i)\left(f(x_{t_i}) - f(\widehat{x}) + k(t_i, x_{t_i})\right) \\ &\leq \lim_{i \in I} \omega(t_i)\left(k(t_i, \widehat{x}) + \varepsilon(t_i)\right) = g(\widehat{x}).\end{aligned}$$

Thus $\limsup_i \omega(t_i)\left(f(x_{t_i}) - f(\widehat{x})\right) \leq 0$. When $\widehat{x} \in S$ we have $f(x_t) - f(\widehat{x}) \geq 0$ hence $\liminf_{t \to 0} \omega(t)\left(f(x_t) - f(\widehat{x})\right) \geq 0$. Thus (3) holds too. □

The following corollaries are immediate consequences.

Corollary 1 *Suppose the assumptions (a), (b), (c) of the preceding theorem are satisfied, f is l.s.c., its set of minimizers S is nonempty and g has a unique minimizer $\widehat{x}$ in S. Suppose there exist a neighborhood V of 0 in P and a relatively compact family $(x_t)_{t \in T \cap V}$ such that x_t is an $\varepsilon(t)$-approximate solution to $(\mathcal{P}_t)$. Then the whole family $(x_t)_{t \in T}$ converges to $\widehat{x}$ and $\widehat{x}$ is a minimizer of f.*

In fact, it suffices that the family $(x_t)_{t \in T}$ is compact in the following sense: for any net $(t_i)_{i \in I} \to 0$ there exists a subnet $(t_j)_{j \in J}$ of $(t_i)_{i \in I}$ such that $(x_{t_j})_{j \in J}$ has a limit (see [30] for such a concept). In the following corollary we say that a family $(g_t)_{t \in T}$ of functions on a normed vector space X is *equi-coercive* if for any $r \in \mathbb{R}$ the union over $t \in T$ of the sublevel sets $[g_t \leq r] := g_t^{-1}(]-\infty, r])$ is bounded in X.

Corollary 2 *Suppose the assumptions (a), (b), (c) of the preceding theorem are satisfied, X is a reflexive Banach space endowed with its weak topology, f is l.s.c., its set of minimizers S is nonempty and g has a unique minimizer $\widehat{x}$ in S. Suppose $(g_t)_{t \in T} := (\omega(t)k_t)_{t \in T}$ is equi-coercive. Then, if x_t is an $\varepsilon(t)$-approximate solution to $(\mathcal{P}_t)$, the family $(x_t)_{t \in T}$ (weakly) converges to a minimizer $\widehat{x}$ of f.*

Proof. For each $x \in S$, and for $t \in T$ close enough to 0, one has, by assumption (b) and relation (5),

$$g_t(x_t) = \omega(t)k(t, x_t) \leq \omega(t)k(t, x) + \omega(t)\varepsilon(t) \leq g(x) + 1,$$

Thus, for a small enough neighborhood V of 0, the family $(x_t)_{t \in T \cap V}$ is bounded, hence weakly relatively compact. Then the preceding corollary applies. □

It has been kindly pointed out by C. Zalinescu that in the references [51] Thms 5.5 and 5.6, [52] Thms 3.7 and 3.8 similar results are displayed; see also [25]. However here f is not supposed to be convex, g is not supposed to be sublinear and no compactness or asymptotic assumption is made. Let us add that after

submission of the present paper we got acquainted with the results of the pioneer paper [22] in which a viscosity selection principle is given under convexity assumptions and conditions on minimizing sequences.
The equi-coercivity assumption is satisfied if the penalization term k is of the decomposed form $k(t,x) = \gamma(t)g(x)$ with g coercive and if $\omega(t) = 1/\gamma(t)$, as in the classical case we record now.

Corollary 3 *Suppose X is a reflexive Banach space, f is weakly l.s.c., its set of minimizers S is nonempty and g is finite, weakly l.s.c., coercive, and has a unique minimizer $\widehat{x}$ in S. Suppose $k(t,x) = \gamma(t)g(x)$ where $\gamma(t) \to 0_+$ as $t \to 0$. Let x_t be an $\varepsilon(t)$-approximate solution to $(\mathcal{P}_t)$, with $\varepsilon(t) \to 0$ and $\varepsilon(t)/\gamma(t) \to 0$ as $t \to 0$ in T. Then the family $(x_t)_{t\in T}$ weakly converges to $\widehat{x}$ as $t \to 0$. Moreover $\widehat{x}$ is a minimizer of f on X and a minimizer of g on S and*

$$g(x_t) \to g(\widehat{x}) \text{ as } t \to 0, \tag{7}$$

$$\frac{f(x_t) - f(\widehat{x})}{\gamma(t)} \to 0 \text{ as } t \to 0. \tag{8}$$

In order to get strong convergence we will suppose that g has the *Kadec-Klee property* in the following sense: a weakly converging net $(x_i)_{i\in I} \to x$ converges in norm whenever $(g(x_i))_{i\in I} \to g(x)$. This property is satisfied if g is of the form $g = \varphi \circ \|\cdot\|$, where the norm $\|\cdot\|$ satisfies the classical Kadec-Klee property and $\varphi : \mathbb{R}_+ \to \mathbb{R}_+$ is increasing with a continuous inverse. In particular, this property is satisfied if X is a Hilbert space and $g(x) = \|x\|^p$.

Corollary 4 *Suppose the assumptions of the preceding corollary are satisfied. Suppose moreover that g satisfies the Kadec-Klee property. Then, if x_t is an $\varepsilon(t)$-approximate solution to $(\mathcal{P}_t)$, the family $(x_t)_{t\in T}$ strongly converges to $\widehat{x}$.*

Similar results can be given in the case X is a dual Banach space endowed with its weak* topology.
The fact that in Theorem 1 we considered a penalization term $k(t,x)$ which is not in a decomposed form enables us to treat the case in which the objective $f+k_t$ of $(\mathcal{P}_t)$ is replaced by f_t+k_t, where f_t an approximation of f. It suffices to replace k_t by k_t+f_t-f and to check the assumptions with this new penalization term. This remark will be applied in the next section.

3 Tikhonov Regularization for Inclusions

Let us illustrate what precedes with the Tikhonov regularization method for the equation

$$(\mathcal{E}) \qquad Ax = b,$$

where X and Y are Banach spaces, A is a continuous linear operator from X into Y, $b \in Y$, or, more generally, with the inclusion

$$(\mathcal{I}) \qquad b \in F(x),$$

where X and Y are topological spaces, $F : X \rightrightarrows Y$ is a multifunction and Y is endowed with a metric d such that $d(b, \cdot)$ is l.s.c. This assumption is satisfied if Y is a dual Banach space endowed with its weak* topology and d is the distance associated with the dual norm.
We denote by C the set of best approximations of b in the closure $\mathrm{cl}(F(X))$ of $F(X)$ for d and we make use of the following assumption for some $x \in X$:

(A_x) the set of best approximations of b in $F(x)$ is nonempty.

This assumption is satisfied if F has compact values, in particular if F is a mapping, or if Y is a dual Banach space endowed with its weak* topology and F has weak* closed bounded values.
We associate to $(\mathcal{I})$ the minimization problem

$$(\mathcal{PI}) \qquad \text{Minimize } f(x) := h\left(d(b, F(x))\right) \qquad x \in X.$$

where $h : \mathbb{R}_+ \to \mathbb{R}_+$ is a locally Lipschitzian increasing function such that $h(0) = 0$; usually one takes $h(r) = r^p$, where $p \geq 1$. As the following simple lemma shows, the minimization problem $(\mathcal{PI})$ is related to $(\mathcal{I})$ and to the family of inclusions

$$(\mathcal{I}_c) \qquad c \in F(x),$$

whose sets of solutions are denoted by $S_c := F^{-1}(c)$.

Lemma 1 *Let S be the set of minimizers of f. Then*

$$S \supset F^{-1}(C) = \bigcup_{c \in C} S_c$$

If assumption (A_x) holds for each $x \in S$, then this inclusion is an equality. In particular, when $b \in F(X)$, S is the set of solutions to $(\mathcal{I})$ and $S = S_b$.

Proof. Let x be in $F^{-1}(c)$, where c is in C. Then, for any $x' \in X$ one has

$$f(x) = h\left(d(b, F(x))\right) \leq h\left(d(b, c)\right) \leq h\left(d(b, F(x'))\right) = f(x'),$$

so that $x \in S$. Conversely, if $x \in S$, and if (A_x) holds, then any best approximation c of b in $F(x)$ is a best approximation of b in $F(X)$, hence is a best approximation of b in $\mathrm{cl}(F(X))$, so that $c \in C$ and $x \in F^{-1}(C)$. The last assertion is obvious. □

Now we suppose b is approximated by some b_t, as in many cases of interest b is known in an approximative way only. Given $\gamma : T \to \mathbb{P} := (0, \infty)$, and a nonnegative, lower semicontinuous function g on X, we associate to $(\mathcal{PI})$ its regularization

$$(\mathcal{PI}_t) \qquad \text{Minimize } f_t(x) := h\left(d(b_t, F(x))\right) + \gamma(t)g(x) \qquad x \in X.$$

We need the following assumption in which $\overline{S}$ denotes again the set of minimizers of the l.s.c. hull $\overline{f}$ of f.

(H) for any $\overline{x} \in \overline{S}$ and any net $(x_i)_{i\in I}$ in X which converges to $\overline{x}$, with $(f(x_i))_{i\in I} \to f(\overline{x})$, there exist $\overline{y} \in F(\overline{x})$, a subnet $(x_j)_{j\in J}$ and a net $(y_j)_{j\in J}$ which converges to $\overline{y}$ such that $y_j \in F(x_j)$ for each $j \in J$.

Assumption (H) is satisfied if X and Y are Banach spaces endowed with their weak topologies and if F is a continuous linear operator A (as A is continuous for the weak topologies). It is also satisfied if Y is a dual Banach space endowed with its weak* topology, if F is locally bounded and if the graph of F is closed, X being an arbitrary topological space.

Proposition 1 *If assumption (H) is satisfied, then S coincides with $\overline{S}$.*

Proof. Let $\overline{x} \in \overline{S}$. There exists a net $(x_i)_{i\in I} \to \overline{x}$ such that $(f(x_i)) \to \overline{f}(\overline{x})$. Let $\overline{y}$, $(x_j)_{j\in J}$ and $(y_j)_{j\in J}$ be as in (H). Then, as $d(b,\cdot)$ is l.s.c. on Y one has

$$f(\overline{x}) \le h\left(d(b,\overline{y})\right) \le h\left(\liminf_{j\in J} d(b,y_j)\right) \le \liminf_{j\in J} h(d(b,y_j)) = \lim_{j\in J} f(x_j) = \overline{f}(\overline{x}).$$

As $\overline{f} \le f$ we get $\overline{f}(\overline{x}) = f(\overline{x})$ and $\overline{x} \in S$.□

Proposition 2 *Let f, g, h, F be as above and such that assumption (H) is satisfied. Let $\gamma, \varepsilon : T \to \mathbb{R}_+$ be such that $\gamma(t) \to 0$, $\varepsilon(t)/\gamma(t) \to 0$ as $t \to 0$ in T and let x_t be an $\varepsilon(t)$-approximate solution to $(\mathcal{PI}_t)$. Suppose $d(b_t, b)/\gamma(t) \to 0$ as $t \to 0$ and $d(b, F(\cdot))$ is locally bounded on X. Then any cluster point $\widehat{x}$ of (x_t) is a minimizer of f and a minimizer of g on S. If moreover the set of best approximations of b in $F(\widehat{x})$ is nonempty, then, for each c in this set, $\widehat{x}$ is a solution to $(\mathcal{I}_c)$.*

Proof. Let us take $k(t,x) = |h\left(d(b_t, F(x))\right) - h\left(d(b, F(x))\right)| + \gamma(t)g(x)$ and $\omega(t) = 1/\gamma(t)$. Since g is lower semicontinuous, $d(b_t, b)/\gamma(t) \to 0$ as $t \to 0$ and since there exists $\kappa > 0$ such that

$$|h\left(d(b_t, F(v))\right) - h\left(d(b, F(v))\right)| \le \kappa d(b_t, b)$$

for t close enough to 0, and v close enough to $x \in X$, we see that assumptions (b), (c) of Theorem 1 are satisfied. □

Corollary 5 *Suppose X is a dual Banach space endowed with its weak* topology and let $f, g, h, F, \gamma, \varepsilon$ be as above with g coercive (and l.s.c.) and such that assumption (H) is satisfied. Then, for any $\varepsilon(t)$-approximate solution x_t to $(\mathcal{PI}_t)$,* *the family (x_t) has a cluster point $\widehat{x}$ as $t \to 0$ and $\widehat{x}$ is a minimizer of f on X and a minimizer of g on S. If moreover g is strictly convex and S is convex, then (x_t) weak* converges to the unique minimizer $\widehat{x}$ of g on S. If moreover g satisfies the Kadec-Klee property, then x_t converges to $\widehat{x}$ in norm.*

The following example is classical. Note that the use of the viscosity selection principle avoids the recourse to spectral methods (as in [23]).

Example 3. Suppose X and Y are Hilbert spaces, A is a continuous linear operator from X into Y, $b \in Y$, $h(r) = r^2$, $T = \mathbb{P}$, $\gamma(t) = t$, $g = \|\cdot\|^2$, $b_t = b \in \mathrm{dom}(A^\dagger) := A(X) + A(X)^\perp$. Then the solution x_t of the minimization problem $(\mathcal{PI}_t)$ satisfies the necessary and sufficient condition

$$A^* A x_t + t x_t = A^* b.$$

Corollary 4 shows that (x_t) converges to the least square solution $\widehat{x}$ of the equation $Ax = b$, i.e. the element $\widehat{x}$ of least norm in the set of solutions to

$$A^* A x = A^* b.$$

When the data b is subject to some noise, one can still get estimates about the rate of convergence (and this topic is of much interest); we just present an illustrative example.

Example 4. ([17]) Suppose X and Y are Hilbert spaces, A is a continuous linear operator from X into Y, $b \in Y$, $h(r) = r^2$, $T = \mathbb{P}$, $\gamma(t) \to 0$ as $t \to 0$, $\delta(t) := d(b_t, b)$ is such that $\delta(t)\,(\gamma(t))^{-1/2} \to 0$ as $t \to 0$. Suppose that for a closed convex subset D of X one has

$$F(x) = \{A(x)\} \text{ if } x \in D,\ F(x) = \emptyset \text{ otherwise.}$$

Suppose b has a best approximation c in $A(D)$, $g(x) := \|x - x_0\|^2$ for some $x_0 \in X$ and the condition

$$x_0 - \widehat{x} \in A^*(Y) + N(D, \widehat{x}) \tag{9}$$

holds, where $\widehat{x}$ is the element of the set S of minimizers of $f := \|A(\cdot) - b\|^2$ on D which is the closest to x_0 and $N(D, \widehat{x})$ is the normal cone to D at $\widehat{x}$. Then one has the estimate

$$\|x_t - \widehat{x}\| \leq \delta(t)\,(\gamma(t))^{\ 1/2} + \alpha\gamma(t)^{1/2}$$

where α is the distance from 0 to the set $(A^*)^{-1}(x_0 - \widehat{x} - N(D, \widehat{x}))$. Moreover

$$\|A(x_t) - c\| \leq 2\delta(t) + 2\alpha\gamma(t).$$

Note that condition (9) means that there exists a Lagrange multiplier for the problem of minimizing g on D under the constraint $A(x) = c$ which defines S. In [17] concrete criteria are presented which ensure (9).

4 Regularization of Functions

In the present section and in the following one, we consider the application of the viscosity principle to the regularization of functions. We focus our attention on the Moreau (-Yosida) regularization, but we also consider other processes and side questions. Recall that the (generalized) *Moreau approximation* (or Moreau envelop [42]) of a function $f : X \to \overline{\mathbb{R}}$ on a metric space X associated to a nondecreasing, nonnegative function $h : \mathbb{R}_+ \to \overline{\mathbb{R}}_+$, such that $h^{-1}(0) = \{0\}$, is given for $t > 0$ by

$$f_t(x) := \inf_{u \in X} \left(f(u) + th(t^{-1}d(x,u))\right), \qquad x \in X. \tag{10}$$

In the classical case, X is a Hilbert space and $h(r) = \frac{1}{2}r^2$. However, more general kernels have to be used in order to ensure that $f_t > -\infty$ when f is not quadratically minorized, as noted in [12], [13], [4]. It has been observed by several authors that f and f_t have the same infima m and, if f is lower semicontinuous, the same minimizer set $S := S_f$ by the very definition of f_t. In the classical case, when f is convex, f_t is differentiable; when f is *paraconvex* (or semiconvex) in the sense that $f + \kappa \left\|\cdot\right\|^2$ is convex and lower semicontinuous for some $\kappa \geq 0$, f_t is differentiable for $t > 0$ small enough ([12], Theorem 4.1). It is worth observing that if $f = \min(f^1, f^2)$, then

$$f_t = \min(f_t^1, f_t^2). \tag{11}$$

Another point of interest from the point of view of optimization is the question of conditioning. It is easy to find examples of well-conditioned functions for which there exist arbitrarily close approximations which are not well-conditioned. Thus, one may wonder whether the classical approximations which are known in optimization and analysis remain well-conditioned when the function is well-conditioned. For the Baire approximation, i.e. the case $h(r) = \kappa\left|r\right|$ for some constant κ, the answer is positive (see [5]). Here we consider the case of the general Moreau approximation defined above in (10). In the proof below we will use the fact that when a function $\gamma : \mathbb{R}_+ \to \overline{\mathbb{R}}_+$ is extended to $\mathbb{R}$ into an even function, its Moreau approximation γ_t given as in (10) satisfies

$$\gamma_t(r) = \inf_{s \geq 0} \left(\gamma(s) + th\left(t^{-1}\left|s - r\right|\right)\right) \qquad \forall t > 0,\ \forall r \geq 0.$$

Proposition 3 *Suppose the function f satisfies the growth estimate*

$$\forall x \in X \quad f(x) \geq m + \gamma(d(x,S)),$$

where $m := \inf f(X)$, and $S := \arg\min f$ is supposed to be nonempty. Given $t > 0$, a nondecreasing, nonnegative function $h : \mathbb{R} \to \mathbb{R}$, such that $h(0) = 0$, its Moreau approximation f_t satisfies $\inf f_t(X) = m$ *and the growth estimate*

$$\forall x \in X \quad f_t(x) \geq m + \gamma_t(d(x,S)).$$

Proof. As $d_S := d(\cdot, S)$ is Lipschitzian with rate 1 and as h is nondecreasing, for any $u, x \in X$ we have

$$\begin{aligned} th\left(t^{-1}\left|d_S(x) - d_S(u)\right|\right) &\leq th\left(t^{-1}d(x,u)\right), \\ m + \gamma(d_S(u)) + th\left(t^{-1}\left|d_S(x) - d_S(u)\right|\right) &\leq f(u) + th\left(t^{-1}d(x,u)\right). \end{aligned}$$

Thus, setting $s := d_S(u)$, $r := d_S(x)$ and taking infima, we get

$$\begin{aligned} m + \gamma_t(r) &\leq m + \inf_{s \geq 0}\left(\gamma(s) + th\left(t^{-1}\left|s - r\right|\right)\right) \\ &\leq \inf_{u \in X}\left(f(u) + th\left(t^{-1}d(x,u)\right)\right) = f_t(x). \square \end{aligned}$$

Let us observe that the conditioning function γ_t of f_t is not as good as the conditioning function γ of f, i.e. $\gamma_t \leq \gamma$. Nonetheless, when γ is a gage (i.e. is nondecreasing and positive on the set of positive numbers) and h is positive on the set of positive numbers, γ_t is a gage.
The following result, which is of general interest, will be used for the generalized Moreau regularization.

Proposition 4 *Let X, P be topological spaces, $T \subset P$, $0 \in \text{cl } T$, and let g, $j : X \to \mathbb{R}$ be lower semicontinuous functions. Suppose j is bounded below by m. Let u_t be a minimizer of $u \mapsto j(u) + \omega(t)g(u)$ on X, where $\omega : T \to]0, \infty[$ is such that $\omega(t) \to \infty$ as $t \to 0$ in T. Suppose (u_t) has a limit $\widehat{u}$ as $t \to 0$ in T. Then $j(u_t) \to j(\widehat{u}), \omega(t)\left(g(u_t) - g(\widehat{u})\right) \to 0$ as $t \to 0$.*

Proof. Let us set $\gamma(t) = 1/\omega(t)$,

$$f_t(u) = g(u) + \gamma(t)\left(j(u) - m\right).$$

Then for each $u \in X$, one has $\gamma(t)\left(j(u) - m\right) \to 0$ as $t \to 0$ in T and $j(u) - m \leq \liminf_{(t,v)\to(0,u)} \omega(t)\gamma(t)\left(j(v) - m\right)$. Thus Theorem 1 applies with f and g replaced by g and $j(u) - m$ respectively, u_t being a minimizer of f_t. Relation (2) yields $j(u_t) \to j(\widehat{u})$ and relation (3) yields $\omega(t)\left(g(u_t) - g(\widehat{u})\right) \to 0$ as $t \to 0$. $\square$
Let us apply this result to the generalized Moreau regularization 10 by taking $j = f$, $g = \overline{h} \circ d(\cdot, x)$, where x is a fixed element of X, $\overline{h}(r) = \frac{1}{p}r^p$, $\omega(t) = t^{1-p}$, X being a normed vector space. We suppose j is bounded below by a continuous affine function $\langle a, \cdot\rangle - b$, with $a \in X^*$, $b \in \mathbb{R}$. Then, if u_t is a minimizer of $u \mapsto j(u) + \omega(t)g(u)$ one has

$$j(u_t) + \frac{1}{p}t^{1-p}\left\|u_t - x\right\|^p \leq j(x),$$

hence

$$\frac{1}{p}t^{1-p}\left\|u_t - x\right\|^p \leq j(x) - \langle a, u_t - x\rangle - \langle a, x\rangle + b.$$

Then, Young's inequality yields

$$\frac{1}{p}\|u_t - x\|^p \leq 2t^{p-1}\left(j(x) - \langle a, x\rangle + b + 2^{q-1}t\|a\|\right),$$

so that $u_t \to x$ as $t \to 0_+$. The preceding statement gives the more precise result that $j(u_t) \to j(x)$, $t^{1-p}\|u_t - x\|^p \to 0$ as $t \to 0_+$.
Another application of the viscosity selection principle to the Moreau regularization of convex functions is described in [2], [3]; let us extend it to the case of paraconvex functions. Here we also consider a kernel which is not necessarily quadratic. Given a function k on a Banach space W, a function j on W is said to be *k-paraconvex*, or in short paraconvex (especially when $k = \frac{1}{2}\|\cdot\|^2$) if $j + k$ is a closed proper convex function. Given $t > 0$ and a nonnegative, coercive, closed, proper, strictly convex function H on the dual X of W, one defines the regularized function j_t of j as

$$j_t(w) := [(j+k)^* + tH]^*(w) - k(w).$$

Here we denote by f^* the Fenchel conjugate of a function $f \in \overline{\mathbb{R}}^X$, given by $f^*(w) = \sup_{x\in X}(\langle w, x\rangle - f(x))$ and we keep the same notation for the conjugate of a function in $\overline{\mathbb{R}}^W$. When the Fenchel conjugate L of H is coercive (this occurs whenever H is bounded above on a ball centered at 0) one can write

$$j_t(w) = ((j+k)\Box(tH)^*)(w) - k(w) := \inf_{v\in W}\left[j(v) + k(v) + tL(\frac{w-v}{t})\right] - k(w),$$

a formula which reduces to the usual Moreau regularization when $k = 0$ and $H = \frac{1}{2}\|\cdot\|^2$. Given $w \in W$, since $((j+k)^* + tH)^{**} = (j+k)^* + tH$, we observe that

$$\begin{aligned} x_t &\in \partial(j_t + k)(w) \Leftrightarrow w \in \partial\left((j+k)^* + tH\right)(x_t) \\ &\Leftrightarrow x_t \in \operatorname{Arg\,min}\left((j+k)^* - \langle w, \cdot\rangle + tH\right). \end{aligned}$$

Since $(j+k)^* - \langle w, \cdot\rangle + tH$ is strictly convex, its minimizer set is a singleton and we get that $j_t + k$ is Gâteaux-differentiable at w; hence j_t is Gâteaux-differentiable when k is Gâteaux-differentiable. Fréchet differentiability results can be given in the line of [13] Prop. 3.6.
Let $f := (j+k)^* - \langle w, \cdot\rangle$ and let $S := \operatorname{Arg\,min} f = \partial(j+k)(w)$. Corollary 2 ensures that x_t weak* converges to the minimizer $\widehat{x}$ of H on S. Thus, when k is differentiable, the derivative $Dj_t(x)$ of j_t at x converges to $\widehat{x} - Dk(x)$ as $t \to 0_+$. Note that when j is differentiable at x one has $\widehat{x} - Dk(x) = Dj(x)$.
On the other hand, by the Young-Fenchel equality, for $x_t \in \partial(j_t + k)(w)$ we have

$$-j_t(w) = (j+k)^*(x_t) + tH(x_t) + k(w) - \langle x_t, w\rangle = f(x_t) + k(w) + tH(x_t).$$

As the Fenchel equality associated with the relation $\widehat{x} \in \partial(j+k)(w)$ can be written

$$f(\widehat{x}) := (j+k)^*(\widehat{x}) - \langle \widehat{x}, w \rangle = -(j+k)(w),$$

we get, by relation (3),

$$0 = \lim_{t \to 0} \frac{1}{t}\left(f(x_t) - f(\widehat{x})\right) = -\lim_{t \to 0} \frac{1}{t}\left(j_t(w) + tH(x_t) - j(w)\right)$$

and since $H(x_t) \to H(\widehat{x})$ by relation (2), we obtain

$$\frac{d}{dt}\Big|_{t=0} j_t(w) + H(\widehat{x}) = 0. \tag{12}$$

In view of the preceding observations, this equation is a kind of Hamilton-Jacobi equation. For $k = 0$, this is exactly the Hamilton-Jacobi equation associated to H, and, as shown in [13], this choice can be made in the case X is a Hilbert space and j is $\|\cdot\|^2$-paraconvex, as j_t is differentiable for $t > 0$ small enough. The links of the Moreau regularization with Hamilton-Jacobi equation are well known. The extensions of these relationships to other kinds of regularizations have been put in light in several recent contributions ([7], [8], [35], [49]).
When H is finitely valued, we see easily that $(j+k)^* + tH$ epi-converges (in the sense of Mosco) to $(j+k)^*$ as $t \to 0$. Thus, the continuity of the Fenchel correspondence ensures that j_t epi-converges to j as in the classical Moreau regularization. Moreover, as $t \to 0_+$ the function j_t increases to its limit j.
The Moreau regularization can be used for nonconvex functions (see [12], [13] where the crucial role of growth conditions has been put in light, and [4] where a further extension is given). However, the Moreau regularization is not adapted to the case of quasiconvex functions, as the sum of a quasiconvex function with a convex function is not always quasiconvex.
Example. Let $f : \mathbb{R}^2 \to \mathbb{R}$ be given by $f(0,0) = 0$, $f(x,y) = 1$ for $(x,y) \in]-\infty, 1] \times \mathbb{R} \backslash \{(0,0)\}$, $f(x,y) = 2$ for $(x,y) \in]1, +\infty[\times \mathbb{R}$. Then, for $t > 0$ one has $f_t(x,y) = \min(\frac{1}{2t}(x^2+y^2), 1 + \frac{1}{2t}(x-1)_+^2, 2)$, with $r_+ = \max(r, 0)$. For $t > \frac{1}{2}$ the sublevel set of height 1 of this function is not convex.
Moreover, the Moreau regularization of a quasiconvex function is not always differentiable.
Example. For f given on $\mathbb{R}$ by $f(x) = \min(x, 0)$, in view of relation (11), one has $f_t(x) = \min(x - \frac{1}{2}t, 0)$.
Thus, for the class of quasiconvex functions another process may be more adapted. It relies on an operation which has attracted much attention during the last few years, the *sublevel-convolution* $h := f \diamond g$ of the two functions f, g given by

$$h(w) := \inf_{x \in X} f(x) \vee g(w - x).$$

This operation, which is the analogue of the infimal convolution of convex analysis, is of fundamental importance for quasiconvex analysis inasmuch as the usual sum does not preserve quasiconvexity in general, whereas supremum does (see [32], [35], [39], [44], [46], [47], [49] ...). Moreover, it can be checked that the *strict sublevel sets* of h are given by

$$[h < s] = [f < s] + [g < s], \tag{13}$$

whereas the sublevel sets satisfy

$$[h \leq s] = [f \leq s] + [g \leq s] \tag{14}$$

whenever the infimum is attained in the formula defining h (then one says that the sublevel-convolution is exact). Let us note the following exactness criterion.

Proposition 5 *([39]) Suppose X is reflexive, f and g are l.s.c., quasiconvex and g is coercive. Then the sublevel-convolution $h := f \diamond g$ is exact: for any $z \in X$ there exists $x, y \in X$ such that $z = x + y$ and $h(z) = f(x) \vee g(y)$. If moreover there is no local minimizer of f in $f^{-1}(r)$ for $r := h(z)$ and if $g = \varphi \circ \|\cdot\|$ with $\varphi : \mathbb{R} \to \mathbb{R}$ increasing and l.s.c., then, for any x, y as above, x is a projection of z on $[f \leq r]$.*

Let us indicate the interest of the operation $\diamond$ for regularization purposes; details will be given elsewhere (see [39]). Let us first observe that when g is everywhere finite and bounded below, for any proper function f, the sublevel-convolution $h := f \diamond g$ is everywhere finite. If moreover g is Lipschitzian, then $h := f \diamond g$ is Lipschitzian ([44] Proposition 3.1). The following assertions refine that result in order it applies to the case $g = \|\cdot\|^p$ with $p > 1$.

Proposition 6 *Suppose g is Lipschitzian on bounded sets and is coercive. Then, for any proper function f, the sublevel-convolution $h := f \diamond g$ is Lipschitzian on bounded sets.*

Proof. Let $x_0 \in \operatorname{dom} f$ and let $r_0 = f(x_0)$. For each $r > 0$ the function $z \mapsto g(z - x_0)$ is bounded above by some a_r on the ball $B(0, r)$. Thus, for $z \in B(0, r)$ one has $h(z) \leq c_r := f(x_0) \vee a_r$. Let $s > 0$ be such that $g(y) > c_r$ for each $y \in X \backslash B(0, s)$. Then, for $z, z' \in B(0, r)$ one has

$$h(z) = \inf_{x \in B(0, r+s)} f(x) \vee g(z - x)$$

hence, if k is the Lipschitz rate of g on $B(0, r + s)$

$$\begin{aligned} h(z') &\leq \inf_{x \in B(0, r+s)} f(x) \vee g(z' - x) \\ &\leq \inf_{x \in B(0, r+s)} f(x) \vee g(z - x) + k \|z - z'\| \\ &\leq h(z) + k \|z - z'\| \end{aligned}$$

and the result follows by symmetry of the roles of z and z'. □

We leave the proof of the following similar assertion to the reader.

Proposition 7 *Suppose g is Lipschitzian on its sublevel sets. Then, for any proper function f, the sublevel-convolution $h := f \diamond g$ is Lipschitzian on its sublevel sets.*

Let us define the *sublevel-regularized function* by

$$f_{[t]}(w) := \inf_{x \in X} f(x) \vee (g(t^{-1}(w-x)) + m) \qquad \text{with } m := \inf f(X).$$

This formula is devised in such a way that when $\inf g = 0$ one has $\inf f_{[t]} = \inf f$. If moreover $g(0) = 0$ one has $f_{[t]}(w) \leq f(w)$ for any $w \in X$ and any $t > 0$. The interest of such a process stems from formulas (13), (14) which entail that when $g := \varphi \circ \|\cdot\|$ where φ is nondecreasing and upper semicontinuous, the strict sublevel set $[f_{[t]} < s]$ of the regularized function $f_{[t]}$ is the strict r-enlargement $U(S,r)$ of $S := [f < s]$ with $r := t \sup\{q : \varphi(q) < s - m\}$. Here, given $r > 0$, the *r-enlargement* of a given nonempty subset S of a normed space X is the set $B(S,r)$ of points of X whose distance to S is not greater than r; the strict r-enlargement of S is the set $U(S,r) := \{x \in X : d_S(x) < r\}$, where $d_S(x) := \inf\{\|x-u\| : u \in S\}$. Its interest stems from the fact that the enlargement of a convex subset is still convex (so that $f_{[t]}$ is quasiconvex when f is quasiconvex) and is usually smoother than the original set, as shown by the next proposition. Our observation about this smoothing effect relies on the following result of independent interest in which the *proximal subdifferential* $\partial^\pi f(x)$ of a function f at x is the set of $x^* \in X^*$ such that for some constants $c, r > 0$ one has

$$f(x+u) - f(x) - \langle x^*, u\rangle \geq -c\,\|u\|^2 \qquad \forall u \in B(0,r).$$

Lemma 2 *([21] Thm 4.1) For a given nonempty subset S of a Hilbert space X and $r > 0$, the following assertions are equivalent:*
(a) for each $x \in U(S,r)$ the proximal subdifferential $\partial^\pi d_S(x)$ of d_S at x is nonempty;
(b) d_S is continuously differentiable on $U(S,r)$.

Proposition 8 *For any convex subset S of a Hilbert space and for each $r > 0$ the enlargement $B(S,r)$ is a smooth submanifold with boundary.*

Corollary 6 *If X is a Hilbert space, if $g := \varphi \circ \|\cdot\|$ where $\varphi : \mathbb{R} \to \mathbb{R} \cup \{\infty\}$ is lower semicontinuous, nondecreasing such that $\varphi(t) \to \infty$ as $t \to \infty$, and if $f : X \to \mathbb{R} \cup \{\infty\}$ is lower semicontinuous, bounded below and quasiconvex, then its sublevel regularized function $f_{[t]} = f \diamond g(t^{-1}\cdot)$ has C^1-smooth sublevel sets.*

More information about the generalized derivatives or subdifferentials of the regularized function $f_{[t]}$ can be derived from results about sublevel-convolutions (see [38], [39], [44] for instance). The preceding proposition implies that when f_t has no local minimizer in $f^{-1}(f(x))$, the subdifferentials of Greensberg-Pierskalla type are contained in a half-line, hence are as small as possible. It is also interesting to compute the lower subdifferential of $f_{[t]}$ where the *lower subdifferential* $\partial^{<} f(x_0)$ of f at x_0 is given by

$$x_0^* \in \partial^{<} f(x_0) \Leftrightarrow \forall x \in [f < f(x_0)] \quad f(x) - f(x_0) \geq \langle x_0^*, x - x_0 \rangle.$$

For such an aim one has to use the operation $\diamond$ (called in [37] the harmonic sum) on the set of nonempty convex subsets of X given by

$$A \diamond B = \mathrm{cl}\left(\bigcup_{\lambda \in]0,1[} \lambda A \cap (1-\lambda) B \right).$$

It is a slight variant of the inverse sum considered in [41], [43] which has been introduced (within a spectrum of operations) by A. Seeger [43] in the case A and B are convex subsets containing 0 (see [43] Prop. 4.2). Since $\partial^{<} f(x_0) = X^*$ whenever 0 belongs to $\partial^{<} f(x_0)$, it is of interest to consider the case of general convex subsets as in [37], [38].

Proposition 9 *([38]) Let* $f_{[t]} = f \diamond t^{-1} \|\cdot\|$*, where* $f : X \to \overline{\mathbb{R}}$ *is such that* $\inf f = 0$*. Consider also* $x_0, y_0, w_0 \in X$ *such that* $w_0 = x_0 + y_0$*. If* x_0 *is not a local minimizer of* f *and* $f_{[t]}(w_0) = f(x_0) = t^{-1} \|y_0\|$ *then*

$$\partial^{<} f_{[t]}(w_0) = \partial^{<} f(x_0) \diamond t^{-1} \partial^{<} \|\cdot\| (y_0). \tag{15}$$

Conversely, if $\partial^{<} f(x_0) \diamond t^{-1} \partial^{<} \|\cdot\| (y_0) \neq \emptyset$*, then* $f_{[t]}(w_0) = t^{-1} \|y_0\| \geq f(x_0)$ *and* $\partial^{<} f(x_0) \diamond t^{-1} \partial^{<} \|\cdot\| (y_0) \subset \partial^{<} f_{[t]}(w_0)$*. Moreover, if* x_0 *is not a local minimizer of* f *and* f *is upper semi-continuous at* x_0 *then* $f(x_0) = t^{-1} \|y_0\|$ *and (15) holds.*

Note that for $y_0 \neq 0$ one has $\partial^{<} \|\cdot\| (y_0) = [1, \infty[J(y_0)$ where

$$J(y_0) = \partial \|\cdot\| (y_0) = \{ y_0^* \in X^* : \|y_0^*\| = 1, \ \langle y_0^*, y_0 \rangle = y_0 \}.$$

5 A General Regularization Method

The preceding two regularization methods are part of a general framework in which duality plays a key role. Given two sets X and Y, a *duality* is a mapping $D : f \mapsto f^D$ from $\overline{\mathbb{R}}^X$ into $\overline{\mathbb{R}}^Y$ such that, for any family $(f_i)_{i \in I}$ of $\overline{\mathbb{R}}^X$ one has

$$(\inf_{i \in I} f_i)^D = \sup_{i \in I} f_i^D \tag{16}$$

The reverse duality $D' : \overline{\mathbb{R}}^Y \to \overline{\mathbb{R}}^X$ is then defined by

$$g^{D'} := \inf \left\{ f \in \overline{\mathbb{R}}^X : \ f^D \leq g \right\}.$$

Then the pair (D, D') is a Galois correspondence between the complete lattices $\overline{\mathbb{R}}^X$ and $\overline{\mathbb{R}}^Y$. This fact has important consequences:

$$\begin{array}{rcll} \forall f & \in & \overline{\mathbb{R}}^X & f^{DD'D} = f^D, \quad f^{DD'} \leq f \\ \forall (f,g) & \in & \overline{\mathbb{R}}^X \times \overline{\mathbb{R}}^Y & f^D \leq g \Longleftrightarrow g^{D'} \leq f \\ \forall f & \in & \overline{\mathbb{R}}^X & f^{DD'} = f \Leftrightarrow \exists g \in \overline{\mathbb{R}}^Y : f = g^{D'}. \end{array}$$

Thus $D'D$ and DD' are closure operations.
Important examples of dualities arise with conjugacies. A duality D is a *conjugacy* if

$$(f+c)^D = -(c - f^D) \text{ for any } f \in \overline{\mathbb{R}}^X, c \in \overline{\mathbb{R}},$$

where the addition of $\mathbb{R}$ is extended to $\overline{\mathbb{R}}$ by setting $r + (+\infty) = +\infty$ for each $r \in \overline{\mathbb{R}}$ and $(-\infty) + (-\infty) = -\infty$ and where $r - s := r + (-s)$. Any coupling function $c : X \times Y \to \overline{\mathbb{R}}$ induces a conjugacy given by

$$f^c(y) := -\inf_{x \in X} (f(x) - c(x,y)).$$

It can be shown that in fact any conjugacy is of this type (called a Fenchel-Moreau conjugacy) for a certain coupling function c. The coupling function c also yields a symmetric conjugacy $\overline{\mathbb{R}}^Y \to \overline{\mathbb{R}}^X$, also denoted by $g \mapsto g^c$ when there is no risk of confusion, given by

$$g^c(x) := -\inf_{y \in Y} (g(y) - c(x,y)).$$

Given a duality D (resp. a conjugacy c) the class of functions $f \in \overline{\mathbb{R}}^X$ such that $f = f^{DD'}$, where $f^{DD'} := (f^D)^{D'}$ will be denoted by $\Gamma_D(X)$ and will be called the class of D-convex functions (or convexoid functions, or D-closed functions as this class has only generalized convexity properties).
Let us suppose X and Y are normed vector spaces, or more generally, metric spaces. Given a function $H : Y \to \mathbb{R}$ which is bounded on bounded sets, for any function $f \in \overline{\mathbb{R}}^X$ the family of functions $(f^D + tH)_{t>0}$ converges to f^D uniformly on bounded subsets of Y as $t \to 0$ and thus in several other convergences in $\overline{\mathbb{R}}^Y$. If the reverse duality $D' : \overline{\mathbb{R}}^Y \to \overline{\mathbb{R}}^X$ is continuous for one such convergence on $\overline{\mathbb{R}}^Y$ and another one on $\overline{\mathbb{R}}^X$ we get that

$$f_{t,D} := (f^D + tH)^{D'} \to f^{DD'} \quad \text{as } t \to 0.$$

When there is in $\overline{\mathbb{R}}^X$ a binary operation denoted by $\Box_D$ such that

$$(f \Box_D g)^D = f^D + g^D \tag{17}$$

and when tH is the conjugate of some function L_t on X, the regularization $f_{t,D}$ can be replaced by the generalized convolution

$$f_{t,D} = f \Box_D L_t.$$

Then, it is of interest to give conditions ensuring that $\left(f^D + tH\right)^{D'} = f \Box_D L_t$. In the case X and Y are normed vector spaces in duality, the Moreau regularization is obtained by taking the infimal convolution with $L_t := tH^*(t^{-1}\cdot)$, H^* being the Fenchel conjugate of H. The sublevel regularization corresponds to the case of the sublevel convolution $\diamond$ since relation (17) is satisfied for the "flat duality" ([7], [8], [35], [49]...) given by

$$f^{\flat}(x^*, r) := \sup\left\{\langle x^*, x\rangle : x \in X,\ f(x) < r\right\}.$$

Since for each $(x^*, r) \in X^* \times \mathbb{R}$, one has $H(x^*, r) := \|\cdot\|^D (x^*, r) = r\|x^*\|$ for $r \geq 0$, $H(x^*, r) = 0$ for $r < 0$, one sees that, for each function f such that $\inf f = 0$,

$$f^{\flat}_{[t]}(x^*, r) = f^{\flat}(x^*, r) + tr\|x^*\| \qquad \forall (x^*, r) \in X^* \times \mathbb{R}.$$

Thus, $\left(f^{\flat}_{[t]}\right) \to f^{\flat}$ uniformly on bounded sets as $t \to 0_+$.
The numerous dualities evoked in [35] suggest to devise other regularization processes.

6 Regularization of Operators

The question arises of how one can regularize an operator F from a Banach space X to its dual X^*. As here F is a multivalued mapping (in short, a multimapping), several meanings can be given to the word "regularize". In the best cases one can expect to approximate F by a family $(F_t)_{t>0}$ of operators which have X as their domains, are single-valued and Lipschitzian. Such a situation occurs when X is a Hilbert space and when F is maximal monotone, taking for F_t the Yosida regularization of F (see [2], [6], [14], [15]...). In [36] X is taken to be a Banach space for which the usual assumptions (in particular strict convexity) are relaxed and a general duality mapping is extensively used, observing that such a notion can be adapted to concrete situations such as L_p or Sobolev spaces. Moreover, the connection with the notion of inverse sum is put in light. Here, we observe that the approach of [36] can be further extended to the case one disposes of a convex coercive differentiable function k on X and one sets

$$F_t = \left(F^{-1} + tK^{-1}\right)^{-1},$$

where $K = \partial k$ is the subdifferential of k. In [36] k is taken to be $\varphi \circ \|\cdot\|$, where φ is an even one-variable convex function of class C^1 such that $\varphi(0) = 0$,

$\varphi(t)/t \to \infty$ as $t \to \infty$. Here, using the analogy with the Moreau approximation of functions as expounded in [2] and recalled above, we observe it is not necessary to suppose k has that special form.
We also note that it may be useful to introduce a regularization process adapted to some classes of generalized monotone multimappings. For such an aim, the harmonic sum is likely to play a role.

7 Some Links with Nonsmooth Analysis

We close our contribution with a short review of recent papers by J. Borwein [10], A. Ioffe [24], M. Lassonde [28], Q.J. Zhu [55] and others about the relationships between various results in nonsmooth analysis among which fuzzy sum rules and mean value theorems are the most important. We point out that penalization techniques are at the core of these results, in connection with decoupling properties; we hope to convince the reader that such methods are not exotic but on the contrary are simple and natural techniques.
Let us start with the notion of *relaxed or fuzzy infimum* (or uniform infimum) of a function $j : X \to \overline{\mathbb{R}}$ over a nonempty closed subset A of a metric space X, which plays a key role in [1], [19], [28]. We suppose $m := \inf_X j > -\infty$ and we set for a tolerance $t > 0$

$$\begin{aligned} r(j,A,t) \quad &: \quad = \inf\{j(x) : x \in X, d(x,A) \leq t\}, \\ r(j,A) \quad &: \quad = \sup_{t>0} r(j,A,t) = \lim_{t \to 0_+} r(j,A,t). \end{aligned}$$

The relaxed infimum $r(j,A)$ can be related to the infima of the function j penalized by the distance function to A, $d(\cdot,A)$ as we show now.

Lemma 3 *Setting* $j_n(x) := j(x) + nd(x,A)$ *for* $n \in \mathbb{N}$, $x \in X$, *one has* $r(j,A) = \sup_{n>0} \inf j_n(X)$.

Proof. We first observe that for each $n > 0$ and for each $t > 0$ we have

$$\inf j_n(X) \leq r(j,A,t) + nt,$$

hence, taking the limit as $t \to 0_+$ and then the supremum on $n \in \mathbb{N}$

$$\begin{aligned} \inf j_n(X) &\leq r(j,A), \\ \sup_{n>0} \inf j_n(X) &\leq r(j,A). \end{aligned}$$

Now, let $s < r(j,A)$. There exists $t > 0$ such that $s < r(j,A,t)$. Taking $n > t^{-1}(r(j,A) - m)$, for each $x \in X \backslash B(A,t)$, where $B(A,t)$ is the enlargement $B(A,t) := \{x \in X : d(x,A) \leq t\}$ of A, we have

$$r(j,A,t) \leq r(j,A) < m + nt \leq j(x) + nd(x,A)$$

while for $x \in B(A,t)$ we have $r(j,A,t) \leq j(x) + nd(x,A)$. Thus

$$r(j,A,t) \leq \inf j_n(X).$$

Taking the suprema over $t > 0$ and $n > 0$ we get $r(j,A) \leq \sup_{n>0} \inf j_n(X)$. □
Let us point out a connection of this classical penalization procedure with the viscosity selection principle.

Proposition 10 *Suppose j is l.s.c. and bounded below on X, $(\varepsilon_n) \to 0_+$, x_n is an ε_n-minimizer of the penalized function j_n and suppose the sequence (x_n) has a cluster point $\widehat{x}$. Then $\widehat{x}$ is a minimizer of j on A, $(j(x_n)) \to j(\widehat{x})$, and $(nd(x_n,A)) \to 0$ as $n \to \infty$.*

Proof. Replacing j by $j - \inf j$, we may suppose j is non negative. Let us set $P = \overline{\mathbb{R}}$, $T = \mathbb{N}$, $f := d(\cdot, A)$, $g = j$, $k(t,x) = t^{-1}j(x)$ so that f and g are l.s.c., x_n is an ε_n/n-minimizer of

$$x \mapsto d(x,A) + \frac{1}{n} j(x) := f(x) + k(n,x).$$

Moreover, the set S of minimizers of f is just A. Since for $\omega(t) := t$ for $t \in T$, we have $\omega(t)\varepsilon_t/t = \varepsilon_t \to 0$ as $t \to \infty$ in T, Theorem 1 yields that $\widehat{x}$ is a minimizer of f, thus belongs to A and

$$(j(x_n)) \to j(\widehat{x}), \ (nd(x_n,A)) \to 0 \text{ as } n \to \infty.$$

□

Let us give a short account of the use of the preceding methods in nonsmooth analysis. Given a finite family $f_1, ..., f_k$ of functions on a Banach space X we set $f(x_1, ..., x_k) := f_1(x_1) + ... + f_k(x_k)$ for $x_1, ..., x_k \in X$, $\Delta_A^k := \{(x, ..., x) \in A^k\}$ and

$$r(f_1, ..., f_k, A) := r(f, \Delta_A^k) := \lim_{t \to 0_+} \inf \left\{ \sum_{i=1}^{k} f_i(x_i) : d((x_1, ..., x_k), \Delta_A^k) < t \right\}.$$

It is easy to see that

$$r(f_1, ..., f_k, A) = \lim_{t \to 0_+} \inf \left\{ \sum_{i=1}^{k} f_i(x_i) : d(x_i, A) < t, \ d(x_i, x_j) < t \right\}.$$

In particular

$$r(f_1, ..., f_k) := r(f_1, ..., f_k, X) = \lim_{t \to 0_+} \inf \left\{ \sum_{i=1}^{k} f_i(x_i) : \sup_{1 \leq i,j \leq n} d(x_i, x_j) < t \right\}.$$

With these notions one has the following theorem ([9]-[11], [24], [28], [55]) in which X is a Banach space with a norm which is Fréchet differentiable off 0 and

$\partial^- f(x)$ is the *Fréchet subdifferential* of a function f on X finite at x, i.e. the set of $x^* \in X^*$ such that

$$f(x+u) - f(x) - \langle x^*, u\rangle \geq -\varepsilon(u)\,\|u\|$$

for some function $\varepsilon(\cdot)$ with limit 0 at 0. This fuzzy sum rule is central in nonsmooth analysis.

Theorem 2 *Let $f_1, ..., f_k$ be a family of lower semicontinuous functions on X such that $r(f_1, ..., f_k)$ is finite. Then, for any $\varepsilon > 0$, and for $i = 1, ..., k$, there exist $x_i \in X$, $x_i^* \in \partial^- f_i(x_i)$ satisfying*

$$\begin{aligned} 0 &\in \sum_{i=1}^{k} x_i^* + \varepsilon B_{X^*}, \\ \|x_i - x_j\| \max(1, \|x_1\|, ..., \|x_k\|) &< \varepsilon, \\ \sum_{i=1}^{k} f_i(x_i) &\leq r(f_1, ..., f_k) + \varepsilon. \end{aligned}$$

A proof of this result involves the penalized functions

$$g_n(x_1, ..., x_k) := \sum_{i=1}^{k} f_i(x_i) + n \sum_{i,j=1}^{k} \|x_i - x_j\|^2 \qquad n \in \mathbb{N},$$

hence is a combination of penalization and decoupling. We refer to the quoted references for more insight about these questions and the relationships with multidirectional mean value theorems.

Acknowledgements. The author is most grateful to an anonymous referee for his careful reading and his criticisms which led to a more precise writing of this paper.

References

[1] D. Aussel, J.-N. Corvellec and M. Lassonde, Nonsmooth constrained optimization and multidirectional mean value inequalities, SIAM J. Optim., to appear.

[2] H. Attouch, Viscosity solutions of minimization problems, epi-convergence and scaling, *Séminaire d'Analyse Convexe, Montpellier*, vol 22 (1992), 8.1-8.48.

[3] H. Attouch, Viscosity solutions of minimization problems, *SIAM J. Optim.* 6 (3) (1996), 769-806.

[4] H. Attouch and D. Azé, Approximation and regularization of arbitrary functions in Hilbert spaces by the Lasry-Lions method, *Ann. Institut H. Poincaré Anal. non linéaire* 10, (3) (1993), 289 - 312.

[5] D. Azé and A. Rahmouni, Intrinsic bounds for Kuhn-Tucker points of perturbed convex programs, in "*Recent developments in optimization, Seventh French-German Conference on Optimization*", R. Durier and C. Michelot (eds), Lecture Notes in Economics and Math. Systems # 429 Springer Verlag, Berlin (1995), 17-35.

[6] V. Barbu, *Nonlinear semigroups and differential equations in Banach spaces*, Ed. Academiei Rep. socialiste romania (Bucharest)-Noordhoff (Leyden) (1976).

[7] E.N. Barron and W. Liu, Calculus of variations in L^∞, *Applied Math. Opt.* 35 (1997), 237-243.

[8] E.N. Barron, R. Jensen and W. Liu, Hopf-Lax formula for $u_t + H(u, Du) = 0$. *J. Differ. Eq.* 126 (1996), 48-61.

[9] J.M. Borwein, J. S. Treiman and Q.J. Zhu, Necessary conditions for constrained optimization problems with semicontinuous and continuous data, *Trans. Amer. Math. Soc.* 350 (1998), 2409-2429.

[10] J.M. Borwein, Q.J. Zhu, Viscosity solutions and viscosity subderivatives in smooth Banach spaces with applications to metric regularity, *SIAM J. Control and Optim.* 34 (5) (1996), 1568-1591.

[11] J.M. Borwein and Q.J. Zhu, A survey of subdifferential calculus with applications, to appear in J. Nonlinear Anal. Th. Meth. Appl.

[12] M. Bougeard, J.-P. Penot, Approximation and decomposition properties of some classes of locally d.c. functions, *Math. Program.* 41 (1988), 195-227.

[13] M. Bougeard, J.-P. Penot and A. Pommellet, Towards minimal assumptions for the infimal convolution regularization, *J. Approx. Th.* 64 (3) (1991), 245-270.

[14] H. Brézis, *Opérateurs maximaux monotones*, North Holland, Amsterdam, 1973.

[15] F.E. Browder, Nonlinear variational inequalities and maximal monotone mappings in Banach spaces, *Math. Ann.* 183 (1969), 213-231.

[16] G. Chavent, Strategies for the regularization of nonlinear least squares problems, in "*Inverse problems in diffusion processes*", H.W. Engl and W. Rundell (eds.), SIAM-GAMM, SIAM Philadelphia (1995), 217-232.

[17] G. Chavent and K. Kunish, Convergence of Tikhonov regularization for constrained ill-posed inverse problems, *Inverse Problems* 10 (1994), 63-76.

[18] G. Chavent and K. Kunish, On weakly nonlinear inverse problems, *SIAM J. Appl. Math.* 56 (2) (1996), 542-572.

[19] F.H. Clarke and Yu. S. Ledyaev, Mean value inequalities in Hilbert spaces, *Trans. Amer. Math. Soc.* 344 (1994), 307-324.

[20] F.H. Clarke, Yu. S. Ledyaev, R.J. Stern, Complements, approximations, smoothings and invariance properties, *J. Convex Anal.* 4 (2) (1997), 189-219.

[21] F.H. Clarke, R.J. Stern, P.R. Wolenski, Proximal smoothness and the lower-C^2 property, *J. Convex Anal.* 2 (1995), 117-144.

[22] A. Gioan, Regularized minimization under weaker hypotheses, *Appl. Math. Optim.* 8 (1981), 59-67.

[23] C.W. Groetsch, *The theory of Tychonov regularization for Fredholm equations of the first kind*, Research Notes in Maths 105, Pitman, Boston, 1984.

[24] A.D. Ioffe, Fuzzy principles and characterization of trustworthiness, *Set-Valued Anal.* 6 (3) (1998), to appear.

[25] G. Isac and M. Théra, Complementary problem and the existence of post-critical equilibrium state of the thin elastic plate, Séminaire d'Analyse numérique, Université Paul Sabatier, Toulouse III (1985-1986), XI-1-XI-27.

[26] A. Kaplan and R. Tichatschke, *Stable methods for ill-posed variational problems. Prox-regularization of elliptic variational inequalities and semi-infinite optimization problems*, Akademie Verlag, Berlin, 1994.

[27] A. Kaplan and R. Tichatschke, Proximal point methods and nonconvex optimization, *J. Global Opt.* 13 (4) (1998),

[28] M. Lassonde, First order rules for nonsmooth constrained optimization, preprint, Univ. Antilles-Guyane, Pointe-à-Pitre, 1998

[29] J.-P. Penot, A characterization of tangential regularity, *Nonlinear Anal. Th. Meth. Appl.* 25 (6) (1981), 625-643.

[30] J.-P. Penot, Compact nets, filters and relations, *J. Math. Anal. Appl.* 93 (2) (1983), 400-417.

[31] J.-P. Penot, Conditioning convex and nonconvex problems, *J. Optim. Th. Appl.* 90 (3) (1995), 539-548.

[32] J.-P. Penot, Are generalized derivatives useful for generalized convex functions? in "*Generalized convexity, generalized monotonicity: recent results*", J.-P. Crouzeix et al (eds), Kluwer, Dordrecht (1998), 3-59.

[33] J.-P. Penot, Proximal mappings, *J. Approx. Theory,* 94 (1998), 203-221.

[34] J.-P. Penot, Well-behavior, well-posedness and nonsmooth analysis, *Pliska Stud. Math. Bulgar.* 12 (1998), 1001-1050.

[35] J.-P. Penot, What is quasiconvex analysis?, *Optimization*, to appear.

[36] J.-P. Penot and R. Ratsimahalo, On the Yosida approximation of operators, preprint, Univ. of Pau, 1997, revised July 1998.

[37] J.-P. Penot and C. Zălinescu, Harmonic sum and duality, preprint, Univ. of Pau, 1999.

[38] J.-P. Penot and C. Zălinescu, Elements of quasiconvex subdifferential calculus, preprint (1998), *J. Convex Anal.*, to appear.

[39] J.-P. Penot and C. Zălinescu, Regularization of quasiconvex functions, in preparation.

[40] F. Plastria, Lower subdifferentiable functions and their minimization by cutting plane, *J. Optim. Th. Appl.* 46 (1) (1994), 37-54.

[41] R.T. Rockafellar, *Convex Analysis*, Princeton Univ. Press, Princeton, N.J., 1970.

[42] R.T. Rockafellar and R. J-B. Wets, *Variational Analysis*, Springer-Verlag, Berlin, 1997.

[43] A. Seeger, Direct and inverse addition in convex analysis and applications, *J. Math. Anal. Appl.* 148 (1990), 317-341.

[44] A. Seeger, M. Volle, On a convolution operation obtained by adding level sets: classical and new results, *Recherche Opérationnelle* 29 (1995), 131-154.

[45] A. Tikhonov et V. Arsénine, *Méthodes de résolution de problèmes mal posés*, Mir, Moscow, (1976).

[46] S. Traore and M. Volle, Quasi-convex conjugation and Mosco convergence, *Richerche di Mat.* 44 (2) (1995), 369-388.

[47] S. Traore and M. Volle, Dualité pour la minimization du maximum de deux fonctions convexes: applications à la somme en niveaux des fonctions convexes, preprint, Univ. of Avignon, (1996).

[48] S. Traoré and M. Volle, Epiconvergence d'une suite de sommes en niveaux de fonctions convexes, preprint, Univ. of Avignon, (1996).

[49] M. Volle, Duality for the level sum of quasiconvex functions and applications, *ESAIM: Control, Optimisation and Calculus of Variations*, 3 (1998), 329-343, http://www.emath.fr/cocv/

[50] C. Zălinescu, On an abstract control problem, *Numer. Funct. Anal. Appl.* 2 (6) (1980), 531-542.

[51] C. Zălinescu, Stability for a class of nonlinear optimization problems and applications, in "*Nonsmooth Optimization and Related Topics*", F.H. Clarke, V.F. Dem'yanov and F. Giannessi, eds, Plenum Press, New York, (1989), 437-458.

[52] C. Zălinescu, A note on d-stability of convex programs and limiting Lagrangians", *Math. Progr.* 53 (1992), 267-277.

[53] C. Zălinescu, *Mathematical programming in infinite dimensional normed linear spaces* (Rumanian) Editura Academiei, Bucharest (1998), French translation to appear.

[54] E. Zeidler, *Nonlinear Functional Analysis and its Applications*, Springer Verlag, Berlin, (1990).

[55] Q.J. Zhu, The equivalence of several basic theorems for subdifferentials, *Set-Valued Anal.* 6 (2) (1998), 171-185.

Variational and Extended Sums of Monotone Operators *

Julian P. Revalski ** Michel Théra ‡

Abstract

In this article we show that the notion of variational sum of maximal monotone operators, introduced by Attouch, Baillon and Théra in [3] in the setting of Hilbert spaces, can be successfuly extended to the case of reflexive Banach spaces, preserving all of its properties. We make then a comparison with the usual pointwise sum and with the notion of extended sum proposed in our paper [26].

Keywords: monotone operators, sums of operators, enlargements, Yosida regularizations, subdifferentials

AMS subject classification: 47H05, 52A41, 26B25

1 Introduction

In [3], Attouch, Baillon and Théra introduced a notion of sum of two monotone operators, called *variational sum*. This was done in the setting of Hilbert spaces and the general idea was to use an appropriate approximation (the Yosida regularization) of a monotone operator in a Hilbert space and to take as a sum of two operators a suitable limit of the pointwise sum of their approximations. In several cases this sum appears to be larger than the usual pointwise sum, as for example in the case of subdifferentials of convex functions.

The need to look for a notion of sum of two monotone operators, which is a generalization of the usual pointwise sum, is inspired by the fact that sometimes (as above in the case of subdifferentials) the study of a problem, with monotone operators involved, leads to an operator that turns out to be larger than the pointwise sum (see e.g. [3, 4], where problems arising in partial differential equations possessing this property are studied). Therefore, different authors have tried to investigate possible generalized notions of sum of monotone operators

*This work was completed while the first author was visiting, during the Fall semester 1998, the LACO (Laboratoire d'Arithmétique, Calcul Formel et Optimisation) at the University of Limoges. The same author was also partially supported by the Bulgarian National Fund for Scientific Research under contract No. MM-701/97.

**Institute of Mathematics and Informatics, Bulgarian Academy of Sciences, Acad. G. Bonchev Street, block 8, 1113 Sofia, Bulgaria; e-mail: revalski@math.bas.bg

‡Université de Limoges, LACO, 123, Avenue Albert Thomas, F-87060 Limoges Cedex, France; e-mail: michel.thera@unilim.fr

like, the variational sum mentioned above, or the parallel sum used in [16] to study electrical networks, or a sum based on the Trotter-Lie formula in [17], or recently an accretive extension of the algebraic sum of two linear operators in [10].

Our aim in this article is twofold. First, to show that the notion of variational sum can be naturally extended to the more general setting of reflexive Banach spaces using the same general idea of approximation. And second, to compare this sum with a notion of extended sum introduced in our paper [26], based on another approximation of a given monotone operator–a suitable enlargement of the operator. Therefore, in the rest of the paper we proceed as follows: after some necessary preliminaries given in Section 2, in Section 3 we present the idea of the extended sum from [26]. Further, in Section 4 we show how the concept of variational sum can be extended to the case of a reflexive Banach space. We see that the generalized notion keeps its properties from Hilbert space setting. We prove also that, when the variational sum of two maximal monotone operators is maximal monotone, it contains the extended (and hence the pointwise) sum of the operators and all the three concepts coincide if we suppose the maximality of the pointwise sum. In the last Section 5, we see that another important property from the Hilbert space setting–the fact that the subdifferential of the sum of two convex proper lower semicontinuous functions is equal to the variational sum of their subdifferentials–is also preserved in the new setting. In particular, we obtain that the two concepts–the variational sum and the extended sum–agree also for the case of subdifferentials of convex functions.

2 Some Preliminaries

Throughout this article we will assume that X is a reflexive real Banach space with continuous dual X^*. The norms in X and X^* will be designated by $\|\cdot\|$, w will mean the weak topology in X and X^*. For the usual pairing between X and X^* we use the symbol $\langle\cdot,\cdot\rangle$.

Due to a result of John and Zizler [14], we may think that X is endowed with a Fréchet differentiable (away from the origin) locally uniformly rotund norm whose dual is also Fréchet differentiable except the origin and locally uniformly rotund. In particular (see e.g. Diestel [11]), these norms are not only strictly convex, but also satisfy the Kadec-Klee property:

$$\text{if } x_n \to x \text{ weakly in } X \text{ and } \|x_n\| \to \|x\|, \text{ then } x_n \to x \text{ strongly in } X,$$

and

$$\text{if } x_n^* \to x^* \text{ weakly in } X^* \text{ and } \|x_n^*\| \to \|x^*\|, \text{ then } x_n^* \to x^* \text{ strongly in } X^*.$$

Evereywhere in the sequel, we will assume that the norms in X and X^* are as above.

Given a (multivalued) operator $A : X \rightrightarrows X^*$, as usual the *graph* of A will be denoted by

$$\mathrm{Gr}(A) := \{(x, x^*) \in X \times X^* : x^* \in Ax\}$$

and the *domain* and the *range* of A by

$$\mathrm{Dom}(A) := \{x \in X : \ Ax \neq \emptyset\}$$

and

$$\mathrm{R}(A) := \bigcup\{Ax : x \in \mathrm{Dom}(A)\}.$$

The inverse operator of $A : X \rightrightarrows X^*$ is $A^{-1} : X^* \rightrightarrows X$:

$$A^{-1}x^* := \{x \in X : x^* \in Ax\}, \ x^* \in X^*.$$

Obviously $\mathrm{Dom}(A^{-1}) = \mathrm{R}(A)$.
The operator A is called *monotone* if the following condition is fulfilled:

$$\langle y - x, y^* - x^*\rangle \geq \ 0 \ \text{ for every two pairs } (x, x^*), (y, y^*) \in \mathrm{Gr}(A).$$

Let us mention that if A is monotone, then its inverse A^{-1} is also monotone from X^* into X.
For an operator $A : X \rightrightarrows X^*$, denote by co$A$ (resp. by $\bar{A}$) the operator $(\mathrm{co}A)x := \mathrm{co}(Ax)$ (resp. $\bar{A}x := \overline{Ax}$), $x \in X$. Here co means convex hull, while the overbar has the meaning of the closure of a set with respect to the norm topology in X^*. Obviously, $\mathrm{Dom}(A) = \mathrm{Dom}(\bar{A}) = \mathrm{Dom}(\mathrm{co}A)$ and if A is monotone then so are $\bar{A}$ and co$A$.
A monotone operator $A : X \rightrightarrows X^*$ is said to be *maximal* if its graph is not contained properly in the graph of any other monotone operator from X to X^*. In other words, A is maximal, if and only if, whenever one has $\langle x-y, x^*-y^*\rangle \geq 0$ for every $(y, y^*) \in \mathrm{Gr}(A)$, it follows that $(x, x^*) \in \mathrm{Gr}(A)$. It is easily seen that, if A is a maximal monotone operator, then the operators $A, \bar{A}, \mathrm{co}A, \overline{\mathrm{co}}A$ coincide. Evidently, if A is maximal monotone then A^{-1} is also a maximal monotone operator between X^* and X. Finally, it is easily checked that if A is a maximal monotone operator then its graph $\mathrm{Gr}(A)$ is a closed subset with respect to the product of the norm topologies in $X \times X^*$.
Among the most important examples of maximal monotone operators are the subdifferentials of convex functions. Let $f : X \to \mathbf{R} \cup \{+\infty\}$ be an extended real-valued proper lower semicontinuous convex function in X (proper means that the *domain* of f, $\mathrm{dom}\, f := \{x \in X : f(x) < +\infty\}$ is nonempty). Given $\varepsilon \geq 0$, the *ε-subdifferential* of f is defined at $x \in \mathrm{dom}f$ by:

$$\partial_\varepsilon f(x) := \{x^* \in X^* : f(y) - f(x) \geq \langle y - x, x^*\rangle - \varepsilon \quad \text{for every } y \in X\},$$

and $\partial_\varepsilon f(x) := \emptyset$, if $x \notin \mathrm{dom}f$. It is known that $\mathrm{Dom}(\partial_\varepsilon f) = \mathrm{dom}f$ for every $\varepsilon > 0$. When $\varepsilon = 0$, $\partial_0 f$ is the *subdifferential* ∂f of f (the latter can be empty at some points of dom$f$). Obviously, one has $\partial f(x) \subset \partial_\varepsilon f(x)$ for every $x \in X$ and $\varepsilon > 0$.
The following result, which has been also a motivation for our investigations, was proved by Hiriart-Urruty and Phelps:

Theorem 2.1 (Hiriart-Urruty and Phelps [12]) *Let $f, g : X \to \mathbf{R} \cup \{+\infty\}$ be two proper lower semicontinuous convex functions. Then for every $x \in$ $\mathrm{dom} f \cap \mathrm{dom} g$ one has:*

$$\partial(f+g)(x) = \bigcap_{\varepsilon>0} \overline{\partial_\varepsilon f(x) + \partial_\varepsilon g(x)}.$$

The result is true in a general Banach space, in which case the closure in the right hand side is taken with respect to the weak star topology in X^*.

3 The Concept of Extended Sum

The notion of approximate subdifferential has inspired the search of similar enlargements for monotone operators. The following one has been studied intensively in the last years:
Given a monotone operator $A : X \rightrightarrows X^*$ and $\varepsilon \geq 0$, the ε-*enlargement* of A is the operator $A^\varepsilon : X \rightrightarrows X^*$, determined by

$$A^\varepsilon x := \{x^* \in X : \langle y - x, y^* - x^* \rangle \geq -\varepsilon \quad \text{for any } (y, y^*) \in \mathrm{Gr}(A)\}.$$

This concept was observed by several authors (see e.g. [19]). But for the first time a detailed study was provided in [8] in finite dimensions, with applications to approximate solutions of variational inequalities, and in [9] with applications to finding a zero of a maximal monotone operator. Similar notions could be found in [18, 21, 22, 25, 34].
It is seen that A^ε is with closed convex images for any $\varepsilon \geq 0$. Due to the monotonicity of A, one has $Ax \subset A^\varepsilon x$ for every $x \in X$ and every $\varepsilon \geq 0$. I.e. A^ε is indeed an enlargement of A. If A and B are two monotone operators such that $A \subset B$ (equivalently $\mathrm{Gr}(A) \subset \mathrm{Gr}(B)$) then $B^\varepsilon \subset A^\varepsilon$ for every $\varepsilon \geq 0$! In the particular case when $A = \partial f$ one has $\partial_\varepsilon f \subset (\partial f)^\varepsilon$ and the inclusion can be strict (for instance for the function $f(x) = x^2$, $x \in \mathbf{R}$, cf. e.g. [19]). Let us mention that a Brøndsted-Rockafellar lemma is true for these enlargements (see Torralba [33] for the case of reflexive spaces (cf. also [9] where the Hilbert space setting is considered) and our paper [26] for a generalization outside these settings). For other properties of this notion we refer to [8, 9, 26].
Let $A, B : X \rightrightarrows X^*$ be two monotone operators. As usual the operator $A+B :$ $X \rightrightarrows X^*$ means the pointwise sum of A and B:

$$(A+B)x = Ax + Bx, \quad x \in X.$$

$A + B$ is a monotone operator with domain $\mathrm{Dom}(A + B) = \mathrm{Dom}A \cap \mathrm{Dom}B$. But it is well known that if A and B are maximal monotone operators, then, in general, their sum $A + B$ need not be a maximal monotone operator. A very well-known case of this phenomenon is the sum of the subdifferentials of two convex proper lower semicontinous functions which can be strictly less than the subdifferential of the sum of the functions. Other counterexamples for this, as

well as sufficient conditions for the maximality of the pointwise sum, could be found in the monographs of Phelps [24] and Simons [28].
As we pointed out, the above phenomenon motivated the study of possible generalized sums of monotone operators. One is the variational sum which will be considered in the next section. Another one was proposed in our paper [26] using the above enlargements:
The *extended sum* of two monotone operators $A, B : X \rightrightarrows X^*$ at the point $x \in X$ is defined in [26] by

$$A \underset{ext}{+} B(x) = \bigcap_{\varepsilon>0} \overline{A^\varepsilon x + B^\varepsilon x}.$$

In the general case of an arbitrary Banach space, the closure on the right hand side is taken in [26] with respect to the weak star topology in X^* (which, of course, in our setting coincides with the norm closure, due to the reflexivity of the space and convexity of the images of the enlargements). Obviously, $A+B \subset \overline{A+B} \subset A \underset{ext}{+} B$ and hence, $\mathrm{Dom}(A) \cap \mathrm{Dom}(B) \subset \mathrm{Dom}(A \underset{ext}{+} B)$. Moreover, the extended sum is commutative. As it is shown in [26], in several important cases the extended sum is a maximal monotone operator: e.g.:

Theorem 3.1 ([26], Theorem 3.1 and Corollary 3.2). *Let $\overline{A+B}$ (resp. $A+B$) be a maximal monotone operator. Then, $\overline{A+B} = A \underset{ext}{+} B$ (resp. $A+B = A \underset{ext}{+} B$).*

Moreover, the subdifferential of the sum of two convex proper lower semicontinuous functions is equal to the extended sum of their subdifferentials ([26], Theorem 3.3; see also the particular case of this result at end of this paper).

4 Variational Sum of Monotone Operators

In this section we see how the notion of variational sum of maximal monotone operators introduced in the setting of Hilbert spaces by Attouch, Baillon and Théra in [3], can be extended to the case of reflexive Banach spaces, keeping its properties. We compare it then with the pointwise and extended sum.
For the extension of the variational sum to the more general setting we will follow the same scheme of approximations used in [3]. For this we need first to introduce the well-known *duality mapping* J between X and X^*, defined by:

$$Jx := \{x^* \in X^* : \langle x, x^* \rangle = \|x\|^2 = \|x^*\|^2\}, \quad x \in X.$$

J is the subdifferential of the convex function $f(x) = (1/2)\|x\|^2, x \in X$ (see e.g. Phelps [24], Example 2.26), i.e. it is a maximal monotone operator with $\mathrm{Dom}(J) = X$. Remember that we consider norms in X and X^* which are Fréchet differentable away from the origin and locally uniformly routund, in particular with the Kadec-Klee property. In such a situation, the duality mapping J between X and X^* is one-to-one, surjective and norm-to-norm continuous.

Before giving the regularization scheme, let us recall a well-known result due to Rockafellar [27], Proposition 1 (which is also a generalization of the classical result of Minty for Hilbert spaces): in our setting of reflexive Banach spaces, with the norms we consider, if $A : X \rightrightarrows X^*$ is a maximal monotone operator then for any $\lambda > 0$ one has $\mathrm{R}(A + \lambda J) = X^*$ and $(A + \lambda J)^{-1}$ is a single-valued maximal monotone operator which is norm-to-weak continuous (the latter notion is termed also *demi-continuous*).
We proceed by introducing the elements of the regularization scheme. Given a maximal monotone operator $A : X \rightrightarrows X^*$ and a positive λ, there are two equivalent ways to define the well-known *Yosida regularization* (or *Yosida approximation*) of A of order λ: the first one is naturally motivated by the special case of subdifferentials of convex functions; one defines first the *resolvent* J_λ^A of A for $\lambda > 0$, as the operator from X into X determined by: for any $x \in X$, $J_\lambda^A x$ is the unique (by the result of Rockafellar above) solution x_λ of the inclusion:

$$0 \in J(x_\lambda - x) + \lambda A x_\lambda. \tag{4.1}$$

Then, the Yosida regularization $A_\lambda : X \rightrightarrows X^*$ of A is defined by

$$A_\lambda x := \frac{1}{\lambda} J(x - x_\lambda), \quad x \in X. \tag{4.2}$$

One easily sees that for each $\lambda > 0$

$$J_\lambda^A x = x - \lambda J^{-1} A_\lambda x \quad \text{for every } x \in X. \tag{4.3}$$

The second equivalent way to define the above regularization is purely analytical:

$$A_\lambda := (A^{-1} + \lambda J^{-1})^{-1}, \ \lambda > 0. \tag{4.4}$$

By the result of Rockafellar above, for any $\lambda > 0$, the Yosida regularization A_λ is an everywhere defined single-valued maximal monotone operator which is demi-continuous.
When X is a Hilbert space, then J is the identity and the above formulae reduce to $A_\lambda = (J - J_\lambda^A)/\lambda$ and $J_\lambda^A = (J + \lambda A)^{-1}$.
Observe that (4.1) and (4.2) above show that, in particular, for any $\lambda > 0$:

$$A_\lambda x \in A(J_\lambda^A x) \text{ for every } x \in X. \tag{4.5}$$

Let us list several well-known properties of the Yosida regularizations due to Brezis, Crandall and Pazy [6]. In what follows, given a maximal monotone operator $A : X \rightrightarrows X^*$ and $x \in \mathrm{Dom}(A)$, by $A^{\min} x$ we denote the unique element of Ax which has minimal norm in Ax (the existence of such an element is guaranteed by the reflexivity of X and the maximality of A (Ax is closed and convex), while its uniqueness follows by the fact that the norm in X^* is strictly convex).

Proposition 4.1 (see [6], Lemma 1.3) *Let A be a maximal monotone operator between the reflexive Banach space X and its dual X^* and the norm in X and the dual norm in X^* are locally uniformly rotund. Then*

(a) *for every* $\lambda > 0$ *the Yosida regularization* A_λ *maps bounded sets into bounded sets;*

(b) *for every* $\lambda > 0$ *and* $x \in \mathrm{Dom}(A)$ *we have* $\|A_\lambda x\| \le \|A^{\min} x\|$*;*

(c) *for every* $x \in \mathrm{Dom}(A)$*,* $J_\lambda^A x$ *strongly converges to* x *as* $\lambda \downarrow 0$ *and* $A_\lambda x$ *strongly converges to* $A^{\min} x$ *as* $\lambda \downarrow 0$*.*

In fact, the second statement in (c) above is not proved in [6], Lemma 1.3, but can be derived from the asertions (d) and (e) of the same Lemma 1.3, using the fact that the norm in X^* satisfies the Kadec-Klee property. Indeed, in order to prove that $A_\lambda x$ strongly converges to $A^{\min} x$ as $\lambda \downarrow 0$, it is enough to show that any subsequence $A_{\lambda_n} x$ such that $\lambda_n \downarrow 0$ has a further subsequence which strongly converges to $A^{\min} x$. Let $A_{\lambda_n} x$ be such that $\lambda_n \downarrow 0$. By (b) above there exists a subsequence $A_{\lambda_{n_k}} x$ which converges weakly, say to y^*. By condition (e) of Lemma 3.1 from [6] $(x, y^*) \in A$ and hence, again by (b) above, $y^* = A^{\min} x$. Now, using (b) above and the (weak) lower semicontinuity of the norm in X^* we have

$$\|A^{\min} x\| \ge \limsup_k \|A_{\lambda_{n_k}} x\| \ge \liminf_k \|A_{\lambda_{n_k}} x\| \ge \|A^{\min} x\|.$$

Therefore, $\lim_k \|A_{\lambda_{n_k}} x\| = \|A^{\min} x\|$ and by the Kadec-Klee property we conclude that $A_{\lambda_{n_k}} x$ strongly converges to $A^{\min} x$.

Let now $\mathcal{I} := \{(\lambda, \mu) \in \mathbf{R}^2 : \lambda \ge 0, \mu \ge 0, \quad \lambda + \mu \ne 0\}$. Given two maximal monotone operators $A, B : X \rightrightarrows X^*$, the idea of the variational sum is to consider the operators of the type $A_\lambda + B_\mu$ for $(\lambda, \mu) \in \mathcal{I}$ and to take as a sum of A and B an appropriate limit of the above perturbations. Here for convention $A_0 = A$. Observe that, since $\lambda + \mu \ne 0$, $A_\lambda + B_\mu$ is always a maximal monotone operator due to a classical result of Browder [7]. Let $\mathcal{F}$ be the filter of all neighborhoods of the zero in $\mathcal{I}$. Then writing $\lim_{\mathcal{F}}$ we will have in mind the limit when $\lambda \to 0$, $\mu \to 0$, $(\lambda, \mu) \in \mathcal{I}$.

A natural idea related to convergence of operators, that has turned out to be useful in different investigations, is the idea of graph convergence (see Attouch [2]). This simply means that one identifies the operators with their graphs in $X \times X^*$ and considers on the latter the Painlevé-Kuratowski convergence determined by an appropriate convergence in the Cartesian product $X \times X^*$.

Given a family of operators $\{C_{\lambda,\mu} : (\lambda, \mu) \in \mathcal{I}\}$ between X and X^*, let us remind the definitions of lower and upper limit of this family in the sense of Painlevé-Kuratowski, when we consider in $X \times X^*$ the product topology generated by the strong topologies in X and X^*.

The lower limit of the family $\{C_{\lambda,\mu} : (\lambda, \mu) \in \mathcal{I}\}$ in the sense of Painlevé-Kuratowski is the following set:

$\|\cdot\| \times \|\cdot\| - \liminf_{\mathcal{F}} C_{\lambda,\mu} := \{(x, x^*) \in X \times X^* :$ for every $\|\cdot\| \times \|\cdot\|$-neighborhood $\mathcal{U}$ of (x, x^*) there exists $F \in \mathcal{F}$ such that $\mathcal{U} \cap \mathrm{Gr}(C_{\lambda,\mu}) \ne \emptyset$ for every $(\lambda, \mu) \in F\}$,

while its upper limit in the same sense is:

$\|\cdot\| \times \|\cdot\| - \limsup_{\mathcal{F}} C_{\lambda,\mu} := \{(x, x^*) \in X \times X^* :$ for every $\|\cdot\| \times \|\cdot\|$-neighborhood $\mathcal{U}$ of (x, x^*) and for every $F \in \mathcal{F}$ there exists $(\lambda, \mu) \in F$ such that $\mathcal{U} \cap \mathrm{Gr}(C_{\lambda,\mu}) \neq \emptyset\}$.

The family of monotone operators $\{C_{\lambda,\mu} : (\lambda, \mu) \in \mathcal{I}\}$ graph-converges to the monotone operator $C : X \rightrightarrows X^*$ if

$$\mathrm{Gr}(C) = \|\cdot\| \times \|\cdot\| - \liminf_{\mathcal{F}} C_{\lambda,\mu} = \|\cdot\| \times \|\cdot\| - \limsup_{\mathcal{F}} C_{\lambda,\mu}.$$

With slight abuse of notation we will omit in the above equality the symbol Gr and will write $C = \|\cdot\| \times \|\cdot\| - \lim_{\mathcal{F}} C_{\lambda,\mu}$.
It is easily seen that equivalent sequential definitions of the above notions are: $(x, x^*) \in \|\cdot\|\times\|\cdot\|-\liminf_{\mathcal{F}} C_{\lambda,\mu}$ exactly when for every sequence $\{(\lambda_n, \mu_n)\} \subset \mathcal{I}$ such that $\lambda_n, \mu_n \to 0$, there exists a sequence $(x_n, x_n^*) \in C_{\lambda_n,\mu_n}$ such that $(x_n, x_n^*) \to (x, x^*)$ strongly; $(x, x^*) \in \|\cdot\| \times \|\cdot\| - \limsup_{\mathcal{F}} C_{\lambda,\mu}$ iff there exists a sequence $\{(\lambda_n, \mu_n)\} \subset \mathcal{I}$ with $\lambda_n, \mu_n \to 0$ and such that there is a sequence $(x_n, x_n^*) \in C_{\lambda_n,\mu_n}$ for which $(x_n, x_n^*) \to (x, x^*)$ strongly.
Using the latter definitions, one can define $\|\cdot\| \times seq.-w-$(lower, upper) limit of a family of operators, when one considers on X^* the sequential weak convergence. In this case, one gets formally bigger sets as limits. But as we will see below, in the case of maximal monotone operators we have the same limits.
The next proposition is well-known (see e.g. [3]).

Proposition 4.2 *Let X be a reflexive Banach space and $C, \{C_{\lambda,\mu} : (\lambda, \mu) \in \mathcal{I}\}$ be maximal monotone operators between X and X^*. Then*

(a) $C = \|\cdot\| \times \|\cdot\| - \lim_{\mathcal{F}} C_{\lambda,\mu}$ *if and only if* $C \subset \|\cdot\| \times \|\cdot\| - \liminf_{\mathcal{F}} C_{\lambda,\mu}$;

(b) $C = \|\cdot\|\times seq.-w-\lim_{\mathcal{F}} C_{\lambda,\mu}$ *if and only if* $C \subset \|\cdot\|\times seq.-w-\liminf_{\mathcal{F}} C_{\lambda,\mu}$.

An obvious corollary from the above is:

Corollary 4.3 *Let X be a reflexive Banach space and $C, \{C_{\lambda,\mu} : (\lambda, \mu) \in \mathcal{I}\}$ be maximal monotone operators between X and X^*. Then $C = \|\cdot\|\times\|\cdot\|-\lim_{\mathcal{F}} C_{\lambda,\mu}$ implies $C = \|\cdot\| \times seq. - w - \lim_{\mathcal{F}} C_{\lambda,\mu}$.*

As it is seen from the following proposition, Yosida approximations of a given maximal monotone operator graph-converge to the operator.

Proposition 4.4 **([2])** *Let X be a reflexive Banach space and $A : X \rightrightarrows X^*$ be a maximal monotone operator. Then*

$$A = \|\cdot\| \times \|\cdot\| - \lim_{\lambda \downarrow 0} A_\lambda = \|\cdot\| \times seq. - w - \lim_{\lambda \downarrow 0} A_\lambda.$$

Now we pass to the definition of the variational sum. The following notion was introduced by Attouch, Baillon and Théra in [3] for the setting of Hilbert spaces. We give here its natural extension for the case of a reflexive Banach space.

Definition 4.5 *Let A and B be two maximal monotone operators in the reflexive Banach space X. The variational sum between A and B denoted by $A \underset{v}{+} B$ is the operator between X and X^* having the following graph:*

$$A \underset{v}{+} B := \|\cdot\| \times \|\cdot\| - \liminf_{\mathcal{F}}(A_\lambda + B_\mu).$$

As it was mentioned, an equivalent working definition is: $x^* \in (A \underset{v}{+} B)(x)$ exactly when for every sequence $\{(\lambda_n, \mu_n)\}_{n=1}^{\infty} \subset \mathcal{I}$ with $(\lambda_n, \mu_n) \to 0$ there are $x_n \in X$ and $x_n^* \in X^*$ such that $x_n \to x$, $x_n^* \to x^*$ and $x_n^* \in A_{\lambda_n} x_n + B_{\mu_n} x_n$ for every $n = 1, 2, \ldots$ It is seen that the definition of the variational sum at a certain point $x \in X$ takes into account the behaviour of the operators also at nearby points. This differs from the definition of extended sum above which involves the values of the approximations of the operators only at the point x. Let us prove some properties of the variational sum which are extensions of the similar ones established in Proposition 4.2, [3], in the setting of Hilbert spaces.

Proposition 4.6 *Let X be a reflexive Banach space and $A, B : X \rightrightarrows X^*$ be maximal monotone operators. Then*

(1) $\mathrm{Dom}(A) \cap \mathrm{Dom}(B) \subset \mathrm{Dom}(A \underset{v}{+} B)$;

(2) $A \underset{v}{+} B$ *is a monotone operator;*

(3) *If $A \underset{v}{+} B$ is a maximal monotone operator then* $A \underset{v}{+} B = \|\cdot\| \times \|\cdot\| - \lim_{\mathcal{F}}(A_\lambda + B_\mu) = \|\cdot\| \times seq. - w - \lim_{\mathcal{F}}(A_\lambda + B_\mu)$;

(4) $A \underset{v}{+} B = B \underset{v}{+} A$.

Proof: (1) Let $x \in \mathrm{Dom}(A) \cap \mathrm{Dom}(B)$. Then by Proposition 4.1 we have that $A_\lambda x$ $\|\cdot\|$-converges to $A^{\min} x$ as $\lambda \downarrow 0$ and $B_\mu x$ $\|\cdot\|$-converges to $B^{\min} x$ as $\mu \downarrow 0$. Hence $(x, A^{\min} x + B^{\min} x) \in \mathrm{Gr}(A \underset{v}{+} B)$. (3) follows by Proposition 4.2 and Corollary 4.3 and (4) is clear by the definition. As to (2), take $(x, x^*), (y, y^*) \in \mathrm{Gr}(A \underset{v}{+} B)$. Let $(\lambda_n, \mu_n) \in \mathcal{I}$ be a sequence so that $\lambda_n, \mu_n \to 0$ and $\lambda_n, \mu_n > 0$. By the definition of the variational sum for every n there are couples $(x_n, x_n^*) \in \mathrm{Gr}(A_{\lambda_n} + B_{\mu_n})$ and $(y_n, y_n^*) \in \mathrm{Gr}(A_{\lambda_n} + B_{\mu_n})$ such that $(x_n, x_n^*) \to (x, x^*)$ and $(y_n, y_n^*) \to (y, y^*)$ strongly. Having in mind this (and that $\lambda_n, \mu_n > 0$) we get

$$\begin{aligned} \langle x - y, x^* - y^* \rangle &= \lim_n \langle x_n - y_n, x_n^* - y_n^* \rangle \\ &= \lim_n \langle x_n - y_n, A_{\lambda_n} x_n + B_{\mu_n} x_n - A_{\lambda_n} y_n - B_{\mu_n} y_n \rangle \\ &\geq 0, \end{aligned}$$

the last inequality being true because of monotonicity of the Yosida approximations. The proof is completed. ■

Before passing to the relation between the variational sum and the extended one, we need an important auxiliary lemma.

Lemma 4.7 *Let A and B be two maximal monotone operators in the reflexive Banach space X such that* $\mathrm{Dom}(A) \cap \mathrm{Dom}(B) \neq \emptyset$. *Suppose that* $(x, x^*) \in \mathrm{Gr}(A \underset{v}{+} B)$ *and that* $(x, x^*) = \|\cdot\| \times \|\cdot\| - \lim_{\lambda,\mu\downarrow 0}(x_{\lambda,\mu}, A_\lambda x_{\lambda,\mu} + B_\mu x_{\lambda,\mu})$. *Then* $\|\cdot\| - \lim_{\lambda,\mu\downarrow 0} \lambda A_\lambda x_{\lambda,\mu} = 0$ *and* $\|\cdot\| - \lim_{\lambda,\mu\downarrow 0} \mu B_\mu x_{\lambda,\mu} = 0$.

Proof: We will use a piece of argument from [3]. Let $y \in \mathrm{Dom}(A) \cap \mathrm{Dom}(B)$ and take $u^* \in Ay$ and $v^* \in By$. Let $\lambda, \mu > 0$ be fixed. Then using (4.5) and the monotonicity of A and B we have the following inequalities:

$$\langle y - J_\lambda^A x_{\lambda,\mu}, u^* - A_\lambda x_{\lambda,\mu}\rangle \geq 0$$

$$\langle y - J_\mu^B x_{\lambda,\mu}, v^* - B_\mu x_{\lambda,\mu}\rangle \geq 0.$$

Now by (4.3) we get

$$\langle y - x_{\lambda,\mu} + \lambda J^{-1}(A_\lambda x_{\lambda,\mu}), u^* - A_\lambda x_{\lambda,\mu}\rangle \geq 0$$

$$\langle y - x_{\lambda,\mu} + \mu J^{-1}(B_\mu x_{\lambda,\mu}), v^* - B_\mu x_{\lambda,\mu}\rangle \geq 0.$$

After adding these two inequalities we obtain:

$$\begin{gathered}\langle y - x_{\lambda,\mu}, u^* + v^* - (A_\lambda x_{\lambda,\mu} + B_\mu x_{\lambda,\mu})\rangle + \\ \langle \lambda J^{-1}(A_\lambda x_{\lambda,\mu}), u^* - A_\lambda x_{\lambda,\mu}\rangle + \langle \mu J^{-1}(B_\mu x_{\lambda,\mu}), v^* - B_\mu x_{\lambda,\mu}\rangle \geq 0.\end{gathered}$$

Further, we use the definition of the duality mapping to get that:

$$\begin{aligned}&\langle y - x_{\lambda,\mu}, u^* + v^* - (A_\lambda x_{\lambda,\mu} + B_\mu x_{\lambda,\mu})\rangle \\ (4.6)\qquad &\quad + \lambda\|u^*\|\,\|A_\lambda x_{\lambda,\mu}\| + \mu\|v^*\|\,\|B_\mu x_{\lambda,\mu}\| \\ &\geq \lambda\|A_\lambda x_{\lambda,\mu}\|^2 + \mu\|B_\mu x_{\lambda,\mu}\|^2.\end{aligned}$$

Observe that, for fixed y, u^*, v^*, when $\lambda, \mu \downarrow 0$, the first term on the left hand side of the above inequality is bounded because of the assumptions of the lemma. Hence one concludes from the last inequality that there is some constant $M > 0$ so that $\lambda^{1/2}\|A_\lambda x_{\lambda,\mu}\| < M$ and $\mu^{1/2}\|B_\mu x_{\lambda,\mu}\| < M$ for sufficiently small $\lambda, \mu > 0$, whence we deduce the assertion of the lemma. The proof is completed. ■

Remark 4.8 Let us stress, in connection with the proof of Theorem 4.12 below, that the inequality (4.6) above is true for every $y \in \mathrm{Dom}(A) \cap \mathrm{Dom}(B)$, every $u^* \in Ay$, $v^* \in By$ and $x_{\lambda,\mu}$, $\lambda, \mu > 0$, without supposing the convergence of $\{(x_{\lambda,\mu}, A_\lambda x_{\lambda,\mu} + B_\mu x_{\lambda,\mu})\}$.

Now we pass to the comparison of the variational sum with the usual pointwise sum and the extended sum.

Theorem 4.9 *Let X be a reflexive Banach space and $A, B : X \rightrightarrows X^*$ be two maximal monotone operators with* $\mathrm{Dom}(A) \cap \mathrm{Dom}(B) \neq \emptyset$ *and such that $A \underset{v}{+} B$ is a maximal monotone operator. Then for every $x \in X$ we have*

$$(A \underset{ext}{+} B)(x) \subset (A \underset{v}{+} B)(x).$$

Proof: Let $x^* \in (A \underset{ext}{+} B)(x) = \bigcap_{\varepsilon>0} \overline{A^\varepsilon(x) + B^\varepsilon(x)}$ for some $x \in X$. We will show that

$$(4.7) \qquad \langle x - y, x^* - y^* \rangle \geq 0 \;\text{ for every } (y, y^*) \in \mathrm{Gr}(A \underset{v}{+} B).$$

This, together with maximality of the variational sum, will imply that $x^* \in (A \underset{v}{+} B)(x)$.

To this end, take an arbitrary $(y, y^*) \in \mathrm{Gr}(A \underset{v}{+} B)$ and fix it. Let $\varepsilon > 0$ be an arbitrary fixed positive number. Since x^* belongs to the norm-closure of $A^\varepsilon(x) + B^\varepsilon(x)$ there exist $u^*_\varepsilon \in A^\varepsilon(x)$ and $v^*_\varepsilon \in B^\varepsilon(x)$ so that

$$(4.8) \qquad \|x - y\| \, \|x^* - u^*_\varepsilon - v^*_\varepsilon\| \leq \varepsilon.$$

Fix these u^*_ε and v^*_ε. Further, for every $n = 1, 2, \ldots$, take $\lambda_n, \mu_n > 0$ so that $\lim_n \lambda_n = \lim_n \mu_n = 0$. Since $(y, y^*) \in \mathrm{Gr}(A \underset{v}{+} B)$ we get the existence of a sequence $\{(y_n, y_n^*)\}_{n=1}^\infty$ such that $y = \|\cdot\| - \lim_n y_n$, $y^* = \|\cdot\| - \lim_n y_n^*$ and $y_n^* = A_{\lambda_n} y_n + B_{\mu_n} y_n$. Let $M > 0$ be an upper bound of the norms of the sequence $\{y_n^*\}_{n=1}^\infty$ and n be so large that:

$$(4.9) \qquad \begin{aligned} &|\langle x - y, y_n^* - y^* \rangle| < \varepsilon \\ &\|y - y_n\| < \frac{\varepsilon}{\|u^*_\varepsilon + v^*_\varepsilon\| + M}. \end{aligned}$$

Finally, using Lemma 4.7 we choose further n to be so large that

$$(4.10) \qquad \begin{aligned} &\|u^*_\varepsilon\| \, \|\lambda_n A_{\lambda_n} y_n\| < \varepsilon \\ &\|v^*_\varepsilon\| \, \|\mu_n B_{\mu_n} y_n\| < \varepsilon. \end{aligned}$$

Now, having in mind (4.8) and (4.9), we obtain the following chain of inequalities:

$$\begin{aligned} \langle x - y, x^* - y^* \rangle &= \langle x - y, u^*_\varepsilon + v^*_\varepsilon - y^* \rangle + \langle x - y, x^* - u^*_\varepsilon - v^*_\varepsilon \rangle \\ &\geq \langle x - y, u^*_\varepsilon + v^*_\varepsilon - y_n^* \rangle + \langle x - y, y_n^* - y^* \rangle \\ &\quad - \|x - y\| \, \|x^* - u^*_\varepsilon - v^*_\varepsilon\| \\ &\geq \langle x - y_n, u^*_\varepsilon + v^*_\varepsilon - y_n^* \rangle \\ &\quad + \langle y_n - y, u^*_\varepsilon + v^*_\varepsilon - y_n^* \rangle - 2\varepsilon \\ &\geq \langle x - y_n, u^*_\varepsilon + v^*_\varepsilon - y_n^* \rangle \\ &\quad - \|y_n - y\| \left(\|u^*_\varepsilon + v^*_\varepsilon\| + M \right) - 2\varepsilon \\ &\geq \langle x - y_n, u^*_\varepsilon + v^*_\varepsilon - A_{\lambda_n} y_n - B_{\mu_n} y_n \rangle - 3\varepsilon. \end{aligned}$$

Next we use (4.3) to continue the above chain of inequalities as follows:

$$\begin{aligned} \langle x - y, x^* - y^* \rangle &\geq \langle x - y_n, u^*_\varepsilon - A_{\lambda_n} y_n \rangle + \langle x - y_n, v^*_\varepsilon - B_{\mu_n} y_n \rangle - 3\varepsilon \\ &= \langle x - J^A_{\lambda_n} y_n - \lambda_n J^{-1}(A_{\lambda_n} y_n), u^*_\varepsilon - A_{\lambda_n} y_n \rangle \\ &\quad + \langle x - J^B_{\mu_n} y_n - \mu_n J^{-1}(B_{\mu_n} y_n), v^*_\varepsilon - B_{\mu_n} y_n \rangle - 3\varepsilon \\ &= \langle x - J^A_{\lambda_n} y_n, u^*_\varepsilon - A_{\lambda_n} y_n \rangle + \langle x - J^B_\mu y_n, v^*_\varepsilon - B_{\lambda_n} y_n \rangle \\ &\quad - \lambda_n \langle J^{-1}(A_{\lambda_n} y_n), u^*_\varepsilon - A_{\lambda_n} y_n \rangle \\ &\quad - \mu_n \langle J^{-1}(B_{\mu_n} y_n), v^*_\varepsilon - B_{\mu_n} y_n \rangle - 3\varepsilon. \end{aligned}$$

Remember now that $u_\varepsilon^* \in A^\varepsilon(x)$ and $v_\varepsilon^* \in B^\varepsilon(x)$. This together with (4.5) show that the first two terms on the right hand side of the last equality in the above chain are greater or equal to $-\varepsilon$. Hence (using also the definition of J^{-1} and (4.10)):

$$\begin{aligned}\langle x-y, x^*-y^*\rangle &\geq -\lambda_n\langle J^{-1}(A_{\lambda_n}y_n), u_\varepsilon^* - A_{\lambda_n}y_n\rangle \\ &\quad -\mu_n\langle J^{-1}(B_{\mu_n}y_n), v_\varepsilon^* - B_{\mu_n}y_n\rangle - 5\varepsilon \\ &= \lambda_n\|A_{\lambda_n}y_n\|^2 - \lambda_n\langle J^{-1}(A_{\lambda_n}y_n), u_\varepsilon^*\rangle \\ &\quad +\mu_n\|B_{\mu_n}y_n\|^2 - \mu_n\langle J^{-1}(B_{\mu_n}y_n), v_\varepsilon^*\rangle - 5\varepsilon \\ &\geq -\lambda_n\|A_{\lambda_n}y_n\|\,\|u_\varepsilon^*\| - \mu_n\|B_{\mu_n}y_n\|\,\|v_\varepsilon^*\| - 5\varepsilon \\ &\geq -7\varepsilon.\end{aligned}$$

Consequently, we have proved that

$$\langle x-y, x^*-y^*\rangle \geq -7\varepsilon,$$

and since $\varepsilon > 0$ was arbitrary, we conclude that

$$\langle x-y, x^*-y^*\rangle \geq 0,$$

i.e. (4.7) is true. The proof is completed. ■

The following corollary is immediate.

Corollary 4.10 *Let A, B be two maximal monotone operators in the reflexive Banach space X such that $A \underset{v}{+} B$ is a maximal monotone operator. Then $A + B \subset A \underset{v}{+} B$.*

On the other hand, the inequality (4.7) was proved for arbitrary pairs from $A \underset{v}{+} B$ and $A \underset{ext}{+} B$. Therefore, we have the following theorem:

Theorem 4.11 *Let A, B be two maximal monotone operators in the reflexive Banach space X with $\mathrm{Dom}(A) \cap \mathrm{Dom}(B) \neq \emptyset$ and such that $A \underset{ext}{+} B$ is a maximal monotone operator. Then $A \underset{v}{+} B \subset A \underset{ext}{+} B$.*

Finally, we prove that when $\overline{A+B}$ or $A+B$ is maximal, then all the three concepts of sum coincide. In particular, with this we extend for reflexive spaces Theorem 6.1 from [3]. Namely, we have:

Theorem 4.12 *Let X be a reflexive Banach space and $A, B : X \rightrightarrows X^*$ be two maximal monotone operators such that $\overline{A+B}$ is a maximal monotone operator. Then*

$$\overline{A+B} = A \underset{ext}{+} B = A \underset{v}{+} B.$$

Proof: Since $\overline{A+B}$ is maximal monotone, the conclusion of the theorem will follow by Theorem 4.9 if we prove the inclusion $\overline{A+B} \subset A \underset{v}{+} B$. For the latter, we will develop further an argument from the setting of Hilbert spaces used in [3].

Let $x^* \in \overline{A+B}(x)$ for some $x \in X$. Take a sequence $\{(\lambda_n, \mu_n)\}_{n=1}^{\infty} \subset \mathcal{I}$, so that $(\lambda_n, \mu_n) \to (0,0)$. We will consider first the case $\lambda_n, \mu_n > 0$ for every $n = 1, 2, \ldots$

For any $n = 1, 2, \ldots$, let x_n be, by virtue of the result of Rockafellar above, the unique solution of the equality:

$$Jx + x^* = Jx_n + A_{\lambda_n} x_n + B_{\mu_n} x_n, \quad n = 1, 2, \ldots \tag{4.11}$$

Take arbitrary $y \in \mathrm{Dom}(A) \cap \mathrm{Dom}(B)$, $u^* \in Ay$ and $v^* \in By$. As it was mentioned in Remark 4.8, the inequality (4.6) is true if we put in the place of $x_{\lambda,\mu}$ the points x_n. Hence, using (4.6), (4.11) and the definition of the duality mapping we have for every $n = 1, 2, \ldots$

$$\begin{aligned} &\langle y - x_n, u^* + v^* - Jx - x^* \rangle + \langle y, Jx_n \rangle \\ &\quad + \lambda_n \|u^*\| \, \|A_{\lambda_n} x_n\| + \mu_n \|v^*\| \, \|B_{\mu_n} x_n\| \\ &\geq \lambda_n \|A_{\lambda_n} x_n\|^2 + \mu_n \|B_{\mu_n} x_n\|^2 + \|x_n\|^2. \end{aligned} \tag{4.12}$$

Using the last inequality, one can see that there exists a constant $M > 0$ so that $\|x_n\| \leq M$, $\sqrt{\lambda_n}\, \|A_{\lambda_n} x_n\| \leq M$ and $\sqrt{\mu_n}\, \|B_{\mu_n} x_n\| \leq M$ for every $n = 1, 2, \ldots$ In particular, $\lim_n \lambda_n \|A_{\lambda_n} x_n\| = \lim_n \mu_n \|B_{\mu_n} x_n\| = 0$.

Since $\{x_n\}$ is bounded, it has at least one weak cluster point, say $\bar{x}$. Now, since for every n, $\|x_n\|^2 = \langle x_n, Jx_n \rangle$ and since by the monotonicity of J we have $\langle x_n - y, Jx_n \rangle \geq \langle x_n - y, Jy \rangle$, we obtain from (4.12) that for every $n = 1, 2, \ldots$

$$\begin{aligned} &\langle y - x_n, u^* + v^* - Jx - x^* \rangle + \\ &\quad + \lambda_n \|u^*\| \, \|A_{\lambda_n} x_n\| + \mu_n \|v^*\| \, \|B_{\mu_n} x_n\| \\ &\geq \langle x_n - y, Jy \rangle. \end{aligned}$$

¿From here, taking a subsequence of $\{x_n\}$ which weakly converges to $\bar{x}$ and passing to the limit we get:

$$\langle y - \bar{x}, u^* + v^* - Jx - x^* \rangle \geq \langle \bar{x} - y, Jy \rangle,$$

or equivalently,

$$\langle y - \bar{x}, Jy + u^* + v^* - Jx - x^* \rangle \geq 0.$$

Since $u^* \in Ay$ and $v^* \in By$ were arbitrary, we conclude that

$$\langle y - \bar{x}, Jy + w^* - Jx - x^* \rangle \geq 0$$

for every $w^* \in \overline{A+B}(y)$. Now, remember that $\overline{A+B}$ was maximal, hence the same is true for $J + \overline{A+B}$. Therefore, since y was also arbitrary, the last inequality entails that $Jx + x^* \in J\bar{x} + \overline{A+B}(\bar{x})$. Finaly, since $x^* \in \overline{A+B}(x)$, the result of Rockafellar above implies that $\bar{x} = x$, from where we conclude that the whole sequence $\{x_n\}$ converges weakly to x.

Further, we show that $||x_n|| \to ||x||$. First, since J is the subdifferential of the function $(1/2)||x||^2$ we have for every $n = 1, 2, \ldots$

$$\langle x, Jx_n\rangle \leq \langle x_n, Jx_n\rangle + \frac{1}{2}||x||^2 - \frac{1}{2}||x_n||^2 = \frac{1}{2}||x||^2 + \frac{1}{2}||x_n||^2.$$

Whence, using (4.12) with $y = x$ we easily get that $\limsup_n ||x_n||^2 \leq ||x||^2$. Therefore, using this and also the weak lower semicontinuity of the norm in X we obtain:

$$||x||^2 \leq \liminf_n ||x_n||^2 \leq \limsup_n ||x_n||^2 \leq ||x||^2,$$

i.e. $||x_n||^2 \to ||x||^2$, whence $||x_n|| \to ||x||$ and by the Kadec-Klee property we obtain that $\{x_n\}$ converges to x strongly. By (4.11) and the norm-to-norm continuity of J we deduce that $A_{\lambda_n}x_n + B_{\mu_n}x_n$ converges strongly to x^*.
In order to conclude that $(x, x^*) \in A \underset{v}{+} B$, it remains also to consider operators of the type $A_{\lambda_n} + B$ or $A + B_{\mu_n}$. But in this last case one uses the well-known results of Brezis, Crandall and Pazy [6], Theorem 2.1, to see that (x, x^*) is a norm-limit of a sequence from $A_{\lambda_n} + B$ or $A + B_{\mu_n}$. The proof is completed. ■

The following theorem is an immediate corollary from the above theorem:

Theorem 4.13 *Let X be a reflexive Banach space and $A, B : X \rightrightarrows X^*$ be two maximal monotone operators such that $A+B$ is a maximal monotone operator. Then*

$$A + B = A \underset{ext}{+} B = A \underset{v}{+} B.$$

5 Lower Semicontinuous Convex Functions and Sums of Operators

In this section we see that the variational sum keeps another important property from the original setting of Hilbert spaces–namely, when the case of subdifferentials of convex functions is considered, we have that the subdifferential of the sum of two convex proper lower semicontinuous functions is equal to the variational sum of their subdifferentials. As a consequence we get that in the case of subdifferentials the extended and the variational sum are the same.
Let $f : X \to \mathbf{R} \cup \{+\infty\}$ be a proper lower semicontinuous convex function. The well-known Moreau-Yosida regularization of f of order $\lambda > 0$ is given by the formula:

$$f_\lambda(x) := \inf_{y\in X}\{f(y) + \frac{1}{2\lambda}||y - x||^2\}, \quad x \in X.$$

It is also well-known that for a given $\lambda > 0$ the relation between the subdifferential of f_λ and the Yosida approximation of this subdifferential of order λ considered in the previous section is:

$$\partial(f_\lambda) = (\partial f)_\lambda. \tag{5.1}$$

Before giving the next result, let us remind the definition of *Mosco convergence* of convex functions ([20]): It is said that the sequence $\{f_n\}_{n=1}^{\infty}$ of proper lower semicontinuous convex functions in X Mosco-converges to f_0 if for every $x \in X$ the following two conditions are fulfilled:

(i) for every sequence $\{x_n\} \subset X$ which converges weakly to x we have $f(x) \leq \liminf_{n\to\infty} f_n(x_n)$;

(ii) there exists a sequence $\{x_n\} \subset X$ which strongly converges to x and such that $\limsup_{n\to\infty} f_n(x_n) \leq f(x)$.

The following fact was observed in [3], Theorem 7.2, for the setting of Hilbert spaces. We give its natural extension for the case of reflexive spaces we consider. In this last setting, the result has also been obtained directly by Jourani [15], using Ekeland variational principle. Other sequential formulae for the subdifferential of the sum of two convex functions can be found in the papers of Thibault [29, 32]. For similar results in the case of a general Banach space the reader is refered to the paper of Hiriart-Urruty and Phelps [12], the survey [13], the papers of Penot [23], Thibault [30, 31] and of the authors [26].

Theorem 5.1 *Let* $f, g : X \to \mathbf{R} \cup \{+\infty\}$ *be two proper convex lower semicontinuous functions in the reflexive Banach space* X *such that* $\mathrm{dom} f \cap \mathrm{dom} g \neq \emptyset$. *Then*

$$\partial(f+g) = \partial f \underset{v}{+} \partial g.$$

Moreover,

$$\partial(f+g) = \|\cdot\| \times \|\cdot\| - \lim_{\mathcal{F}} (\partial f_\lambda + \partial g_\mu).$$

Proof: Since $\mathrm{dom} f \cap \mathrm{dom} g \neq \emptyset$ we see that $\partial(f+g)$ is a maximal monotone operator. The function f_λ for $\lambda > 0$ (or g_μ for $\mu > 0$) is everywhere defined and continuous. Hence, by the classical result of Moreau-Rockafellar $\partial f_\lambda + \partial g_\mu = \partial(f_\lambda + g_\mu)$ for every $(\lambda, \mu) \in \mathcal{I}$. Therefore, by (5.1) we conclude that

$$(\partial f)_\lambda + (\partial g)_\mu = \partial(f_\lambda + g_\mu) \quad \text{for all } (\lambda, \mu) \in \mathcal{I}. \tag{5.2}$$

On the other hand, by a result from [5], Theorem 3.20, we deduce that the family $\{f_\lambda + g_\mu : (\lambda, \mu) \in \mathcal{I}\}$ Mosco converges to $f + g$ as (λ, μ) converges to $(0, 0)$. Now, an already classical result of Attouch [1] allows to derive from the latter that $\|\cdot\| \times \|\cdot\| - \lim_{\mathcal{F}} \partial(f_\lambda + g_\mu) = \partial(f+g)$. It remains to take into account (5.2) above and the definition of the variational sum. ■

The following is an immediate corollary from the above theorem, Theorem 4.9 and the formula of Hiriart-Urruty and Phelps (Theorem 2.1). It shows that in the case of subdifferentials of convex functions the variational and the extended sum coincide.

Corollary 5.2 *Let* $f, g : X \to \mathbf{R} \cup \{+\infty\}$ *be two proper convex and lower semicontinuous functions in the reflexive Banach space* X *such that* $\mathrm{dom} f \cap \mathrm{dom} g \neq \emptyset$. *Then:*

$$\partial(f+g) = \partial f \underset{v}{+} \partial g = \partial f \underset{ext}{+} \partial g.$$

In fact, the subdifferential of the sum of two convex proper lower semicontinuous functions is equal to the extended sum of their subdifferentials in the more general setting of an arbitrary Banach space as shown in our paper [26].

Acknowledgements. We would like to thank one anonymous referee whose detailed remarks helped us to improve both the presentation of the ideas and of the results.

References

[1] H. Attouch, "On the maximality of the sum of two maximal monotone operators", *Nonlinear Anal., TMA* 5 (1981) 143–147.

[2] H. Attouch, *Variational convergence for functions and operators*, (Applicable Math. Series, Pitman, London, 1984).

[3] H. Attouch, J.-B. Baillon and M. Théra, "Variational sum of monotone operators", *J. Convex Anal.* 1 (1994) 1–29.

[4] H. Attouch, J.-B. Baillon and M. Théra, "Weak solutions of evolution equations and variational sum of maximal monotone operators", *SEA Bull. Math.* 19 (1995) 117–126.

[5] H. Attouch and M. Théra, "Convergences en analyse multivoque et unilatérale", *MATAPLI, Bulletin de liaison* 36 (1993) 23–40.

[6] H. Brezis, M.G. Crandall and A. Pazy, "Perturbations of nonlinear maximal monotone sets in Banach space", *Comm. Pure Appl. Math.* XXIII (1970) 123–144.

[7] F.E. Browder, "Nonlinear maximal monotone operators in Banach space", *Math. Ann.* 175 (1968) 89–113.

[8] R.S. Burachik, A.N. Iusem and B.F. Svaiter, "Enlargements of maximal monotone operators with applications to variational inequalities", *Set-valued Anal.* 5 (1997) 159–180.

[9] R.S. Burachik, C.A. Sagastizábal and B.F. Svaiter, *ε-enlargements of maximal monotone operators: Theory and Applications".* in: Nonsmooth, Piecewise Smooth, Semismooth and Smoothing Methods, M. Fukushima and L.Qi (eds.), Kluwer Academic Publishers, Dordrecht, 1998, pp. 25–43.

[10] T. Diagana, *Sommes d'opérateurs et conjecture de Kato-McIntosh*, (Thèse de l'Université Lyon I, 1999).

[11] J. Diestel, *Geometry of Banach spaces, Selected Topics*, Lecture Notes in Math., Springer-Verlag, Berlin, 1975.

[12] J.-B. Hiriart-Urruty and R.R. Phelps, "Subdifferential calculus using ε-subdifferentials", *J. Funct. Anal.* 118 (1993) 154–166.

[13] J.-B. Hiriart-Urruty, M. Moussaoui, A. Seeger and M. Volle, "Subdifferential calculus, without qualification hypothesis", *Nonlinear Anal. TMA* 24 (1995) 1727–1754.

[14] K. John and V. Zizler, "A renorming of a dual space", *Israel J. Math.* 12 (1972) 331–336.

[15] A. Jourani, *Variational sum of subdifferentials of convex functions*, in: Proc. of the Fourth Catalan Days on Applied Mathematics, C. Garcia, C. Olive, M. Sanroma (eds.), Tarragona Press University, Tarragona, 1998, pp. 71–80.

[16] F. Kubo, "Conditional expectations and operations derived from network connections, *J. Math. Anal. Appl.* 80 (1981) 477–489.

[17] M. Lapidus, *Formules de Trotter et calcul opérationnnel de Feynman*, (Thèse d'Etat, Université Paris VI, 1986).

[18] R. Lucchetti and F. Patrone, "A characterization of Tykhonov well-posedness for minimum problems with applications to variational inequalities", *Numer. Funct. Anal. Optim.* 3(4) (1981) 461–476.

[19] J.E. Martinez-Legaz and M. Théra, "ε-subdifferentials in terms of subdifferentials", *Set-valued Anal.* 4 (1996) 327–332.

[20] U. Mosco, "Convergence of convex sets and of solutions of variational inequalities", *Adv. Math.* 3 (1969) 510–585.

[21] M. Nisipeanu, *Somme variationnelle d'opérateurs et applications*, (Thèse de l'Université de Limoges, 1997).

[22] M. Nisipeanu and M. Théra, "Variational sum of maximal monotone operators: approach via saddle functions". In preparation.

[23] J.-P. Penot, "Subdifferential calculus without qualification conditions", *J. Convex Anal.* 3 (1996) 1–13.

[24] R.R. Phelps, *Convex Functions, Monotone Operators and Differentiability*, Lect. Notes in Math., Springer Verlag, Berlin, # 1364, 1989.

[25] J.P. Revalski, *Variational inequalities with unique solution*, in: Mathematics and Education in Mathematics, Proc. of the 14-th Spring Conf. of the Union of the Bulgarian Mathematicians, 1985, pp. 534–541.

[26] J.P. Revalski and M. Théra, "Enlargements and sums of monotone operators", Preprint (1999), Submitted.

[27] R.T. Rockafellar, "On the maximality of sums of nonlinear monotone operators", *Trans. Amer. Math. Soc.* 149 (1970) 75–88.

[28] S. Simons, *Minimax and Monotonicity*, Lect. Notes in Math., Springer Verlag, Berlin, #1693, 1998.

[29] L. Thibault, "A direct proof of a sequential formula for the subdifferential of the sum of two convex functions", Unpublished paper, (1994).

[30] L. Thibault, *A general sequential formula for subdifferentials of sums of convex functions defined on Banach spaces*, in: Recent Developments in Optimization, R. Durier and C. Michelot (eds.), Lect. Notes in Economics and Mathematical Systems, Springer-Verlag, Berlin, 429, 1995, pp. 340–345.

[31] L. Thibault, "A short note on sequential convex subdifferential calculus", Unpublished paper, (1995).

[32] L. Thibault, "Sequential convex subdifferential calculus and sequential Lagrange multipliers", *SIAM J. Contr. Optim.* 35 (1997) 1434–1444.

[33] D. Torralba, *Convergence épigraphique et changements d'échelle en analyse variationnelle et optimisation*, (Thèse de l'Université de Montpellier II, 1996).

[34] L. Veselý, "Local uniform boundedness principle for families of ε-monotone operators", *Nonlinear Anal., TMA* 24 (1994) 1299–1304.

Error Bounds for Regularized Complementarity Problems *

Paul Tseng **

Abstract

An approach to solving a complementarity problem entails regularizing/perturbing the problem by adding to the given mapping another mapping multiplied by a small positive parameter. We study properties of the limit point of the solution to the regularized problem. We also derive local error bounds on the distance from the solution to its limit point, expressed in terms of the regularization parameter.

Keywords: Regularization, complementarity problem, optimization, error bound

AMS subject classification: 49M39, 90C25, 90C31, 90C34, 90C48

1 Introduction

Consider the complementarity problem (CP) of finding an $x \in \Re^n$ satisfying

$$x \geq 0, \quad F(x) \geq 0, \quad F(x)^T x = 0, \tag{1}$$

where $F : \Re^n_+ \mapsto \Re^n$ is a given continuous mapping. This is a well-known problem in optimization, with many applications [12, 24]. In various regularization/continuation/smoothing approaches to solving this problem, one adds to the mapping F another mapping $G : \Re^n_{++} \mapsto \Re^n$, multiplied by a small positive scalar ϵ, and computes (possibly inexactly) an $x^\epsilon \in \Re^n$ satisfying

$$x^\epsilon \geq 0, \quad F(x^\epsilon) + \epsilon G(x^\epsilon) \geq 0, \quad (F(x^\epsilon) + \epsilon G(x^\epsilon))^T x^\epsilon = 0. \tag{2}$$

Then, one may decrease ϵ and update x^ϵ accordingly. Our interests are in properties of any limit point of x^ϵ (along some sequence of $\epsilon \to 0$) and the distance from x^ϵ to this limit point. There are also the related issues of existence/uniqueness/boundedness of x^ϵ as $\epsilon \to 0$, which we will not focus on.

*This research is supported by National Science Foundation Grant CCR-9731273.

**Department of Mathematics, University of Washington, Seattle, Washington 98195, U.S.A. (tseng@math.washington.edu)

The regularized CP (2) is closely linked to a regularized smooth optimization problem of the form

$$\text{minimize } f_0(u)+\epsilon g_0(u) \text{ subject to } u \geq 0,\ f_i(u)+\epsilon g_i(u) \leq 0, i=1,...,m, \quad (3)$$

where $f_0, f_1, ..., f_m$ are continuously differentiable functions defined on some open set containing $\Re^l_+$ and $g_0, g_1, ..., g_m$ are continuously differentiable functions defined on $\Re^l_{++}$ $(m \geq 0, l \geq 1)$. In particular, it is well known that the associated Karush-Kuhn-Tucker condition is exactly (2) with $x = \left[\begin{smallmatrix} u \\ v \end{smallmatrix}\right]$ and

$$F(x) = \begin{bmatrix} \nabla f_0(u) + \sum_{i=1}^m v_i \nabla f_i(u) \\ -f_i(u)_{i=1}^m \end{bmatrix}, \quad G(x) = \begin{bmatrix} \nabla g_0(u) + \sum_{i=1}^m v_i \nabla g_i(u) \\ -g_i(u)_{i=1}^m \end{bmatrix}.$$

Moreover, if $f_0, f_1, ..., f_m$ are convex (respectively, quadratic) on their respective domains, then this F is monotone (respectively, affine) and continuous on $\Re^{l+m}_+$ [42, Example 8], and similarly for G. A well studied case in this optimization setting is when $g_i \equiv 0$ for $i = 1, ..., m$, i.e., constraint functions are unregularized. For the regularized CP (2), one popular choice of G is the identity mapping

$$G(x) = x, \quad (4)$$

corresponding to the well-known Tikhonov regularization technique. This choice has been much studied [6, 12, 14, 15, 26, 49, 50, 51, 53], including in the general setting of finding a zero of a maximal monotone operator [2, page 62], [8, Chapter II], [32]. The analogous choice of

$$g_0(u) = \|u\|^2/2$$

for (3) has been considered by Karlin [27], Mangasarian [37, 38, 39] and others [36, 44] in the context of linear programs (LP) and by Tikhonov and various others in the general optimization setting (see [1, 13, 32] and references therein). It was shown in [8, Proposition 2.6(iii)] (also see [12, Theorem 5.6.2(b)],[32, Proposition 6.1], [49], [51, Theorem 2]) that, if F is monotone, then each limit point of x^ϵ (as $\epsilon \to 0$) is the least 2-norm solution of CP. Analogous results were obtained by Mangasarian [38] in the context of LP (also see [19, 40] for extensions to other choices of g_0 in this context) and by Levitin and Polyak and others in the general optimization setting (see [1],[13, pages 30, 37],[32, Proposition 6.1] and references therein). If F is only a P_0-function, Sznader and Gowda [51, Theorem 3] showed that any limit point is weak-Pareto-minimal in the sense that no other solution is componentwise strictly less (so any nonpositive solution is weak-Pareto-minimal). A second popular choice of G is the inverse function

$$G(x) = [-1/x_j]_{j=1}^n, \quad (5)$$

corresponding to log-barrier methods and interior-point methods. This choice has been considered Kojima et al. [28, 29, 30, 31] and Güler [23] and, in the general setting of finding a zero of a maximal monotone operator, by McLinden [42]. The analogous choice of

$$g_0(u) = -\sum_{j=1}^{l} \ln(u_j)$$

for (3) has been much studied in the context of LP (see [20, 41, 43, 54] and references therein). It was shown by McLinden [42, Corollary 2] that if F is monotone and CP has a strictly complementary solution, then any limit point of x^ϵ is a least weighted $-\ln(u)$ solution of CP. Analogus results were obtained by McLinden [41, Theorem 9] and Megiddo [43] in the context of LP. A third choice is the logarithm function

$$G(x) = [\ln(x_j) + 1]_{j=1}^n. \tag{6}$$

The analogous choice of

$$g_0(u) = \sum_{j=1}^{l} u_j \ln(u_j) \tag{7}$$

for (3) was considered in the context of LP by Fang et al. [16, 17, 18, 45] and, from a dual exponential penalty view, by Cominetti et al. [10, 11]. It was shown in [11] that any limit point of the solution of the regularized LP is the least $u\ln(u)$-entropy solution of the LP. This result was generalized recently by Auslender et al. [4] to convex programs, with g_0 being a certain kind of separable strictly convex essentially smooth function. A similar result was shown in [52] for the LP case, without the convexity and smoothness assumption. Related results in the general optimization setting are given in [1, 13] and references therein. These results do not assume g_0 to be separable or even continuous, but they do need g_0 to be lower semicontinuous and real-valued at the limit point to be meaningful.

As the preceding discussion shows, there have been many studies of the properties of a limit point $\bar{x}$ of x^ϵ, with particular focus on the cases of G given by (4) or (5) or (6). However, there have been relatively few studies of the distance from x^ϵ to $\bar{x}$, estimated in terms of ϵ. In the context of LP with g_0 given by (7), this distance is known to be in the order of $e^{-\rho/\epsilon}$ for some constant $\rho > 0$ [11, Theorem 5.8]. The same reference also gives distance estimates for the dual LP. If g_0 is more generally a separable strictly convex essentially smooth function, this distance can be estimated in terms of $(\nabla g_0)^{-1}$ and ϵ [52]. If the LP has a multicommodity network flow structure and g_0 is a weighted inverse barrier function whose weights are affine functions of a nonnegative variable, this distance is known to be in the order of $\sqrt{\epsilon}$ [7].

In this paper, we study the above questions in the context of CP and its regularization (2). In particular, we show that if F is pseudo-monotone on $\Re^n_+$ and G is continuous at $\bar{x}$, then $\bar{x}$ solves the variational inequality problem with mapping

G over the solution set. Moreover, if F is analytic on an open set containing $\Re^n_+$, then the generalized distance $(G(x^\epsilon) - G(\bar{x}))^T(x^\epsilon - \bar{x})$ is in the order of ϵ^γ for some $\gamma > 0$, with $\gamma = 1$ if F is affine (see (16)). Alternatively, if

$$G(x) = [G_j(x_j)]_{j=1}^n, \tag{8}$$

where each G_j is strictly increasing and continuous on $\Re_{++}$ but may tend to $-\infty$ at 0 (e.g., G given by (5) or (6)), we show that in each coordinate subspace over which F is pseudo-monotone, $\bar{x}$ solves the variational inequality problem with mapping G over the solution set (see Proposition 3(a)). Moreover, under the assumption that either (i) F is pseudo-monotone on $\Re^n_+$ and $\lim_{t\to 0} tG_j(t) = 0$ for $j \notin J$ (e.g., G given by (6)) or (ii) F is affine with certain principal submatrices of its Jacobian positive semidefinite and spanning the corresponding rows or (iii) F is affine with certain principal submatrix of its Jacobian positive semidefinite and $\limsup_{t\to 0} tG_j(t) < 0$ for $j \notin J$ (e.g., G given by (5)), we estimate $(|x_j^\epsilon - \bar{x}_j|)_{j\in J}$ in terms of $(x_j^\epsilon)_{j\notin J}$ and, in the case where F is affine, we estimate the latter in terms of ϵ, where J is the set of indices j with $G_j(\bar{x}_j) > -\infty$ (see Proposition 3(c1)–(c4)). Thus, our results may be applied to analyze regularization of a convex quadratic program of the form (3). Our study is motivated by a related work in the context of LP [52], although our results and our proofs are quite different from those in [52] due to the different problem structure and regularization.

In our notation, $\Re^n$ denotes the space of n–dimensional real column vectors, $\Re^n_+$ and $\Re^n_{++}$ denote the nonnegative orthant and the positive orthant in $\Re^n$, respectively, and ^{T} denotes transpose. For any $x \in \Re^n$, we denote by x_i the ith component of x, and, for any $I \subseteq N := \{1, ..., n\}$, by x_I the vector obtained by removing from x those x_i with $i \notin I$, and by $(x_I, 0)$ the vector in $\Re^n$ whose ith component is x_i if $i \in I$ and is zero otherwise. [Here and throughout, := means "define".] We denote by $|I|$ the cardinality of I, and denote $I^c := N\backslash I$, $\|x\| := \sqrt{x^T x}$, $\|x\|_\infty := \max_{i\in N} |x_i|$. For any $M \in \Re^{n\times n}$ and any $I, J \subset N$, we denote by M_I the submatrix of M obtained by removing all rows of M with indices outside of I and by M_{IJ} the submatrix of M_I obtained by removing all columns of M_I with indices outside of J. For any $F : \Re^n_+ \mapsto \Re^n$ and any nonempty closed convex set $\Sigma \subset \Re^n_+$, we denote

$$\mathrm{VI}(\Sigma, F) := \{x \in \Sigma : F(x)^T(y - x) \geq 0 \ \forall y \in \Sigma\}.$$

[Thus x satisfies (1) if and only if $x \in \mathrm{VI}(\Re^n_+, F)$ and x^ϵ satsifies (2) if and only if $x^\epsilon \in \mathrm{VI}(\Re^n_+, F + \epsilon G)$.] We denote by F_i the ith component of F and, for any $I \subseteq N$, by F_I the mapping obtained by removing from F those F_i with $i \notin I$. We say F is pseudo-monotone on $\Re^n_+$ [5, page 121] if

$$x, y \in \Re^n_+ \text{ and } F(y)^T(x - y) \geq 0 \quad \Longrightarrow \quad F(x)^T(x - y) \geq 0. \tag{9}$$

For any $x \in \Re^n$ and any nonempty closed set $\Sigma \subset \Re^n$, we denote $\mathrm{dist}(x, \Sigma) := \min_{y\in\Sigma} \|y - x\|$.

2 Error Bounds on Distance to Limiting Solution

First, we have the following bound on the distance from x^ϵ to the solution set of (1) in terms of the regularization $G^\epsilon(x^\epsilon)$. This is a simple consequence of an error bound result for analytic systems [33].

Proposition 1 *Consider an open set $\Omega \subset \Re^n$ containing $\Re^n_+$ and an analytic $F : \Omega \mapsto \Re^n$ with $\Sigma := \mathrm{VI}(\Re^n_+, F)$ nonempty. Then, for every bounded $\Xi \subset \Re^n$, there exist $\tau > 0$ and $\gamma > 0$ such that*

$$\mathrm{dist}(x^\epsilon, \Sigma) \leq \tau \left(\|\max\{0, G^\epsilon(x^\epsilon)\}\| + |G^\epsilon(x^\epsilon)^T x^\epsilon|\right)^\gamma \tag{10}$$

for all $G^\epsilon : \Re^n_+ \mapsto \Re^n$ ($\epsilon > 0$) and all $x^\epsilon \in \Xi \cap \mathrm{VI}(\Re^n_+, F + G^\epsilon)$.

Proof. We have that an $x \in \Sigma$ satisfies

$$-x \leq 0, \quad -F(x) \leq 0, \quad F(x)^T x = 0$$

and that an $x^\epsilon \in \mathrm{VI}(\Re^n_+, F + G^\epsilon)$ satisfies

$$-x^\epsilon \leq 0, \quad -F(x^\epsilon) \leq G^\epsilon(x^\epsilon), \quad F(x^\epsilon)^T x^\epsilon = -G^\epsilon(x^\epsilon)^T x^\epsilon.$$

So if x^ϵ is also in the bounded set Ξ, then since F is analytic on an open set containing $\Re^n_+$, an error bound result of Lojasiewicz, as extended by Luo and Pang for analytic systems [33, Theorem 2.2], yields (10) with $\tau > 0$ and $\gamma > 0$ some constants. ∎

Note 1. Proposition 1 does not say anything about the existence or uniqueness or boundedness of $x^\epsilon \in \mathrm{VI}(\Re^n_+, F + G^\epsilon)$. In the case where F is monotone and affine and $G^\epsilon(x) = M^\epsilon x + q^\epsilon$, Robinson [47, Theorem 2] showed that Σ being nonempty and bounded is both necessary and sufficient for the existence of x^ϵ satisfying $\mathrm{dist}(x^\epsilon, \Sigma) \to 0$ as $\|M^\epsilon\| + \|q^\epsilon\| \to 0$. If F is a continuously differentiable P_0-function, a result of Facchinei [14, Theorem 4.4] implies that Σ being nonempty and bounded is sufficient for the existence of x^ϵ satisfying $\mathrm{dist}(x^\epsilon, \Sigma) \to 0$ as $\epsilon \to 0$, where G^ϵ is continuous and satisfies $\lim_{\epsilon \to 0} \sup_{x:\mathrm{dist}(x,\Sigma) \leq \sigma} \|G^\epsilon(x)\| = 0$ for some $\sigma > 0$. If F is a continuously differentiable P_0-function and $G^\epsilon(x) = \epsilon x$, Facchinei and Kanzow [15, Theorem 3.5] showed existence and uniqueness of x^ϵ for all $\epsilon > 0$ and, if in addition Σ is nonempty and bounded, then x^ϵ is bounded and $\mathrm{dist}(x^\epsilon, \Sigma) \to 0$ as $\epsilon \to 0$. Ravindran and Gowda [46] extended the preceding two results to CP with bound constraints, and they weakened the differentiability assumption on F to continuity. In the case where F is a polynomial P_0-function and $G^\epsilon(x) = \epsilon x$, Sznajder and Gowda [51, Theorem 5] showed that x^ϵ either converges or diverges in norm.

Proposition 1 does not give an estimate of the Hölder constant γ. [In the case where F is affine and monotone, a bound with $\gamma = 1/2$ can be shown [33, Theorem 5.4].] By adding a mild assumption on G^ϵ, we derive below a second distance bound in terms of $G^\epsilon(x^\epsilon)$, with $\gamma = 1$ when F is affine. Moreover, in the case where F is pseudo-monotone on $\Re^n_+$ and $G^\epsilon(x^\epsilon)/\epsilon$ converges, we derive a bound on the distance from x^ϵ to its limit point in terms of ϵ.

Proposition 2 *Consider a continuous $F : \Re^n_+ \mapsto \Re^n$, a sequence of positive scalars $\Upsilon = \{\epsilon^1, \epsilon^2, ...\}$ tending to zero and, for each $\epsilon \in \Upsilon$, a $G^\epsilon : \Re^n_+ \mapsto \Re^n$ and an $x^\epsilon \in \Sigma^\epsilon := \mathrm{VI}(\Re^n_+, F + G^\epsilon)$ such that x^ϵ converges to some $\bar{x}$ and $G^\epsilon(x^\epsilon)/\epsilon$ converges to some g as $\epsilon \in \Upsilon \to 0$. Then $\bar{x} \in \Sigma := \mathrm{VI}(\Re^n_+, F)$, and the following hold.*

(a). *If F is pseudo-monotone on $\Re^n_+$, then $\bar{x} \in \mathrm{VI}(\Sigma, g)$.*

(b). *If F is analytic on an open set containing $\Re^n_+$, then there exist $\tau > 0$ and $\gamma > 0$ such that*

$$\mathrm{dist}(x^\epsilon, \Sigma) \leq \tau \|G^\epsilon(x^\epsilon)\|^\gamma \tag{11}$$

for all $\epsilon \in \Upsilon$, with $\gamma = 1$ whenever F is affine.

(c). *If F is pseudo-monotone on $\Re^n_+$ and is analytic on an open set containing $\Re^n_+$, then there exist $\tau > 0$ and $\gamma > 0$ such that*

$$(G^\epsilon(x^\epsilon)/\epsilon - g)^T(x^\epsilon - x^*) \leq \tau \|g\| \|G^\epsilon(x^\epsilon)\|^\gamma \tag{12}$$

for all $x^* \in \mathrm{VI}(\Sigma, g)$ and all $\epsilon \in \Upsilon$, with $\gamma = 1$ whenever F is affine.

Proof. Since $x^\epsilon \in \Sigma^\epsilon$ so that $x^\epsilon \geq 0$, $F(x^\epsilon) + G^\epsilon(x^\epsilon) \geq 0$, $(F(x^\epsilon) + G^\epsilon(x^\epsilon))^T x^\epsilon = 0$ for all $\epsilon \in \Upsilon$, we have in the limit (also using $G^\epsilon(x^\epsilon) \to 0$) that $\bar{x} \geq 0$, $F(\bar{x}) \geq 0$, $F(\bar{x})^T \bar{x} = 0$. Thus $\bar{x} \in \Sigma$.

(a). Assume F is pseudo-monotone on $\Re^n_+$. Then Σ is closed convex [5, page 121]. Moreover, for any $y \in \Sigma$, (9) and the fact that $F(y)^T(x^\epsilon - y) \geq 0$ imply

$$0 \leq F(x^\epsilon)^T(x^\epsilon - y) \leq G^\epsilon(x^\epsilon)^T(y - x^\epsilon),$$

where the second inequality uses $x^\epsilon \in \Sigma^\epsilon$ and $y \in \Re^n_+$. Dividing both sides by ϵ yields in the limit that $0 \leq g^T(y - \bar{x})$.

(b). For each $\epsilon > 0$, since $x^\epsilon \in \Sigma^\epsilon$, we have

$$F(x^\epsilon)_I = -G^\epsilon(x^\epsilon)_I, \quad F(x^\epsilon)_{I^c} \geq -G^\epsilon(x^\epsilon)_{I^c}, \quad x^\epsilon_I \geq 0, \quad x^\epsilon_{I^c} = 0, \tag{13}$$

for some $I \subset N$. Let $\Upsilon_I := \{\epsilon \in \Upsilon : (13) \text{ holds}\}$. Consider any $I \subset N$ such that $|\Upsilon_I| = \infty$. Since $G^\epsilon(x^\epsilon) \to 0$ as $\epsilon \in \Upsilon_I \to 0$, then any cluster point x of x^ϵ satisfies

$$F(x)_I = 0, \quad F(x)_{I^c} \geq 0, \quad x_I \geq 0, \quad x_{I^c} = 0. \tag{14}$$

Assume F is analytic on an open set containing $\Re^n_+$. Then an error bound result of Lojasiewicz, as extended by Luo and Pang to analytic systems [33, Theorem 2.2], implies the nonlinear system (14) has a solution y^ϵ satisfying

$$\|x^\epsilon - y^\epsilon\| \le \tau_I \|G^\epsilon(x^\epsilon)\|^{\gamma_I}, \tag{15}$$

where $\tau_I > 0$ and $\gamma_I > 0$ are constants depending on F and I and and $\sup_{\epsilon \in \Upsilon_I} \|x^\epsilon\|$ only. Thus, $y^\epsilon \in \Sigma$ and, moreover, in the case where F is affine, a lemma of Hoffman [25] implies that $\gamma_I = 1$. For any I with $|\Upsilon_I| < \infty$, let y^ϵ be any fixed element of Σ for all $\epsilon \in \Upsilon_I$ and then, for any $\gamma_I \ge 1$, (15) would hold for a suitable τ_I (since its left-hand side is bounded and its right-hand side is bounded away from zero). Taking $\gamma := \min_I \gamma_I$ and $\tau := \max_I \{\sup_{\epsilon \in \Upsilon_I} \tau_I \|G^\epsilon(x^\epsilon)\|^{\gamma_I - \gamma}\}$ yields (11) for all $\epsilon \in \Upsilon = \cup_I \Upsilon_I$, with $\gamma = 1$ whenever F is affine.
(c) Assume F is pseudo-monotone on $\Re^n_+$ and is analytic on an open set containing $\Re^n_+$. Fix any $x^* \in \mathrm{VI}(\Sigma, g)$ and any $\epsilon \in \Upsilon$. Letting $y^\epsilon \in \Sigma$ satisfy $\|x^\epsilon - y^\epsilon\| = \mathrm{dist}(x^\epsilon, \Sigma)$, we have together with (11) in part (b) that

$$\begin{aligned} 0 \le g^T(y^\epsilon - x^*) &= g^T(y^\epsilon - x^\epsilon) + g^T(x^\epsilon - x^*) \\ &\le \tau\|g\|\|G^\epsilon(x^\epsilon)\|^\gamma + g^T(x^\epsilon - x^*) \end{aligned}$$

for some constants $\tau > 0$ and $\gamma > 0$. Also, since $x^* \in \Sigma$, we have from (9) and $x^\epsilon \in \Sigma^\epsilon$ that

$$0 \le F(x^\epsilon)^T(x^\epsilon - x^*), \qquad 0 \le (F(x^\epsilon) + G^\epsilon(x^\epsilon))^T(x^* - x^\epsilon).$$

Adding the above two inequalities to the previous inequality multiplied by ϵ, we obtain

$$0 \le \epsilon\tau\|g\|\|G^\epsilon(x^\epsilon)\|^\gamma + (\epsilon g - G^\epsilon(x^\epsilon))^T(x^\epsilon - x^*).$$

Rearranging terms yields (12). ■

Note 2. Notice that Proposition 2 is stated in the setting of ϵ along a sequence, rather than ϵ in a continuum as in Proposition 1. Although for practical purposes such as analyzing the convergence of an iterative method, the former setting is sufficient, it is nevertheless possible to extend Proposition 2 to the latter setting, provided $\|G^\epsilon(x^\epsilon)\|$ is bounded away from zero whenever ϵ in the continuum is bounded away from zero. Also, Propositions 1 and 2 may possibly be extended to F being piecewise-analytic and, more generally, "subanalytic" [34].

Note 3. In the case where F is an analytic P_0-function and the solution set Σ is nonempty and bounded, [14, Theorem 4.4] implies that x^ϵ is defined and bounded as $\epsilon \to 0$ and so Proposition 2(b) yields that, for any sequence of ϵ along which x^ϵ converges, (11) holds for all ϵ in this sequence, where τ, γ depend on the limit point ($\gamma = 1$ if F is affine). A similar result was shown earlier by Robinson [47, Theorem 2] in the case of F being monotone and affine. If in addition F is polynomial and $G^\epsilon(x) = \epsilon x$, then x^ϵ converges [51]. Bounds of the type (11) were also derived by Fischer [21, Section 3.2] under similar, though not identical, assumptions on F and G^ϵ. Fischer derived his bounds by applying a stability result of Klatte for parametric optimization. In the case where the set-valued mapping $F \rightrightarrows VI(\Re^n_+, F)$ has the *Aubin property* relative to $\{F + G : G : \Re^n_+ \mapsto \Re^n, \sup_{x\in\Re^n_+ : \|x-\bar{x}\|\le 1} \|G(x)\| < 1\}$ at F for $\bar{x}$, a bound similar to (11) with $\gamma = 1$ holds [48, 9F]. However, verifying the Aubin property may be difficult. In the optimization setting, an analogous Lipschitzian property can be shown, under very mild assumptions, for the set of ϵ-approximate solutions [3, Theorem 4.3].

Note 4. In the case where F is pseudo-monotone on $\Re^n_+$ and $G^\epsilon(x) = \epsilon G(x)$ with G continuous at $\bar{x}$, Proposition 2(a),(c) imply $g = G(\bar{x})$ and $\bar{x} \in \mathrm{VI}(\Sigma, G)$. This extends previous results [6, Theorem 2.3], [12, Theorem 5.6.2(b)], [49] for the case of F being monotone or pseudo-monotone and $G(x) = x$ (also see [2, page 63], [8, Proposition 2.6(iii)], [32, Proposition 6.1] for analogous results in the context of maximal monotone operators in an infinite-dimensional space). If F is also analytic on an open set containing $\Re^n_+$, taking $x^* = \bar{x}$ in (12) yields

$$(G(x^\epsilon) - G(\bar{x}))^T(x^\epsilon - \bar{x}) \le (\tau\|G(\bar{x})\|\|G(x^\epsilon)\|^\gamma)\,\epsilon^\gamma. \tag{16}$$

Thus, if in addition G is strictly monotone at $\bar{x}$ in the sense that there exist $\sigma > 0, \varrho > 0, \delta > 0$ such that

$$(G(x) - G(\bar{x}))^T(x - \bar{x}) \ge \sigma\|x - \bar{x}\|^\varrho \qquad \forall x \ge 0 \text{ with } \|x - \bar{x}\| \le \delta, \tag{17}$$

then (16) would yield the error bound that $\|x^\epsilon - \bar{x}\|$ is in the order of $(\epsilon)^{\gamma/\varrho}$ whenever $\|x^\epsilon - \bar{x}\| \le \delta$. Notice that G essentially needs to be continuous at $\bar{x}$ in order to satisfy the assumption that $G^\epsilon(x^\epsilon)/\epsilon$ converges as $x^\epsilon \to \bar{x}$.

In deriving the error bound in Proposition 2(c), we have required $G^\epsilon(x^\epsilon) \to 0$ as $\epsilon \to 0$. This rules out the important case of $G^\epsilon(x) = \epsilon G(x)$, where G is given by (5) or (6) or, more generally, (8) with possibly $\lim_{t\downarrow 0} G_j(t) = -\infty$. In Proposition 3 below, we consider this case and we study properties of any limit point $\bar{x}$ of x^ϵ (see part (a)) and derive error bounds on the distance from x^ϵ to $\bar{x}$ (see parts (c1)–(c4)). In particular, parts (c1), (c2), (c4) of this proposition estimate, under various assumptions on F and $G_1, ..., G_n$, the distance $\|(x-\bar{x})_J\|$ in terms of $\|x_{J^c}\|$, and parts (c3) and (c4) estimate, in the case where F is affine, the latter in terms of ϵ, with J being the set of indices j with $G_j(\bar{x}_j) > -\infty$. While these error bounds may be complex, Example 1 below suggests that this complexity is needed to account for the different (relative) growth rates of $G_1, ..., G_n$ near zero and the linkage among the components of x^ϵ as imposed by the complementarity condition (2).

Proposition 3 *Consider a continuous* $F : \Re^n_+ \mapsto \Re^n$ *and a continuous* $G : \Re^n_{++} \mapsto \Re^n$ *given by* (8), *where* $\lim_{t\downarrow 0} G_j(t) = G_j(0) \in [-\infty, \infty)$ *for all* $j \in N$. *Consider a sequence of positive scalars* $\Upsilon = \{\epsilon^1, \epsilon^2, ...\}$ *tending to zero and, for each* $\epsilon \in \Upsilon$, *an* $x^\epsilon \in \Sigma^\epsilon := \mathrm{VI}(\Re^n_+, F + \epsilon G)$ *converging to some* $\bar{x}$ *as* $\epsilon \in \Upsilon \to 0$. *Then* $\bar{x} \in \Sigma := \mathrm{VI}(\Re^n_+, F)$ *and the following hold with* $J := \{j \in N : G_j(\bar{x}_j) > -\infty\}$, $\bar{\Sigma}$ *being the path-connected component of* Σ *containing* $\bar{x}$, $\bar{J} := \{j \in N : G_j(x_j) > -\infty$ *for some* $x \in \bar{\Sigma}\}$ *and, for each* $I \subset J$, $\Upsilon_I := \{\epsilon \in \Upsilon : F(x^\epsilon)_I + \epsilon G(x^\epsilon)_I = 0, x^\epsilon_{J\setminus I} = 0\}$.

(a). *For each* $J \subset H \subset \bar{J}$ *such that* $x_H \mapsto F(x_H, 0)$ *is pseudo-monotone on* $\Re^{|H|}_+$, *and* $\Sigma_H := \{x_H : (x_H, 0) \in \bar{\Sigma}\}$ *is convex and has an element* $y_H = (y_j)_{j\in H}$ *with* $G_j(y_j) > -\infty$ *for all* $j \in H$, *we have* $\bar{x}_H \in \mathrm{VI}(\Sigma_H, p_H)$ *(respectively,* $H = J$ *and* $\bar{x}_H \in \mathrm{VI}(\Sigma_H, \zeta p_H + G_H)$*) if* $\|q^\epsilon_H\|/\epsilon \to \infty$ *(respectively,* $\|q^\epsilon_H\|/\epsilon \to$ *some* $\zeta \in \Re_+$*) as* $\epsilon \to 0$ *along some subsequence of* Υ, *where* p_H *is any cluster point of* $q^\epsilon_H/\|q^\epsilon_H\|$ *along this subsequence and* $q^\epsilon := F(x^\epsilon) - F(x^\epsilon_H, 0)$.

(b). *If* F *is analytic on an open set containing* $\Re^n_+$, *then there exist* $\tau > 0$ *and* $\gamma > 0$ *such that*

$$\mathrm{dist}(x^\epsilon, \Sigma) \le \tau(\epsilon\|G(x^\epsilon)_J\| + \|x^\epsilon_{J^c}\|)^\gamma \tag{18}$$

for all $\epsilon \in \Upsilon$, *with* $\gamma = 1$ *whenever* F *is affine.*

(c1). *If* F *is pseudo-monotone on* $\Re^n_+$ *and* $\lim_{t\to 0} tG_j(t) = 0$ *for all* $j \in \bar{J}^c$, *then* $\bar{J} = J$ *and* $\bar{x} \in \mathrm{VI}(\Sigma, G)$. *If in addition* F *is analytic on an open set containing* $\Re^n_+$, *then there exist* $\tau > 0$ *and* $\gamma > 0$ *such that*

$$(G(x^\epsilon) - G(\bar{x}))^T_J(x^\epsilon - \bar{x})_J \le \tau(\epsilon\|G(x^\epsilon)_J\| + \|x^\epsilon_{J^c}\|)^\gamma - G(x^\epsilon)^T_{J^c}x^\epsilon_{J^c} \tag{19}$$

for all $\epsilon \in \Upsilon$, *with* $\gamma = 1$ *whenever* F *is affine.*

(c2). *If* $F(x) = Mx + q$ *for some* $M \in \Re^{n\times n}, q \in \Re^n$, *with* $M_{\bar{J}\bar{J}^c} \in M_{\bar{J}\bar{J}}(\Re^{|\bar{J}|\times|\bar{J}^c|})$ *and* $M_{\bar{J}\bar{J}} \in M^T_{\bar{J}\bar{J}}(\Re^{|\bar{J}|\times|\bar{J}|})$, *and if* $\bar{\Sigma}$ *is convex, then* $\bar{J} = J$ *and* $\bar{x} \in \mathrm{VI}(\bar{\Sigma}, G)$. *Moreover, for each* $I \subset J$ *with* $|\Upsilon_I| = \infty$ *and* M_{II} *positive semidefinite and* $M_{IJ^c} = M_{II}N_{IJ^c}$ *for some* $N_{IJ^c} \in \Re^{|I|\times|J^c|}$, *there exists* $\tau_I > 0$ *such that*

$$(G(x^\epsilon) - G(\bar{x}))^T_J(x^\epsilon - \bar{x})_J \le \tau_I(\epsilon\|G(x^\epsilon)_J\| + \|x^\epsilon_{J^c}\|) - G(x^\epsilon)^T_I N_{IJ^c}x^\epsilon_{J^c} \tag{20}$$

for all $\epsilon \in \Upsilon_I$.

(c3). *If* $F(x) = Mx + q$ *for some* $M \subset \Re^{n\times n}$, $q \in \Re^n$, *and if* $\bar{\Sigma}$ *is convex and* G_j *is strictly increasing for all* $j \in J^c$, *then there exist* $\rho > 0$ *and* $\tau > 0$ *such that*

$$x^\epsilon_j \le G_j^{-1}(-\rho/\epsilon)\ \forall j \in K, \quad \|x^\epsilon_{L\setminus\bar{J}}\| \le \tau\left(\|x^\epsilon_K\| + \sum_{j\in L} h_j(\epsilon) + \epsilon\right) \tag{21}$$

for all $\epsilon \in \Upsilon$ *sufficiently small, where* $K \subset J^c$ *(*K *may depend on* ϵ*),* $L := J^c \backslash K$*, and* $h_j(\epsilon)$ *is the unique* $\eta > 0$ *satisfying* $G_j^{-1}(-\eta/\epsilon) = \eta$ *for* $j \in J^c$*. If* $M = -M^T$*, then (21) remains true without the* h_j *terms.*

(c4). *If* $F(x) = Mx + q$ *for some* $M \subset \Re^{n \times n}$*,* $q \in \Re^n$*, with* $M_{\bar{J}\bar{J}}$ *positive semidefinite, and if* G_j *is strictly increasing with* $\limsup_{t \to 0} tG_j(t) < 0$ *for all* $j \in J^c$*, then* Υ *is the union of a countable collection of subsequences* $\tilde{\Upsilon}$ *for each of which there exist* $\tau > 0, \rho > 0$*, properly nested* $\bar{J} = H_1 \supset \cdots \supset H_{r-1} \supset H_r = J$*, and* $\lambda_l^\epsilon > 0$ *depending on* $M_{N(H_l)^c} x^\epsilon_{(H_l)^c}$ *and* ϵ *only (*$l = 1, .., r$*,* $\epsilon \in \tilde{\Upsilon}$*) such that* $\lambda_l^\epsilon \to 0$ *as* $\epsilon \in \tilde{\Upsilon} \to 0$ *and, for each* $l = 1, .., r-1$*,*

$$\text{either} \quad x_j^\epsilon \le G_j^{-1}(-\rho/\lambda_l^\epsilon) \quad \text{or} \quad x_j^\epsilon \le \tau \lambda_l^\epsilon \quad \forall j \in H_l \backslash H_{l+1} \tag{22}$$

for all $\epsilon \in \tilde{\Upsilon}$ *sufficiently small. And if in addition* G_J *is locally Lipschitzian at* $\bar{x}_J$ *with constant* $\kappa > 0$ *and satisfies (17) with* $\sigma > 0, \delta > 0, \varrho = 1$*, then there exist* $\tau' > 0$ *(independent of* κ, σ*) such that*

$$\|(x^\epsilon - \bar{x})_J\| \le \tau'(\kappa/\sigma + 1 + 1/\sigma)\lambda_r^\epsilon \tag{23}$$

for all $\epsilon \in \tilde{\Upsilon}$ *sufficiently small.*

Proof. For each $j \in J$, we have $G_j(x_j^\epsilon) \to G_j(\bar{x}_j)$ as $\epsilon \in \Upsilon \to 0$ so the fact x^ϵ satisfies (2) for all $\epsilon \in \Upsilon$ yields in the limit that $\bar{x}_j \ge 0$, $F(\bar{x})_j \ge 0$, $F(\bar{x})_j \bar{x}_j = 0$. For each $j \in J^c$, we have $\bar{x}_j = 0$ and $F(x^\epsilon)_j = -\epsilon G_j(x_j^\epsilon) \ge 0$ for all $\epsilon \in \Upsilon$ sufficiently small. The latter yields in the limit $F(\bar{x})_j \ge 0$. Thus $\bar{x}$ satisfies (1) and hence $\bar{x} \in \Sigma$.

(a). Consider any $J \subset H \subset \bar{J}$ such that $x_H \mapsto F(x_H, 0)$ is pseudo-monotone on $\Re_+^{|H|}$, and Σ_H is convex and contains an element $y_H = (y_j)_{j \in H}$ with $G_j(y_j) > -\infty$ for all $j \in H$. Then, we have from $(y_H, 0) \in \Sigma$ and $x^\epsilon \in \Sigma^\epsilon$ that

$$0 \le F(y_H, 0)_H^T (x_H^\epsilon - y_H), \quad 0 \le (F(x_H^\epsilon, 0)_H + q_H^\epsilon + \epsilon G(x^\epsilon)_H)^T (y_H - x_H^\epsilon),$$

with $q^\epsilon := F(x^\epsilon) - F(x_H^\epsilon, 0)$. Since $x_H^\epsilon \in \Re_+^{|H|}$, the first inequality and the pseudo-monotonicity of $x_H \mapsto F(x_H, 0)$ on $\Re_+^{|H|}$ imply $0 \le F(x_H^\epsilon, 0)_H^T (x_H^\epsilon - y_H)$, which when added to the second inequality yields

$$0 \le (q_H^\epsilon + \epsilon G(x^\epsilon)_H)^T (y_H - x_H^\epsilon). \tag{24}$$

Consider any subsequence of Υ along which either (i) $\|q_H^\epsilon\|/\epsilon \to \infty$ or (ii) $\|q_H^\epsilon\|/\epsilon \to \zeta \in \Re_+$, and let p_H be any cluster point of $q_H^\epsilon / \|q_H^\epsilon\|$ along this subsequence. In case (i), dividing both sides of (24) by $\|q_H^\epsilon\|$ and using $y_j > 0 = \bar{x}_j$ and $G_j(x_j^\epsilon) \to -\infty$ for all $j \in J^c$ yield in the limit that

$$0 \le p_H^T (y_H - \bar{x}_H).$$

Since Σ_H is convex, this holds for all $y_H \in \Sigma_H$, so $\bar{x}_H \in \mathrm{VI}(\Sigma_H, p_H)$. In case (ii), dividing both sides of (24) by ϵ and arguing as in case (i) yield in the limit that $H = J$ and

$$0 \le (\zeta p_H + G(\bar{x})_H)^T (y_H - \bar{x}_H).$$

Since Σ_H is convex, this holds for all $y_H \in \Sigma_H$, so $\bar{x}_H \in \mathrm{VI}(\Sigma_H, \zeta p_H + G_H)$.
(b). For each $j \in J^c$, we have $x_j^\epsilon > 0$ for all $\epsilon \in \Upsilon$ and $G_j(x_j^\epsilon) < 0$ for all $\epsilon \in \Upsilon$ below some $\bar{\epsilon}$. Consider any $I \subset J$ such that $|\Upsilon_I| = \infty$. For each $\epsilon \in \Upsilon_I$ below $\bar{\epsilon}$, since $x^\epsilon \in \Sigma^\epsilon$, we have

$$\begin{array}{lll} F(x^\epsilon)_I = -\epsilon G(x^\epsilon)_I, & F(x^\epsilon)_{J\setminus I} \ge -\epsilon G(x^\epsilon)_{J\setminus I}, & F(x^\epsilon)_{J^c} \ge 0, \\ x_I^\epsilon \ge 0, & x_{J\setminus I}^\epsilon = 0, & x_{J^c}^\epsilon = x_{J^c}^\epsilon. \end{array} \tag{25}$$

Since $F(x^\epsilon) \to F(\bar{x})$ and $G(x^\epsilon)_J \to G(\bar{x})_J$ as $\epsilon \in \Upsilon_I \to 0$, (25) yields in the limit that $\bar{x}$ satisfies

$$F(x)_I = 0, \quad F(x)_{I^c} \ge 0, \quad x_I \ge 0, \quad x_{I^c} = 0. \tag{26}$$

Assume F is analytic on an open set containing $\Re^n_+$. Then an error bound result of Lojasiewicz, as extended by Luo and Pang to analytic systems [33, Theorem 2.2], implies the nonlinear system (26) has a solution y^ϵ satisfying

$$\|y^\epsilon - x^\epsilon\| \le \tau_I(\epsilon\|G(x^\epsilon)_J\| + \|x_{J^c}^\epsilon\|)^{\gamma_I}, \tag{27}$$

where $\tau_I > 0$ and $\gamma_I > 0$ are constants depending on F and I and $\sup_{\epsilon\in\Upsilon_I} \|x^\epsilon\|$ only. Thus, $y^\epsilon \in \Sigma$ and, in the case where F is affine, a lemma of Hoffman [25] implies $\gamma_I = 1$. If Υ_I is finite but nonempty, let y^ϵ be any fixed element of Σ for all $\epsilon \in \Upsilon_I$ and then (27) would hold for any $\gamma_I \ge 1$ and a sufficiently large τ_I. Taking $\gamma := \min_I \gamma_I$ and

$$\tau := \max_I \left\{ \sup_{\epsilon\in\Upsilon_I} \tau_I(\epsilon\|G(x^\epsilon)_J\| + \|x_{J^c}^\epsilon\|)^{\gamma_I - \gamma} \right\}$$

yields (18) for all $\epsilon \in \Upsilon = \cup_I \Upsilon_I$.
(c1). Assume F is pseudo-monotone on $\Re^n_+$ and $\lim_{t\to 0} tG_j(t) = 0$ for all $j \in \bar{J}^c$. Then Σ is closed convex [5, page 121] so $\Sigma = \bar{\Sigma}$ and there exists $y \in \Sigma$ such that $G_j(y_j) > -\infty$ for all $j \in \bar{J}$. For each $\epsilon \in \Upsilon$, since $y \in \Sigma$, (9) and $x^\epsilon \in \Sigma^\epsilon$ imply

$$0 \le F(x^\epsilon)^T(x^\epsilon - y) \le \epsilon G(x^\epsilon)^T(y - x^\epsilon). \tag{28}$$

Also, by (a), $\bar{x} \in \Sigma$, so $J \subset \bar{J}$. For each $j \in J$, we have $G_j(x_j^\epsilon)(y_j - x_j^\epsilon)$ converges as $\epsilon \in \Upsilon \to 0$. For each $j \in \bar{J}\setminus J$, we have $y_j > 0$ and $G_j(x_j^\epsilon) \to -\infty$, so $G_j(x_j^\epsilon)(y_j - x_j^\epsilon) \to -\infty$ as $\epsilon \in \Upsilon \to 0$. For each $j \in \bar{J}^c$, we have $x_j^\epsilon \to y_j = 0$ so our assumption on G_j yields $G_j(x_j^\epsilon)(y_j - x_j^\epsilon) \to 0$ as $\epsilon \in \Upsilon \to 0$. Hence, (28) implies $\bar{J}\setminus J = \emptyset$, i.e., $J = \bar{J}$. Now, for any $y \in \Sigma$, (9) and $x^\epsilon \in \Sigma^\epsilon$ imply (28) holds. Dividing both sides of (28) by ϵ and using $y_j = 0$ and $G_j(x_j^\epsilon)x_j^\epsilon \to 0$ for all $j \in \bar{J}^c = J^c$ yields in the limit that

$$0 \le G(\bar{x})_J^T(y - \bar{x})_J = G(\bar{x})^T(y - \bar{x})$$

(here $\infty \cdot 0 = 0$), so $\bar{x} \in \mathrm{VI}(\Sigma, G)$.
Assume in addition F is analytic on an open set containing $\Re_+^n$, so that, by part (a), there exist $\tau > 0$ and $\gamma > 0$ such that (18) holds for all $\epsilon \in \Upsilon$. Let $y^\epsilon \in \Sigma$ satisfy $\|x^\epsilon - y^\epsilon\| = \mathrm{dist}(x^\epsilon, \Sigma)$. Now, for each $\epsilon \in \Upsilon$, since $\bar{x} \in \Sigma$, (9) and $x^\epsilon \in \Sigma^\epsilon$ imply

$$0 \le F(x^\epsilon)^T(x^\epsilon - \bar{x}), \qquad 0 \le (F(x^\epsilon) + \epsilon G(x^\epsilon))^T(\bar{x} - x^\epsilon).$$

Adding these two inequalities and dividing by ϵ gives $0 \le G(x^\epsilon)^T(\bar{x} - x^\epsilon)$. Also, $\bar{x} \in \mathrm{VI}(\Sigma, G)$ and $y^\epsilon \in \Sigma$ imply $0 \le G(\bar{x})_J^T(y^\epsilon - \bar{x})_J$ (since $\bar{J} = J$). Adding these two inequalities and using $\bar{x}_{J^c} = 0$ gives

$$0 \le G(\bar{x})_J^T(y^\epsilon - x^\epsilon)_J + (G(\bar{x}) - G(x^\epsilon))_J^T(x^\epsilon - \bar{x})_J - G(x^\epsilon)_{J^c}^T x_{J^c}^\epsilon.$$

Combining this with (18) and renaming $\tau\|G(\bar{x})_J\|$ as τ yields (19).
(c2). Assume $F(x) = Mx + q$ for some $M \in \Re^{n \times n}, q \in \Re^n$, and assume $\bar{\Sigma}$ is convex. Consider any $I \subset J$ such that $|\Upsilon_I| = \infty$ and $M_{\bar{J}\bar{J}^c} = M_{\bar{J}\bar{J}}N_{\bar{J}\bar{J}^c}$ and $M_{\bar{J}\bar{J}} = M_{\bar{J}\bar{J}}^T N_{\bar{J}\bar{J}}$ for some $N_{\bar{J}\bar{J}^c} \in \Re^{|\bar{J}| \times |\bar{J}^c|}$ and some $N_{\bar{J}\bar{J}} \in \Re^{|\bar{J}| \times |\bar{J}|}$. First, we have $M_{\bar{J}}y + q_{\bar{J}} = 0$ for all $y \in \bar{\Sigma}$. [If $M_i y + q_i > 0$ for some $i \in N$ and some $y \in \bar{\Sigma}$, then the convexity of $\bar{\Sigma}$ would imply $x_i = 0$ for all $x \in \bar{\Sigma}$, so $i \in \bar{J}^c$.] Fix any $y \in \bar{\Sigma}$. Since $\bar{x} \in \bar{\Sigma}$, then $d = y - \bar{x}$ satisfies $d_{\bar{J}^c} = 0$ and $M_{\bar{J}\bar{J}}d_{\bar{J}} = 0$. Moreover, for each $i \in \bar{J}$ with $\bar{x}_i = 0$ we have $d_i \ge 0$. Thus, there exists $\alpha > 0$ such that $z^\epsilon := x^\epsilon + \alpha d \ge 0$ for all $\epsilon \in \Upsilon$ sufficiently small. Then, $x^\epsilon \in \Sigma^\epsilon$ and $\bar{x} \in \bar{\Sigma}$ imply

$$\begin{aligned}
0 &\le (Mx^\epsilon + q + \epsilon G(x^\epsilon))^T(z^\epsilon - x^\epsilon) \\
&= \alpha(M_{\bar{J}}x^\epsilon + q_{\bar{J}} + \epsilon G(x^\epsilon)_{\bar{J}})^T d_{\bar{J}} \\
&= \alpha(M_{\bar{J}}(x^\epsilon - \bar{x}) + \epsilon G(x^\epsilon)_{\bar{J}})^T d_{\bar{J}} \\
&= \alpha\left(M_{\bar{J}\bar{J}}((x^\epsilon - \bar{x})_{\bar{J}} + N_{\bar{J}\bar{J}^c}(x^\epsilon - \bar{x})_{\bar{J}^c})) + \epsilon G(x^\epsilon)_{\bar{J}}\right)^T d_{\bar{J}} \\
&= \epsilon\alpha G(x^\epsilon)_{\bar{J}}^T d_{\bar{J}},
\end{aligned}$$

where the last equality uses $M_{\bar{J}\bar{J}}^T d_{\bar{J}} = N_{\bar{J}\bar{J}}^T M_{\bar{J}\bar{J}} d_{\bar{J}} = 0$. This shows that $G(x^\epsilon)_{\bar{J}}^T d_{\bar{J}} \ge 0$. Since $\bar{x} \in \bar{\Sigma}$ so that $\bar{J} \supset J$, if $\bar{J} \ne J$, then the convexity of $\bar{\Sigma}$ would imply the existence of $y \in \bar{\Sigma}$ with $y_{\bar{J}\setminus J} > 0$. Using this y in the above argument would yield $d_{\bar{J}\setminus J} > 0$ and hence $G(x^\epsilon)_{\bar{J}}^T d_{\bar{J}} \to -\infty$, a contradiction. Thus $\bar{J} = J$. Then, G_j is continuous at $\bar{x}_j$ for all $j \in \bar{J}$ and the above inequality yields in the limit as $\epsilon \to 0$ that

$$0 \le G(\bar{x})_{\bar{J}}^T d_{\bar{J}} = G(\bar{x})^T(y - \bar{x})$$

(here $\infty \cdot 0 = 0$), so $\bar{x} \in \mathrm{VI}(\bar{\Sigma}, G)$.
Consider any $I \subset J$ with $|\Upsilon_I| = \infty$, M_{II} positive semidefinite and $M_{IJ^c} = M_{II}N_{IJ^c}$ for some $N_{IJ^c} \in \Re^{|I|\times|J^c|}$. Then $\bar{x}$ satisfies (26). For each $\epsilon \in \Upsilon_I$, Hoffman's lemma implies (26) has a solution y^ϵ satisfying (27), with $\tau_I > 0$ and $\gamma_I = 1$. Then $y^\epsilon \in \bar{\Sigma}$ (since the line segment joining $\bar{x}$ and y^ϵ lies in Σ) so the fact $\bar{x} \in \mathrm{VI}(\bar{\Sigma}, G)$ implies

$$\begin{aligned} 0 &\le G(\bar{x})^T(y^\epsilon - \bar{x}) \\ &= G(\bar{x})_I^T(y^\epsilon - x^\epsilon)_I + G(\bar{x})_I^T(x^\epsilon - \bar{x})_I \\ &\le \tau_I\|G(\bar{x})_I\|(\epsilon\|G(x^\epsilon)_J\| + \|x^\epsilon_{J^c}\|) + G(\bar{x})_I^T(x^\epsilon - \bar{x})_I. \end{aligned} \tag{29}$$

Also, we have from (25) and $\bar{x}$ satisfying (26) that

$$0 = (M\bar{x} + q)_I, \qquad 0 = (Mx^\epsilon + q + \epsilon G(x^\epsilon))_I,$$

which when subtracted and using $x^\epsilon_{J\setminus I} = 0$ and $\bar{x}_{I^c} = 0$ yields $0 = M_{II}((x^\epsilon - \bar{x})_I + N_{IJ^c}x^\epsilon_{J^c}) + \epsilon G(x^\epsilon)_I$. This and the positive semidefinite property of M_{II} yield

$$\begin{aligned} 0 &\le ((x^\epsilon - \bar{x})_I + N_{IJ^c}x^\epsilon_{J^c})^T M_{II}((x^\epsilon - \bar{x})_I + N_{IJ^c}x^\epsilon_{J^c}) \\ &= -\epsilon((x^\epsilon - \bar{x})_I + N_{IJ^c}x^\epsilon_{J^c})^T G(x^\epsilon)_I. \end{aligned}$$

Dividing the above inequality by ϵ and adding it to the inequality (29), we obtain

$$0 \le \tau_I\|G(\bar{x})_I\|(\epsilon\|G(x^\epsilon)_J\| + \|x^\epsilon_{J^c}\|) + (G(\bar{x}) - G(x^\epsilon))_I^T(x^\epsilon - \bar{x})_I - G(x^\epsilon)_I^T N_{IJ^c}x^\epsilon_{J^c}.$$

Using $(x^\epsilon - \bar{x})_{J\setminus I} = 0$ and renaming $\tau_I\|G(\bar{x})_I\|$ as τ_I yield (20).
(c3). Assume $F(x) = Mx + q$ for some $M \subset \Re^{n\times n}$, $q \in \Re^n$, and assume $\bar{\Sigma}$ is convex and each G_j is strictly increasing. Fix any $I \subset J$ such that $|\Upsilon_I| = \infty$. For each $\epsilon \in \Upsilon_I$, Hoffman's lemma implies (26) has a solution y^ϵ satisfying (27) with $\tau_I > 0$ and $\gamma_I = 1$. Let Ψ denote the set of $x \in \Re^n$ satisfying (26) and $\|x\| \le \sup_{\epsilon\in\Upsilon_I}\|y^\epsilon\|$. We claim that there exists a scalar $\rho > 0$ such that, for every $y \in \Psi$ there exists a $K \subset J^c$ such that

$$M_j y + q_j > 2\rho \ \forall j \in K \quad \text{and} \quad M_L\psi + q_L = 0 \text{ for some } \psi \in \Psi, \tag{30}$$

where $L := J^c\setminus K$ (cf. proof of [52, Proposition 2]). If not, then for every sequence of scalars $\rho^k > 0$, $k = 1, 2, ...$, tending to zero, there would exist a $\nu^k \in \Psi$ such that, for every $K \subset J^c$ we have

$$M_j\nu^k + q_j \le 2\rho^k \ \text{ for some } j \in K \quad \text{or} \quad M_L\psi + q_L \ne 0 \ \forall \psi \in \Psi,$$

where $L := J^c\setminus K$. Since Ψ is bounded and closed, then ν^k, $k = 1, 2, ...$, has a cluster point $\nu \in \Psi$ such that, for every $K \subset J^c$ we have

$$M_j\nu + q_j = 0 \ \text{ for some } j \in K \quad \text{or} \quad M_L\psi + q_L \ne 0 \ \forall \psi \in \Psi,$$

where $L := J^c \backslash K$. However, this cannot be true since the above relations fail to hold for $K = \{j \in J^c : M_j \nu + q_j > 0\}$ and $\psi = \nu$.
For each $\epsilon \in \Upsilon_I$, we have $y^\epsilon \in \Psi$ and hence there exists a $K \subset J^c$ such that (30) holds with $y = y^\epsilon$ and $L := J^c \backslash K$. Since the number of such subset K is finite, by passing into a subsequence if necessary, we can assume it is the same K for all $\epsilon \in \Upsilon_I$. Since (27) and $x^\epsilon \to \bar{x}$ imply $x^\epsilon - y^\epsilon \to 0$ as $\epsilon \in \Upsilon_I \to 0$, we have $M_j x^\epsilon + q_j \geq \rho$ for all $j \in K$ and sufficiently small $\epsilon \in \Upsilon_I$, in which case $M_j x^\epsilon + q_j + \epsilon G_j(x_j^\epsilon) = 0$ and the strictly increasing property of G_j would imply

$$x_j^\epsilon = G_j^{-1}(-(M_j x^\epsilon + q_j)/\epsilon) \leq G_j^{-1}(-\rho/\epsilon) \quad \forall j \in K. \tag{31}$$

For each $\epsilon \in \Upsilon_I$, let $L_1 := \{j \in L : M_j x^\epsilon + q_j \geq h_j(\epsilon)\}$ and let $L_2 := L \backslash L_1$. Since there is only a finite number of different L_1 and L_2, by passing to a subsequence if necessary, we can assume that L_1 and L_2 are the same for all $\epsilon \in \Upsilon_I$. Then, we have as argued above that

$$x_j^\epsilon \leq G_j^{-1}(-h_j(\epsilon)/\epsilon) = h_j(\epsilon) \quad \forall j \in L_1. \tag{32}$$

We claim there exists constant $\tau_1 > 0$ such that

$$\|x^\epsilon_{L_2 \backslash \bar{J}}\| \leq \tau_1 (\|x^\epsilon_{K \cup L_1}\| + \sum_{j \in L_2} h_j(\epsilon) + \epsilon) \tag{33}$$

for all $\epsilon \in \Upsilon_I$. If not, then there would exist a subsequence of $\epsilon \in \Upsilon_I$ along which $(\|x^\epsilon_{K \cup L_1}\| + \sum_{j \in L_2} h_j(\epsilon) + \epsilon)/\|x^\epsilon_{L_2 \backslash \bar{J}}\| \to 0$. By (30), there exists $\psi \in \Psi$ satisfying $M_L \psi + q_L = 0$. Then ψ would satisfy (26), which together with (25) implies

$$\begin{aligned}
M_I(x^\epsilon - \psi) &= -\epsilon G(x^\epsilon)_I, & (x^\epsilon - \psi)_{I'} &\geq 0, \\
M_{J'}(x^\epsilon - \psi) &\geq -\epsilon G(x^\epsilon)_{J'}, & (x^\epsilon - \psi)_{J \backslash I} &= 0, \\
M_{K' \cup L_1}(x^\epsilon - \psi) &\geq 0, & (x^\epsilon - \psi)_{K \cup L_1} &= x^\epsilon_{K \cup L_1}, \\
M_{L_2}(x^\epsilon - \psi) &= M_{L_2} x^\epsilon + q_{L_2}, & (x^\epsilon - \psi)_{L_2} &\geq 0,
\end{aligned}$$

where $I' := \{i \in I : \psi_i = 0\}$, $J' := \{i \in J \backslash I : M_i \psi + q_i = 0\}$, and $K' := \{i \in K : M_i \psi + q_i = 0\}$. Dividing both sides by $\|x^\epsilon_{L_2 \backslash \bar{J}}\|$ and using $0 \leq M_j x^\epsilon + q_j < h_j(\epsilon)$ for $j \in L_2$ would yield in the limit that

$$\begin{aligned}
M_I u &= 0, & u_{I'} &\geq 0, \\
M_{J'} u &\geq 0, & u_{J \backslash I} &= 0, \\
M_{K' \cup L_1} u &\geq 0, & u_{K \cup L_1} &= 0, \\
M_{L_2} u &= 0, & u_{L_2} &\geq 0,
\end{aligned}$$

for some $u \in \Re^n$ with $u_{L_2 \backslash \bar{J}} \neq 0$. Then, since $\psi \in \Psi$ and $M_L \psi + q_L = 0$, the vector $\psi + \alpha u$ would be in Σ for all $\alpha > 0$ sufficiently small. Since ψ and $\bar{x}$ satisfy (26) so that $\psi \in \bar{\Sigma}$, this vector is also in $\bar{\Sigma}$. Since $u_{L_2 \backslash \bar{J}} \neq 0$, this would contradict the fact that $x_{\bar{J}^c} = 0$ for all $x \in \bar{\Sigma}$.

Assume $M = -M^T$. We claim that there exists constant $\tau_1 > 0$ such that

$$\|x^\epsilon_{L\setminus \bar{J}}\| \leq \tau_1(\|x^\epsilon_K\| + \epsilon) \tag{34}$$

for all $\epsilon \in \Upsilon_I$. If not, then there would exist a subsequence of $\epsilon \in \Upsilon_I$ along which $(\|x^\epsilon_K\| + \epsilon)/\|x^\epsilon_{L\setminus \bar{J}}\| \to 0$. By (30), there exists $\psi \in \Psi$ satisfying $M_L\psi + q_L = 0$. Then ψ would satisfy (26), which together with (25) implies

$$\begin{array}{rlrl} M_I(x^\epsilon - \psi) & = -\epsilon G(x^\epsilon)_I, & (x^\epsilon - \psi)_{I'} & \geq 0, \\ M_{J'}(x^\epsilon - \psi) & \geq -\epsilon G(x^\epsilon)_{J'}, & (x^\epsilon - \psi)_{J\setminus I} & = 0, \\ M_{K'}(x^\epsilon - \psi) & \geq 0, & (x^\epsilon - \psi)_K & = x^\epsilon_K, \\ M_L(x^\epsilon - \psi) & \geq 0 & (x^\epsilon - \psi)_L & \geq 0. \end{array}$$

where $I' := \{i \in I : \psi_i = 0\}$, $J' := \{i \in J\setminus I : M_i\psi + q_i = 0\}$, and $K' = \{i \in K : M_i\psi + q_i = 0\}$. Dividing both sides by $\|x^\epsilon_{L\setminus \bar{J}}\|$ would yield in the limit that

$$\begin{array}{rlrl} M_I u & = 0, & u_{I'} & \geq 0, \\ M_{J'} u & \geq 0, & u_{J\setminus I} & = 0, \\ M_{K'} u & \geq 0, & u_K & = 0, \\ M_L u & \geq 0, & u_L & \geq 0, \end{array}$$

for some $u \in \Re^n$ with $u_{L\setminus \bar{J}} \neq 0$. Since $M = -M^T$ so that $u^T M u = 0$, the above implies $u_j(Mu)_j = 0$ for all $j \in L$. Then, since $\psi \in \Psi$ and $M_L\psi + q_L = 0$, the vector $\psi + \alpha u$ would be in Σ for all $\alpha > 0$ sufficiently small and hence, as argued earlier, would be in $\bar{\Sigma}$. Since $u_{L\setminus \bar{J}} \neq 0$, this would contradict the fact that $x_{\bar{J}^c} = 0$ for all $x \in \bar{\Sigma}$.

(c4). Assume $F(x) = Mx + q$ for some $M \subset \Re^{n\times n}$, $q \in \Re^n$, with $M_{\bar{J}\bar{J}}$ positive semidefinite. Also assume G_j is strictly increasing with $\limsup_{t\to 0} tG_j(t) < 0$ for all $j \in J^c$. Fix any $I \subset J$ such that $|\Upsilon_I| = \infty$. Let $H_1 := \bar{J}$. Initialize $\tilde{\Upsilon}$ to comprise all sufficiently small $\epsilon \in \Upsilon_I$ so that $G_j(x^\epsilon_j) < 0$ for all $j \in J^c$.

Given H_l for some $l \geq 1$, we construct below (by passing to a subsequence of $\tilde{\Upsilon}$ if necessary) a proper subset H_{l+1} of H_l having the desired properties (22), until $H_l = J$. For notational simplicity, we will write H_l and H_{l+1} as H and H^{new} respectively, dropping the subscript l. First, by passing to a subsequence if necessary, we assume there exist $q_h \in \Re^n$ and $\alpha^\epsilon_h > 0$ ($h = 1, ..., n$, $\epsilon \in \tilde{\Upsilon}$) satisfying

$$M_{NH^c}x^\epsilon_{H^c} = \sum_{h=1}^n \alpha^\epsilon_h q_h \ \forall \epsilon \in \tilde{\Upsilon}, \quad \alpha^\epsilon_1 \to 0, \ (\frac{\alpha^\epsilon_{h+1}}{\alpha^\epsilon_h})_{h=1,\dots,n-1} \to 0 \text{ as } \epsilon \in \tilde{\Upsilon} \to 0. \tag{35}$$

[To see that such a decomposition exists, let $q^\epsilon := M_{NH^c}x^\epsilon_{H^c}$. If $q^\epsilon = 0$ for all $\epsilon \in \tilde{\Upsilon}$ small, then choose $q_h = 0$ and $\alpha^\epsilon_h = (\epsilon)^h$ for $h = 1, ..., n$. Otherwise take any subsequence of $\tilde{\Upsilon}$ along which $q^\epsilon \neq 0$ and $q^\epsilon/\|q^\epsilon\|$ converges. Let q_1 be its limit and let $\tilde{q}^\epsilon$ be the orthogonal projection of q^ϵ onto the subspace orthogonal to q_1. Then $\tilde{q}^\epsilon = q^\epsilon - \alpha_1^\epsilon q_1$ for some α_1^ϵ satisfying $\|q^\epsilon\|/\alpha_1^\epsilon \to 1$ along the subsequence. Apply the above construction inductively to $\tilde{q}^\epsilon$ (restricted to the above subspace) yields (35).] For each $\epsilon \in \tilde{\Upsilon}$, we have from $\epsilon \in \Upsilon_I$ and $G_j(x_j^\epsilon) < 0$ for all $j \in J^c$ and $J \subset H$ that x^ϵ satisfies (cf. (25))

$$\begin{array}{rlrl} M_{LH}x_H^\epsilon + M_{LH^c}x_{H^c}^\epsilon + q_L & = -\epsilon G(x^\epsilon)_L, & x_I^\epsilon & \geq 0, \\ M_{J\backslash IH}x_H^\epsilon + M_{J\backslash IH^c}x_{H^c}^\epsilon + q_{J\backslash I} & \geq -\epsilon G(x^\epsilon)_{J\backslash I}, & x_{J\backslash I}^\epsilon & = 0, \\ M_{H^cH}x_H^\epsilon + M_{H^cH^c}x_{H^c}^\epsilon + q_{H^c} & \geq 0, & x_{H\backslash J}^\epsilon & > 0, \end{array}$$

where for convenience we let $L := I \cup (H\backslash J)$. Letting $\alpha_0^\epsilon := 1$, $q_0 := q$, $I_0 := J\backslash I$, $K_0 := I$ and $L_0 := H^c$ and $y_0^\epsilon := x^\epsilon$, we see from the above relations and (35) that the following holds with $k = 0$:

$$\begin{array}{ll} M_{LH}(y_k^\epsilon)_H + \sum_{h=k}^n \alpha_h^\epsilon (q_h)_L = -\epsilon G(x^\epsilon)_L, & (y_k^\epsilon)_{K_k} \geq 0, \\ M_{I_kH}(y_k^\epsilon)_H + \sum_{h=k}^n \alpha_h^\epsilon (q_h)_{I_k} \geq -\epsilon G(x^\epsilon)_{I_k}, & (y_k^\epsilon)_{J\backslash I} = 0, \\ M_{L_kH}(y_k^\epsilon)_H + \sum_{h=k}^n \alpha_h^\epsilon (q_h)_{L_k} \geq 0, & (y_k^\epsilon)_{H\backslash J} > 0, \end{array} \tag{36}$$

$$y_k^\epsilon = x^\epsilon - \sum_{h=0}^{k-1} \alpha_h^\epsilon u_h, \qquad I_k \subset J, \tag{37}$$

for all $\epsilon \in \tilde{\Upsilon}$. Now, suppose that (36)–(37) hold for some $k \geq 0$ for all $\epsilon \in \tilde{\Upsilon}$. By further passing to a subsequence if necessary, we can assume one of the following two cases occurs.

Case 1. There exist $j \in H\backslash J$ and $\rho > 0$ such that $\epsilon G_j(x_j^\epsilon)/\alpha_k^\epsilon \leq -\rho$ for all $\epsilon \in \tilde{\Upsilon}$.
In this case, let $H^{new} := H\backslash\{j\}$ and we have that H^{new} is a proper subset of H and contains J. Moreover, the strictly increasing property of G_j implies

$$x_j^\epsilon \leq G_j^{-1}(-\rho\alpha_k^\epsilon/\epsilon). \tag{38}$$

Case 2. For all $j \in H\backslash J$, $\epsilon G_j(x_j^\epsilon)/\alpha_k^\epsilon \to 0$ as $\epsilon \in \tilde{\Upsilon} \to 0$.
In this case, by further passing to a subsequence, we can assume either $\alpha_k^\epsilon/\epsilon \to \infty$ or $\alpha_k^\epsilon/\epsilon$ converges, as $\epsilon \in \tilde{\Upsilon} \to 0$.
Suppose $\alpha_k^\epsilon/\epsilon \to \infty$ as $\epsilon \in \tilde{\Upsilon} \to 0$. Since (36) holds for all $\epsilon \in \tilde{\Upsilon}$, dividing all sides by α_k^ϵ and using $L = I \cup (H\backslash J)$ and the fact we are in Case 2 yield in the limit

$$\begin{array}{rlrl} M_{LH}(u_k)_H + (q_k)_L & = 0, & (u_k)_{K_k} & \geq 0, \\ M_{I_kH}(u_k)_H + (q_k)_{I_k} & \geq 0, & (u_k)_{J\backslash I} & = 0, \\ M_{L_kH}(u_k)_H + (q_k)_{L_k} & \geq 0, & (u_k)_{H\backslash J} & \geq 0, \end{array} \tag{39}$$

for some $u_k \in \Re^n$. [Notice that $u_0 = \bar{x}$, so $(u_0)_{J^c} = 0$.] By further passing to a subsequence if necessary, we can assume one of the following two subcases occurs.

Subcase 2a. There exist $j \in H\backslash J$ and $\tau > 0$ such that $(y_k^\epsilon - \alpha_k^\epsilon u_k)_j/\alpha_1^\epsilon \leq \tau$ for all $\epsilon \in \tilde{\Upsilon}$.

In this subcase, let $H^{new} := H\backslash\{j\}$ and we have that H^{new} is a proper subset of H and contains J. Moreover, $(u_0)_j = \bar{x}_j = 0$, so (37) yields

$$\tau\alpha_1^\epsilon \geq (y_k^\epsilon - \alpha_k^\epsilon u_k)_j = (x^\epsilon - \sum_{h=1}^{k} \alpha_h^\epsilon u_h)_j. \tag{40}$$

Subcase 2b. For all $j \in H\backslash J$, $(y_k^\epsilon - \alpha_k^\epsilon u_k)_j/\alpha_1^\epsilon \to \infty$ as $\epsilon \in \tilde{\Upsilon} \to 0$.

In this subcase, let $I_{k+1} := \{i \in I_k : M_{iH}(u_k)_H + (q_k)_i = 0\}$ and $K_{k+1} := \{i \in K_k : (u_k)_i = 0\}$ and $L_{k+1} := \{i \in L_k : M_{iH}(u_k)_H + (q_k)_i = 0\}$. Then (36) and (39) yield

$$\begin{array}{ll} M_{LH}(y_k^\epsilon - \alpha_k^\epsilon u_k)_H + \sum_{h=k+1}^n \alpha_h^\epsilon (q_h)_L &= -\epsilon G(x^\epsilon)_L, \quad (y_k^\epsilon - \alpha_k^\epsilon u_k)_{K_{k+1}} \geq 0, \\ M_{I_{k+1}H}(y_k^\epsilon - \alpha_k^\epsilon u_k)_H + \sum_{h=k+1}^n \alpha_h^\epsilon (q_h)_{I_{k+1}} &\geq -\epsilon G(x^\epsilon)_{I_{k+1}}, \quad (y_k^\epsilon - \alpha_k^\epsilon u_k)_{J\backslash I} = 0, \\ M_{L_{k+1}H}(y_k^\epsilon - \alpha_k^\epsilon u_k)_H + \sum_{h=k+1}^n \alpha_h^\epsilon (q_h)_{L_{k+1}} &\geq 0, \qquad (y_k^\epsilon - \alpha_k^\epsilon u_k)_{H\backslash J} > 0, \end{array} \tag{41}$$

for all $\epsilon \in \tilde{\Upsilon}$ sufficiently small. Letting $y_{k+1}^\epsilon := y_k^\epsilon - \alpha_k^\epsilon u_k$ and we see that (36)–(37) hold with k replaced by $k+1$. Below we show that $k < n$ so that we can repeat the above construction with k replaced by $k+1$. Suppose not, so that $k = n$. Then, dividing all sides of (41) by $\min_{j\in H\backslash J}(y_{n+1}^\epsilon)_j$ and using the fact that we are in Subcase 2b (so that $\epsilon G_j(x_j^\epsilon)/\alpha_1^\epsilon \to 0$ for all $j \in L$ and $\alpha_1^\epsilon/(y_{n+1}^\epsilon)_j \to 0$ for all $j \in H\backslash J$ as $\epsilon \in \tilde{\Upsilon} \to 0$) yield in the limit that

$$M_{LH}(u_{n+1})_H = 0, \quad (u_{n+1})_{J\backslash I} = 0, \quad (u_{n+1})_j \geq 1 \; \forall j \in H\backslash J,$$

for some $u_{n+1} \in \Re^n$. Then, using this and (41) and $H = L \cup (J\backslash I)$, we see that $y^\epsilon := y_{n+1}^\epsilon - 2\|(y_{n+1}^\epsilon)_{H\backslash J}\|_\infty u_{n+1}$ satisfies

$$M_{LL} y_L^\epsilon = -\epsilon G(x^\epsilon)_L.$$

Also, $y_{n+1}^\epsilon = x^\epsilon - \bar{x} - \sum_{h=1}^n \alpha_h^\epsilon u_h \to 0$ so that $y^\epsilon \to 0$. Multiplying the above equation on the left by $(y_L^\epsilon)^T$ and using the positive semidefinite property of M_{LL} (since M_{JJ} is positive semidefinite and $L \subset \bar{J}$) yields $0 \leq -\epsilon (y_L^\epsilon)^T G(x^\epsilon)_L$. Dividing both sides by ϵ and using $L = I \cup (H\backslash J)$ gives

$$\sum_{j\in H\backslash J} y_j^\epsilon G_j(x_j^\epsilon) \leq -\sum_{j\in I} y_j^\epsilon G_j(x_j^\epsilon). \tag{42}$$

For each $j \in H\backslash J$, we have $y_j^\epsilon = (y_{n+1}^\epsilon)_j - 2\|(y_{n+1}^\epsilon)_{H\backslash J}\|_\infty (u_{n+1})_j \leq -(y_{n+1}^\epsilon)_j = -x_j^\epsilon - \sum_{h=1}^n \alpha_h^\epsilon (u_h)_j$ so that (also using $G_j(x_j^\epsilon) < 0$)

$$y_j^\epsilon G_j(x_j^\epsilon) \geq -\Big(1 + \sum_{h=1}^n (\alpha_h^\epsilon / x_j^\epsilon)(u_h)_j\Big) x_j^\epsilon G_j(x_j^\epsilon).$$

Since we are in Subcase 2b with $k = n$, then for each $j \in H\backslash J$ we have $(y^\epsilon_{n+1})_j/\alpha^\epsilon_1 \to \infty$ as $\epsilon \in \tilde{\Upsilon} \to 0$, so that $(y^\epsilon_{n+1})_j = x^\epsilon_j + \sum_{h=1}^n \alpha^\epsilon_h (u_h)_j$ yields $x^\epsilon_j/\alpha^\epsilon_1 \to \infty$. This together with the above inequality and $\limsup_{t\to 0} tG_j(t) < 0$ implies the left-hand side of (42) is positive and bounded away from zero. On the other hand, we have $G(x^\epsilon)_I \to G(\bar{x})_I$ and $y^\epsilon \to 0$ as $\epsilon \in \tilde{\Upsilon} \to 0$, so the right-hand side of (42) tends to zero, a contradiction.

Suppose instead $\alpha^\epsilon_k/\epsilon$ converges to some $c \in \Re_+$ as $\epsilon \in \tilde{\Upsilon} \to 0$. Then, $k \geq 1$ and, since we are in Case 2 (and $G_j(x^\epsilon_j) \to -\infty$ for $j \in H\backslash J$), it must be that $H\backslash J = \emptyset$, i.e., $H = J$. The first equation in (36) can then be written using $(y^\epsilon_k)_{J\backslash I} = 0$ as

$$M_{II}(y^\epsilon_k)_I + \sum_{h=k}^n \alpha^\epsilon_h (q_h)_I = -\epsilon G(x^\epsilon)_I.$$

Dividing this by ϵ yields in the limit

$$M_{II}(u_k)_I + c(q_k)_I = -G(\bar{x})_I,$$

for some $u_k \in \Re^n$. Combining the above two equations yields

$$-M_{II}\Delta_I + (\alpha^\epsilon_k - c\epsilon)(q_k)_I + \sum_{h=k+1}^n \alpha^\epsilon_h (q_h)_I = \epsilon(G(\bar{x})_I - G(x^\epsilon)_I),$$

where $\Delta := \epsilon u_k - y^\epsilon_k = \bar{x} - x^\epsilon + \sum_{h=1}^{k-1} \alpha^\epsilon_h u_h + \epsilon u_k$. Multiplying the left-hand side by Δ_I^T and using the positive semidefinite property of M_{II} (since $M_{\bar{J}\bar{J}}$ is positive semidefinite and $I \subset \bar{J}$) yields

$$\Delta_I^T \left((\alpha^\epsilon_k - c\epsilon)(q_k)_I + \sum_{h=k+1}^n \alpha^\epsilon_h (q_h)_I \right) \geq \epsilon \Delta_I^T (G(\bar{x})_I - G(x^\epsilon)_I).$$

Thus, dividing both sides by ϵ and expanding Δ_I yields

$$\begin{aligned}
&(\bar{x} - x^\epsilon)_I^T (G(\bar{x})_I - G(x^\epsilon)_I) \\
\leq\; & - \left(\sum_{h=1}^{k-1} \alpha^\epsilon_h (u_h)_I + \epsilon (u_k)_I \right) (G(\bar{x})_I - G(x^\epsilon)_I) \\
+\; & \left((\bar{x} - x^\epsilon)_I + \sum_{h=1}^{k-1} \alpha^\epsilon_h (u_h)_I + \epsilon (u_k)_I \right)^T \left((\frac{\alpha^\epsilon_k}{\epsilon} - c)(q_k)_I + \sum_{h=k+1}^n \frac{\alpha^\epsilon_h}{\epsilon} (q_h)_I \right) \\
\leq\; & \left\| \sum_{h=1}^{k-1} \alpha^\epsilon_h (u_h)_I + \epsilon (u_k)_I \right\| \|G(\bar{x})_I - G(x^\epsilon)_I\| \\
+\; & \left(\|(\bar{x} - x^\epsilon)_I\| + \left\| \sum_{h=1}^{k-1} \alpha^\epsilon_h (u_h)_I + \epsilon (u_k)_I \right\| \right) \left\| (\frac{\alpha^\epsilon_k}{\epsilon} - c)(q_k)_I + \sum_{h=k+1}^n \frac{\alpha^\epsilon_h}{\epsilon} (q_h)_I \right\|
\end{aligned}$$

Suppose in addition G_J is locally Lipschitzian at $\bar{x}_J$ with constant $\kappa > 0$ and satisfies (17) with $\sigma > 0, \delta > 0, \varrho = 1$. Then for all $\epsilon \in \tilde{\Upsilon}$ sufficiently small so that $\|(\bar{x} - x^\epsilon)_I\| \leq \delta$ and $\|G(\bar{x})_I - G(x^\epsilon)_I\| \leq \kappa\|(\bar{x} - x^\epsilon)_I\|$, the preceding inequality yields

$$\sigma\xi^2 \leq (r\kappa + s)\xi + rs,$$

where $r := \|\sum_{h=1}^{k-1} \alpha_h^\epsilon (u_h)_I + \epsilon(u_k)_I\|$, $s := \left\|\left(\frac{\alpha_k^\epsilon}{\epsilon} - c\right)(q_k)_I + \sum_{h=k+1}^{n} \frac{\alpha_h^\epsilon}{\epsilon}(q_h)_I\right\|$, and $\xi := \|(\bar{x} - x^\epsilon)_I\|$. Solving this using the quadratic formula yields

$$\xi \leq \frac{(r\kappa + s) + \sqrt{(r\kappa + s)^2 + 4\sigma rs}}{2\sigma} \leq \frac{r\kappa + s}{\sigma} + \sqrt{r}\sqrt{\frac{s}{\sigma}} \leq \frac{r\kappa + s}{\sigma} + \frac{r + s/\sigma}{2},$$

where the last two inequalities use the identities $\sqrt{a+b} \leq \sqrt{a} + \sqrt{b}$ and $ab \leq (a^2 + b^2)/2$. This and $x_{J\setminus I}^\epsilon = 0$ and, by taking $\tau' > 0$ sufficiently large so that $r \leq \tau'(\alpha_1^\epsilon + \epsilon)$ for all $\epsilon \in \tilde{\Upsilon}$, yields

$$\|(x^\epsilon - \bar{x})_J\| \leq \tau'\left(\frac{\kappa}{\sigma} + \frac{1}{2}\right)(\alpha_1^\epsilon + \epsilon) + \frac{3}{2\sigma}\left|\frac{\alpha_k^\epsilon}{\epsilon} - c\right|\|(q_k)_J\| + \sum_{h=k+1}^{n} \frac{\alpha_h^\epsilon}{\epsilon}\|(q_h)_J\|. \quad (43)$$

Thus, letting $\lambda^\epsilon := \epsilon/\alpha_k^\epsilon$ in case (38) and letting $\lambda^\epsilon := \alpha_1^\epsilon + \sum_{h=1}^{k} \alpha_h^\epsilon (u_h)_j$ in case (40) yield (22). Similarly, (43) yields (23) for suitable choice of λ^ϵ.
By repeating the above argument with $\Upsilon_I \setminus \tilde{\Upsilon}$ in place of Υ_I, we can extract another subsequence of Υ_I having the same properties as $\tilde{\Upsilon}$ and so on. We do this for all $I \subset J$ with $|\Upsilon_I| = \infty$, thus yielding a countable collection of subsequences whose union is Υ. ■

Note 5. A few words about the assumptions in Proposition 3 are in order. First, the assumptions on $F(x_H, 0)$ and Σ_H in part (a) are satisfied by $H = \bar{J}$ if F is pseudo-monotone on $\Re_+^n$ (since $\Sigma = \bar{\Sigma}$ is convex in this case). Second, the assumption of $\lim_{t\to 0} tG_j(t) = 0$ in part (c1) is equivalent to $\lim_{\epsilon\to 0} G_j^{-1}(-1/\epsilon)/\epsilon = 0$ if G_j is strictly increasing. This is because $G_j^{-1}(-1/\epsilon)/\epsilon \geq c > 0$ implies, by G_j being strictly increasing, that $-c \geq c\epsilon G_j(c\epsilon)$. Conversely, $tG_j(t) \leq -c$ for some $c > 0$ implies, by G_j^{-1} being strictly increasing, that $c \leq G_j^{-1}(-c/t)/(t/c)$. Third, the assumptions on M in part (c2) are satisfied by any $I \subset \bar{J} \subset N$ if M is symmetric positive semidefinite (see, e.g., [35, Lemma 5]) or if M is symmetric nondegenerate (i.e., M_{II} is nonsingular for all $I \subset N$). It is also satisfied by any $I \subset \bar{J} \subset N$ if $M = \begin{bmatrix} 1 & 1 & 2 \\ 1 & 1 & 2 \\ 0 & 0 & 1 \end{bmatrix}$, which is neither symmetric nor nondegenerate. Fourth, for the h_j defined in part (c3), direct calculation finds that, for G given by (5), $h_j(\epsilon) = \sqrt{\epsilon}$ and, for G given by (6), $h_j(\epsilon)$ is the uniqe η satisfying $-\eta/\ln(\eta) = \epsilon$, so that $h_j(\epsilon) = o(\epsilon^\beta)$ for any fixed $\beta \in (0,1)$. To see that the bound (21) is reasonable, notice that for $n = 1$ and $F(x) = 2$ and $G(x) = -1/x$, we have $2 + \epsilon G(x^\epsilon) = 0$ so that $x^\epsilon = G^{-1}(-2/\epsilon) = \epsilon/2$. Similarly, for $n = 1$ and $F(x) = x$ and $G(x) = -1/x$, we have $x^\epsilon + \epsilon G(x^\epsilon) = 0$ so that $x^\epsilon = \sqrt{\epsilon}$. Notice that the skew symmetry assumption $M = -M^T$ is satisfied when an LP is formulated as a CP. The dependence of K on ϵ cannot be removed, as is shown by an example in [52] in the context of LP. Fifth, the nesting of index sets in part (c4) reflects a nested dependence of the convergence rate of some components of x^ϵ (indexed by H_l) on the remaining components. Intuitively, if x_j^ϵ converges more slowly than x_i^ϵ, then the term $M_{ij}x_j^\epsilon$ can influence what the limit $\bar{x}_i$ will be and the rate at which x_i^ϵ converges to this limit.

Note 6. If $\bar{x}$ in Proposition 3 satisfies strict complementarity, i.e., $\bar{x}+F(\bar{x}) > 0$, then parts (c3) and (c4) of this proposition simplify considerably. In particular, we have $F(\bar{x})_j > 0$ as well as $F(x^\epsilon)_j + \epsilon G_j(x_j^\epsilon) = 0$ for all $j \in J^c$, so that $F(x^\epsilon)_j \to F(\bar{x})_j$ and the strictly increasing property of G_j yield $x_j^\epsilon \leq G_j^{-1}(-F(x^\epsilon)_j/\epsilon) \leq G_j^{-1}(-\rho/\epsilon)$ for all $\epsilon \in \Upsilon$ sufficiently small, where $\rho := \min_{j \in J} F(\bar{x})_j/2$.

Note 7. If F is affine and pseudo-monotone on $\Re^n_+$ and G_j is strictly increasing with $\lim_{t\to 0} tG_j(t) = 0$ for all $j \in J^c$, then $\Sigma = \bar{\Sigma}$ and Proposition 3(c1),(c3) yield the error bound (19) and (21) for all $\epsilon \in \Upsilon$ sufficiently small, with $\bar{J} = J$, $\gamma = 1$ and $\tau > 0, \rho > 0$ some constants, and with $K \subset J^c$ depending on ϵ and $L := J^c \backslash K$. Similarly, if F is affine and monotone on $\Re^n_+$ and G_j is strictly increasing with $\limsup_{t\to 0} tG_j(t) < 0$ for all $j \in J^c$ and G_j is Lipschitz continuous and strongly monotone near $\bar{x}_j$ for all $j \in J$, then $\Sigma = \bar{\Sigma}$ and Proposition 3(c1),(c4) yield the error bound (21), (22), (23) for all sufficiently small ϵ along some subsequence $\tilde{\Upsilon}$, etc. Moreover, there exists a $c > 0$ such that $tG_j(t) \leq -c$ for all $j \in J^c$ and all $t > 0$ sufficiently small, implying $t \leq G_j(-c/t)$. Thus, the second case in (22) implies the first case.

Note 8. Proposition 3 does not say anything about existence or uniqueness or boundedness of $x^\epsilon \in \text{VI}(\Re^n_+, F + \epsilon G)$. In the case where G is given by (5), it was shown by Kojima et al. [29, Theorem 4.4] that F being a continuous P_0-function and satisfying strict feasibility (i.e., $x > 0, F(x) > 0$ has a solution) and a boundedness condition implies the existence and uniqueness of x^ϵ for all ϵ (also see [30, Theorem 4.4] for the case of affine F and see [28] for extensions to other types of F). Analogous results were shown earlier by McLinden in the context of convex programs [41] and, more generally, when F is a maximal monotone operator [42]. These results were further improved and extended by Kojima et al. [31] and Güler [23]. Recently, Chen et al. [9, Corollary 3.14] showed that F being a continuously differentiable P_0-function and Σ being nonempty and bounded is sufficient for the existence and uniqueness of x^ϵ for all ϵ sufficiently small. Subsequently, Gowda and Tawhid [22, Theorems 8 and 9] weakened the differentiability assumption on F to continuity and considered more general regularizations on F.

We illustrate Proposition 3 with the following example with $n = 3$ variables.

Example 1. Consider

$$F(x) = Mx+q,\ M = \begin{bmatrix} 1 & 1 & 2 \\ 1 & 1 & 1 \\ 0 & 1 & 1 \end{bmatrix},\ q = \begin{bmatrix} -1 \\ -1 \\ 0 \end{bmatrix},\ G(x) = \begin{bmatrix} G_1(x_1) \\ G_2(x_2) \\ G_3(x_3) \end{bmatrix} = \begin{bmatrix} x_1 - .5 \\ x_2 - .5 \\ -1/x_3^\beta \end{bmatrix}$$

with $\beta > 0$. Notice that M is positive semidefinite and it can be checked that $\Sigma = \{(t, 1-t, 0) : 0 \le t \le 1\}$.

(i). Suppose $\beta < 1$ so that $\lim_{t \to 0} tG_3(t) = 0$. Then either by direct calculation or by using Proposition 3(c1), we find that $x^\epsilon \to \bar{x} = (.5, .5, 0) \in \mathrm{VI}(\Sigma, G)$ as $\epsilon \to 0$. Thus $J = \bar{J} = \{1, 2\}$ and, for ϵ sufficiently small, we have $x^\epsilon > 0$ and hence $F(x^\epsilon) + \epsilon G(x^\epsilon) = 0$. Then, direct calculation yields

$$x_3^\epsilon = G_3^{-1}(-(x_2^\epsilon + x_3^\epsilon)/\epsilon) \approx G_3^{-1}(-.5/\epsilon) = (2\epsilon)^{1/\beta}$$

and $(x_1^\epsilon - .5, x_2^\epsilon - .5) = -x_3^\epsilon(1+2\epsilon, \epsilon-1)/(2\epsilon+\epsilon^2) = O(\epsilon^{1/\beta - 1})$. This illustrates parts (c1) and (c3) of Proposition 3.

(ii). Suppose $\beta > 1$ so that $\lim_{t \to 0} tG_3(t) = -\infty$. Then direct calculation finds that, for all ϵ sufficiently small, we have $x_1^\epsilon = 0$ and $x_2^\epsilon = (1 + .5\epsilon - x_3^\epsilon)/(1+\epsilon)$ with x_3^ϵ satisfying $x_3^\epsilon + (1+\epsilon)G_3(x_3^\epsilon) = -(1+.5\epsilon)/\epsilon$. Thus $x^\epsilon \to \bar{x} = (0, 1, 0)$ and $J = \bar{J} = \{1, 2\}$. Moreover, $\bar{x}_J \in \mathrm{VI}(\Sigma_J, p_J)$ with $\Sigma_J = \{(t, 1-t) : 0 \le t \le 1\}$ and $p_J = \left[\begin{smallmatrix} 2 \\ 1 \end{smallmatrix}\right] / \sqrt{5}$. Lastly, we have

$$x_3^\epsilon = G_3^{-1}\left(-(1 + .5\epsilon + \epsilon x_3^\epsilon)/\epsilon(1+\epsilon)\right) = \left(\epsilon(1+\epsilon)/(1 + .5\epsilon + \epsilon x_3^\epsilon)\right)^{1/\beta} \approx \epsilon^{1/\beta}$$

and hence $x_2^\epsilon - 1 \approx -x_3^\epsilon \approx -\epsilon^{1/\beta}$. This illustrates parts (a) and (c3) of Proposition 3.

(iii). Suppose $\beta > 1$ and G_1, G_2 are changed to $G_1(t) = G_2(t) = -1/t$. First, we claim that, for each $\epsilon > 0$, x^ϵ exists and is unique. To see this, let $I := \{1, 2\}$ and note that $M_{II} = \left[\begin{smallmatrix} 1 & 1 \\ 1 & 1 \end{smallmatrix}\right]$ is positive semidefinite and

$$M_{I3}x_3 + q_I = \begin{bmatrix} 2x_3 - 1 \\ x_3 - 1 \end{bmatrix} \in \left\{b_I \in \Re^2 : b_I = y_I - M_{II}x_I \text{ for some } x_I, y_I \in \Re^2_{++}\right\}$$

for all $x_3 \ge 0$, so a result of Kojima et al. [31, Corollary 1.2, Theorem 3.3] implies that, for each $x_3 \ge 0$, the equation $M_{II}x_I + M_{I3}x_3 + q_I + \epsilon G(x)_I = 0$ has a unique solution $x_I(x_3) > 0$ which is continuous in x_3 and is bounded as $x_3 \to 0$. Then the equation

$$M_{3I}x_I(x_3) + M_{33}x_3 + q_3 + \epsilon G_3(x_3) = x_2(x_3) + x_3 - \epsilon/(x_3)^\beta = 0$$

has a solution $x_3^\epsilon > 0$ since the left-hand side is continuous in $x_3 > 0$ and tends to $-\infty$ as $x_3 \to 0$ and tends to ∞ as $x_3 \to \infty$. Then $x^\epsilon := [x_I(x_3^\epsilon)\ x_3^\epsilon]^T > 0$

satisfies

$$F(x^\epsilon) + \epsilon G(x^\epsilon) = \begin{bmatrix} x_1^\epsilon + x_2^\epsilon + 2x_3^\epsilon - 1 + \epsilon G_1(x_1^\epsilon) \\ x_1^\epsilon + x_2^\epsilon + \ x_3^\epsilon - 1 + \epsilon G_2(x_2^\epsilon) \\ x_2^\epsilon + \ x_3^\epsilon \qquad + \epsilon G_3(x_3^\epsilon) \end{bmatrix} = 0. \tag{44}$$

Uniqueness of x^ϵ follows from $F + \epsilon G$ being strictly monotone on $\Re^n_{++}$. Now, (44) and $x^\epsilon > 0$ imply $x_1^\epsilon < 1 + \epsilon/x_1^\epsilon$, $x_2^\epsilon < 1 + \epsilon/x_2^\epsilon$, $x_3^\epsilon < \epsilon/(x_3^\epsilon)^\beta$ for all $\epsilon > 0$, so x^ϵ is bounded as $\epsilon \to 0$. Then, x^ϵ has a cluster point $\bar{x}$ which, by Proposition 3 is in Σ. Since M is positive semidefinite so that $\Sigma = \bar{\Sigma}$ is convex and $\bar{J} = \{1, 2\}$, Proposition 3(a) with $H = \bar{J}$ implies either $\bar{x}_{\bar{J}} \in \mathrm{VI}(\Sigma_{\bar{J}}, p_{\bar{J}})$ or $\bar{x}_{\bar{J}} \in \mathrm{VI}(\Sigma_{\bar{J}}, \zeta p_{\bar{J}} + G_{\bar{J}})$ for some $\zeta \in \Re_+$, where $\Sigma_{\bar{J}} = \{(t, 1-t) : 0 \le t \le 1\}$ and $p_{\bar{J}} = \left[\begin{smallmatrix}2\\1\end{smallmatrix}\right]/\sqrt{5}$. In either case, we have $\bar{x}_2 > 0$, so that the third equation in (44) yields

$$x_3^\epsilon = G_3^{-1}(-(x_2^\epsilon + x_3^\epsilon)/\epsilon) = (\epsilon/(x_2^\epsilon + x_3^\epsilon))^{1/\beta} \approx (\epsilon/\bar{x}_2)^{1/\beta}.$$

Since $\beta > 1$, this shows $x_3^\epsilon/\epsilon \to \infty$ so we are in the case of $\bar{x}_{\bar{J}} \in \mathrm{VI}(\Sigma_{\bar{J}}, p_{\bar{J}})$, yielding $\bar{x} = (0, 1, 0)$ and $J = \{2\}$. Thus $\bar{J} \backslash J = \{1\}$. Now, subtracting the second equation in (44) from the first equation and using $G_2(t) = -1/t$ yields $x_3^\epsilon + \epsilon/x_2^\epsilon + \epsilon G_1(x_1^\epsilon) = 0$, so that (cf. (22))

$$x_1^\epsilon = G_1^{-1}(-(x_3^\epsilon + \epsilon/x_2^\epsilon)/\epsilon) \approx G_1^{-1}(-(\epsilon/\bar{x}_2)^{1/\beta}/\epsilon) = \epsilon^{1-1/\beta}/(\bar{x}_2)^{1/\beta}.$$

Finally, the second equation in (44) implies

$$x_2^\epsilon - 1 = -x_1^\epsilon - x_3^\epsilon + \epsilon/x_2^\epsilon \approx -\epsilon^{\min\{1/\beta, 1-1/\beta\}}/(\bar{x}_2)^{1/\beta}.$$

Notice that G_2 is locally Lipschitzian at $\bar{x}_2 = 1$ and satisfies (17) with some $\sigma > 0, \delta > 0, \varrho = 1$. This illustrates parts (a), (c3) and (c4) of Proposition 3. For part (c4), we have $H_1 = \{1, 2\}$, $H_2 = \{2\}$. Correspondingly, for $l = 1$, the decomposition (35) (with the subscript l restored) holds with $q_{1,1} = M_{N3} = [2\ 1\ 1]^T$, $\alpha_{1,1}^\epsilon = x_3^\epsilon$, yielding $\lambda_1^\epsilon = \epsilon/x_3^\epsilon$. For $l = 2$, (35) holds with $q_{1,2} = M_{N1} = [1\ 1\ 0]^T$, $\alpha_{1,1}^\epsilon = x_1^\epsilon$ if $\beta \in (1, 2)$; or $q_{1,2} = M_{N3} = [2\ 1\ 1]^T$, $\alpha_{1,1}^\epsilon = x_3^\epsilon$ if $\beta > 2$, etc.

(iv). Suppose $\beta > 1$ and G_1, G_2 are changed to $G_1(t) = G_2(t) = -1/t$ as in (iii). Suppose we also change M to $M = \begin{bmatrix} 1 & 1 & 2 \\ 1 & 1 & 2 \\ 0 & 0 & 1 \end{bmatrix}$. It can be seen that this does not change Σ. Moreover, M is positive semidefinite, so $\Sigma = \bar{\Sigma}$ is convex and $\bar{J} = \{1, 2\}$. Using an argument analogous to that used in (iii), we have that x^ϵ exists and is unique for all $\epsilon > 0$, and x^ϵ is bounded as $\epsilon \to 0$. Also, M satisfies the assumptions in Proposition 3(c2) for any $I \subset N$, so it follows that any cluster point of x^ϵ is in $\mathrm{VI}(\Sigma, G) = \{(.5, .5, 0)\}$. Thus $x^\epsilon \to \bar{x} = (.5, .5, 0)$, with $J = \bar{J}$. Moreover, $F(x^\epsilon) + \epsilon G(x^\epsilon) = 0$ yields

$$x_3^\epsilon = h_3(\epsilon) = \epsilon^{1/(\beta+1)}$$

(recall $h_3(\epsilon)$ is the unique $\eta > 0$ satisfying $-\eta/\epsilon = G_3(\eta) = -1/\eta^\beta$) and, using symmetry, $x_1^\epsilon - .5 = x_2^\epsilon - .5 = O(x_3^\epsilon) = O(\epsilon^{1/(\beta+1)})$. This illustrates parts (c2) and (c3) of Proposition 3. Compared to (iii), we see that changing M changes both the limit point $\bar{x}$ and the convergence rate, even when G and the solution set Σ are unchanged!

3 Summary and Open Questions

We have considered regularizing the mapping F in a complementarity problem by another mapping and we studied properties of any limit point of the solution of the regularized problem. We have also derived error bounds on the distance from the solution of the regularized problem to its limit point. These error bounds are fairly complex, reflecting both the local growth rate of the regularization mapping and the linkage among solution components through the complementarity condition.

There remain many open questions to be answered. We list a few below.

Q1. Can parts (c3) and (c4) of Proposition 3 be simplified/strengthened in the case of $G_1 = \cdots = G_n$?

Q2. For the G given by (5), the convergence result of McLinden [42] requires F to be monotone and continuous, whereas our error bound result requires F to be affine and satisfying the assumptions of either part (c2) or part (c4) of Proposition 3. For this particular choice of G, can an error bound result analogous to Proposition 3(c2)–(c4) be obtained for non-affine F?

Q3. Consider higher-order regularization of the form $F^\epsilon(x) := F(x) + \epsilon G^1(x) + \epsilon^2 G^2(x) + \cdots + \epsilon^p G^p(x)$, where $p \geq 1$ and $G^1, ..., G^p$ are suitable mappings. What can we say about any limit point of $x^\epsilon \in \mathrm{VI}(\Re^n_+, F^\epsilon)$ as $\epsilon \to 0$? [See [1, Section 4] for discussions in the optimization setting.] What kind of error bounds can be derived?

Q4. Here we have considered the CP where the feasible set is $\Re^n_+$. Can our results be extended to variational inequality problems where the feasible set is a polyhedral set or, more generally, a nonempty closed convex set of $\Re^n$? How about extension to spaces other than $\Re^n$, such as the space of $n \times n$ symmetric matrices (with $\Re^n_+$ replaced by the convex cone of $n \times n$ symmetric positive semidefinite matrices) or an infinite-dimensional space?

References

[1] Attouch, H., Viscosity Solutions of Minimization Problems, *SIAM Journal on Optimization*, Vol. 6, pp. 769–806, 1996.

[2] Attouch, H., Chbani, Z., and Moudafi, A., Recession Operators and Solvability of Variational Problems in Reflexive Banach Spaces, in *Calculus of Variations, Homogenization and Continuum Mechanics*, edited by G. Bouchitte, G. Buttazzo, and P. Suquet, pp. 51–67, World Scientific, Teaneck, 1993.

[3] Attouch, H. and Wets, R. J.-B., Quantitative Stability of Variational Systems: III. ϵ-approximate Solutions, *Mathematical Programming*, Vol. 61, pp. 197–214, 1993.

[4] Auslender, A., Cominetti, R., and Haddou, M., Asymptotic Analysis for Penalty and Barrier Methods in Convex and Linear Programming, *Mathematics of Operations Research*, Vol. 22, pp. 43–62, 1997.

[5] Auslender, A., *Optimisation: Méthodes Numériques*, Masson, Paris, 1976.

[6] Billups, S. C. and Ferris, M. C., QPCOMP: A Quadratic Program Based Solver for Mixed Complementarity Problems", *Mathematical Programming*, Vol. 76, pp. 533–562, 1997.

[7] Bonnans, J. F. and Haddou, M., Asymptotic Analysis of Congested Communication Networks, Research Report No. 3133, Institut National de Recherche en Informatique et en Automatique, Rocquencourt, March 1997.

[8] Brézis, H., *Opérateurs Maximaux Monotones et Semi-Groupes de Contractions dans les Espaces de Hilbert*, North-Holland, Amsterdam, 1973.

[9] Chen, B., Chen, X.-J., and Kanzow, C., A Penalized Fischer-Burmeister NCP-Function: Theoretical Investigation and Numerical Results, Department of Management and Systems, Washington State University, Pullman, September 1997.

[10] Cominetti, R. and Dussault, J. P., Stable Exponential-Penalty Algorithm with Superlinear Convergence, *Journal of Optimization Theory and Applications*, Vol. 83, pp. 285–309, 1994.

[11] Cominetti, R. and San Martín, J., Asymptotic Analysis of the Exponential Penalty Trajectory in Linear Programming, *Mathematical Programming*, Vol. 67, pp. 169–187, 1994.

[12] Cottle, R. W., Pang, J.-S., and Stone, R. E., *The Linear Complementarity Problem*, Academic Press, New York, New York, 1992.

[13] Dontchev, A. L. and Zolezzi, T., *Well-Posed Optimization Problems*, Springer-Verlag, Berlin, 1993.

[14] Facchinei, F., Structural and Stability Properties of P_0 Nonlinear Complementarity Problems, Dipartimento di Informatica e Sistemistica, Universitá di Roma "La Sapienza", Rome, April 1997.

[15] Facchinei, F. and Kanzow, C., Beyond Monotonicity in Regularization Methods for Nonlinear Complementarity Problems, Dipartimento di Informatica e Sistemistica, Universitá di Roma "La Sapienza", Rome, May 1997.

[16] Fang, S. C., An Unconstrained Convex Programming View of Linear Programming, *Zeitschrift für Operations Research – Methods and Models of Operations Research*, Vol. 36, pp. 149–161, 1992.

[17] Fang, S. C., Rajasekera, J. R., and Tsao, H.-S. J., *Entropy Optimization and Mathematical Programming*, Kluwer, Norwell, 1997.

[18] Fang, S. C., and Tsao, H.-S. J., Linear Programming with Entropic Perturbation, *Zeitschrift für Operations Research – Methods and Models of Operations Research*, Vol. 37, pp. 171–186, 1993.

[19] Ferris, M. C. and Mangasarian, O. L., Finite Perturbation of Convex Programs, *Applied Mathematics and Optimization*, Vol. 23, pp. 263–273, 1991.

[20] Fiacco, A. V. and McCormick, G. P., *Nonlinear Programming: Sequential Unconstrained Minimization Techniques*, Wiley, New York, 1968; reprinted by SIAM, Philadelphia, 1990.

[21] Fischer, A., Merit Functions and Stability for Complementarity Problems, Institute for Numerical Mathematics, Technical University of Dresden, Dresden, December 1997.

[22] Gowda, M. S. and Tawhid, M. A., Existence and Limiting Behavior of Trajectories Associated with P_0-Equations, Department of Mathematics and Statistics, University of Maryland, Baltimore, November 1997.

[23] Güler, O., Existence of Interior Points and Interior Paths in Nonlinear Monotone Complementarity Problems, *Mathematics of Operations Research*, Vol. 18, pp. 128–147, 1993.

[24] Harker, P., and Pang, J.-S., Finite-Dimensional Variational Inequality and Nonlinear Complementarity Problems: a Survey of Theory, Algorithms and Applications, *Mathematical Programming*, Vol. 48, pp. 161–220, 1990.

[25] Hoffman, A. J., On Approximate Solutions of Systems of Linear Inequalities, *Journal of Research of the National Bureau of Standards*, Vol. 49, pp. 263–265, 1952.

[26] Isac, G., Tikhonov's Regularization and the Complementarity Problem in Hilbert Space, *Journal of Mathematical Analysis and Applications*, Vol. 174, pp. 53-66, 1993.

[27] Karlin, S., *Mathematical Methods and Theory in Games, Programming, and Economics*, Vol. I, Addison-Wesley, Reading, 1959.

[28] Kojima, M., Megiddo, N., and Mizuno, S. A General Framework of Continuation Methods for Complementarity Problems, *Mathematics of Operations Research*, Vol. 18, 945–963, 1993.

[29] Kojima, M., Megiddo, N., and Noma T., Homotopy Continuation Methods for Nonlinear Complementarity Problems, *Mathematics of Operations Research*, Vol. 16, pp. 754–774, 1991.

[30] Kojima, M., Megiddo, N., Noma, T. and Yoshise, A., *A Unified Approach to Interior Point Algorithms for Linear Complementarity Problems*, Lecture Notes in Computer Science 538, Springer-Verlag, Berlin, 1991.

[31] Kojima, M., Mizuno, M., and Noma T., Limiting Behavior of Trajectories Generated by a Continuation Methods for Monotone Complementarity Problems, *Mathematics of Operations Research*, Vol. 4, pp. 662–675, 1990.

[32] Lemaire, B., Well-Posedness, Conditioning and Regularization of Minimization, Inclusion, and Fixed-Point Problems, Report, Department of Mathematics, University of Montpellier 2, Montpellier, November 1997; to appear in *Pliska Studia Mathematica.*

[33] Luo, Z.-Q., and Pang, J.-S., Error Bounds for Analytic Systems and Their Applications, *Mathematical Programming*, Vol. 67, pp. 1–28, 1994.

[34] Luo, Z.-Q., and Sturm, J. F., Error Bounds for Quadratic Systems, Department of Electrical and Computer Engineering, McMaster University, Hamilton, January 1998.

[35] Luo, Z.-Q., and Tseng, P., On the Convergence of a Matrix Splitting Algorithm for the Symmetric Monotone Linear Complementarity Problem, *SIAM Journal on Control and Optimization*, Vol. 29, pp. 1037-1060, 1991.

[36] Madsen, K., Nielsen, H. B., and Pinar, M. Ç., A New Finite Continuation Algorithm for Linear Programming, *SIAM Journal on Optimization*, Vol. 6, pp. 600–616, 1996.

[37] Mangasarian, O. L., Iterative Solution of Linear Programs, *SIAM Journal on Numerical Analysis*, Vol. 18, pp. 606–614, 1981.

[38] Mangasarian, O. L., Normal Solution of Linear Programs, *Mathematical Programming Study*, Vol. 22, pp. 206-216, 1984.

[39] Mangasarian, O. L., Some Applications of Penalty Functions in Mathematical Programming, in *Optimization and Related Fields*, edited by R. Conti, E. De Giorgi, and F. Giannessi, pp. 307-329, Springer-Verlag, Heidelberg, 1986.

[40] Mangasarian, O. L. and Meyer, R. R., Nonlinear Perturbation of Linear Programs, *SIAM Journal on Control and Optimization*, Vol. 17, pp. 745-752, 1979.

[41] McLinden, L., An Analogue of Moreau's Proximation Theorem, *Pacific Journal of Mathematics*, Vol. 88, pp. 101–161, 1980.

[42] McLinden, L., The Complementarity Problem for Maximal Monotone Multifunctions, in *Variational Inequalities and Complementarity Problems: Theory and Applications*, edited by R. W. Cottle, F. Giannessi and J.-L. Lions, pp. 251–270, John Wiley & Sons, New York, 1980.

[43] Megiddo, N., Pathways to the Optimal Set in Linear Programming, in *Progress in Mathematical Programming, Interior-Point and Related Methods*, edited by N. Megiddo, pp. 131–158, Springer-Verlag, 1989.

[44] Pinar, M. Ç., Piecewise-Linear Pathways to the Optimal Solution Set in Linear Programming, *Journal of Optimization Theory and Applications*, to appear.

[45] Rajasekera, J. R., and Fang, S. C., Deriving an Unconstrained Convex Program for Linear Programming, *Journal of Optimization Theory and Applications*, Vol. 75, pp. 603–612, 1992.

[46] Ravindran, G. and Gowda, M. S., Regularization of P_0-Functions in Box Variational Inequality Problems, Department of Mathematics and Statistics, University of Maryland, Baltimore, August 1997.

[47] Robinson, S. M., Generalized Equations and Their Solutions, Part I: Basic Theory, *Mathematical Programming Study*, Vol. 10, pp. 128-141, 1979.

[48] Rockafellar, R. T. and Wets, R. J.-B., *Variational Analysis*, Springer-Verlag, Berlin, 1997.

[49] Subramanian, P. K., A Note on Least Two Norm Solutions of Monotone Complementarity Problems, *Applied Mathematics Letters*, Vol. 1, pp. 395-397, 1988.

[50] Sun, D., A Regularization Newton Method for Solving Nonlinear Complementarity Problems, Applied Mathematics Report AMR 97/15, School of Mathematics, the University of New South Wales, Sydney, July 1997; revised February, 1998; to appear in *Applied Mathematics and Optimization*.

[51] Sznajder, R. and Gowda, M. S., On the Limiting Behavior of the Trajectory of Regularized Solutions of a P_0-Complementarity Problem, Department of Mathematics and Statistics, University of Maryland, Baltimore, October 1997.

[52] Tseng, P., Convergence and Error Bound for Perturbation of Linear Programs, Department of Mathematics, University of Washington, Seattle, June 1997; revised May 1998; to appear in *Computational Optimization and Applications*.

[53] Venkateswaran, V., An Algorithm for the Linear Complementarity Problem with a P_0-Matrix, *SIAM Journal on Matrix Analysis and Applications*, Vol. 14, pp. 967-977, 1993.

[54] Wright, S. J., *Primal-Dual Interior-Point Methods*, SIAM Publications, Philadelphia, 1997.

Lecture Notes in Economics and Mathematical Systems

For information about Vols. 1–290 please contact your bookseller or Springer-Verlag

Vol. 291: N. Takahashi, Design of Adaptive Organizations. VI, 140 pages. 1987.

Vol. 292: I. Tchijov, L. Tomaszewicz (Eds.), Input-Output Modeling. Proceedings, 1985. VI, 195 pages. 1987.

Vol. 293: D. Batten, J. Casti, B. Johansson (Eds.), Economic Evolution and Structural Adjustment. Proceedings, 1985. VI, 382 pages.

Vol. 294: J. Jahn, W. Knabs (Eds.), Recent Advances and Historical Development of Vector Optimization. VII, 405 pages. 1987.

Vol. 295. H. Meister, The Purification Problem for Constrained Games with Incomplete Information. X, 127 pages. 1987.

Vol. 296: A. Börsch-Supan, Econometric Analysis of Discrete Choice. VIII, 211 pages. 1987.

Vol. 297: V. Fedorov, H. Läuter (Eds.), Model-Oriented Data Analysis. Proceedings, 1987. VI, 239 pages. 1988.

Vol. 298: S.H. Chew, Q. Zheng, Integral Global Optimization. VII, 179 pages. 1988.

Vol. 299: K. Marti, Descent Directions and Efficient Solutions in Discretely Distributed Stochastic Programs. XIV, 178 pages. 1988.

Vol. 300: U. Derigs, Programming in Networks and Graphs. XI, 315 pages. 1988.

Vol. 301: J. Kacprzyk, M. Roubens (Eds.), Non-Conventional Preference Relations in Decision Making. VII, 155 pages. 1988.

Vol. 302: H.A. Eiselt, G. Pederzoli (Eds.), Advances in Optimization and Control. Proceedings, 1986. VIII, 372 pages. 1988.

Vol. 303: F.X. Diebold, Empirical Modeling of Exchange Rate Dynamics. VII, 143 pages. 1988.

Vol. 304: A. Kurzhanski, K. Neumann, D. Pallaschke (Eds.), Optimization, Parallel Processing and Applications. Proceedings, 1987. VI, 292 pages. 1988.

Vol. 305: G.-J.C.Th. van Schijndel, Dynamic Firm and Investor Behaviour under Progressive Personal Taxation. X, 215 pages.1988.

Vol. 306: Ch. Klein, A Static Microeconomic Model of Pure Competition. VIII, 139 pages. 1988.

Vol. 307: T.K. Dijkstra (Ed.), On Model Uncertainty and its Statistical Implications. VII, 138 pages. 1988.

Vol. 308: J.R. Daduna, A. Wren (Eds.), Computer-Aided Transit Scheduling. VIII, 339 pages. 1988.

Vol. 309: G. Ricci, K. Velupillai (Eds.), Growth Cycles and Multisectoral Economics: The Goodwin Tradition. III, 126 pages. 1988.

Vol. 310: J. Kacprzyk, M. Fedrizzi (Eds.), Combining Fuzzy Imprecision with Probabilistic Uncertainty in Decision Making. IX, 399 pages. 1988.

Vol. 311: R. Färe, Fundamentals of Production Theory. IX, 163 pages. 1988.

Vol. 312: J. Krishnakumar, Estimation of Simultaneous Equation Models with Error Components Structure. X, 357 pages. 1988.

Vol. 313: W. Jammernegg, Sequential Binary Investment Decisions. VI, 156 pages. 1988.

Vol. 314: R. Tietz, W. Albers, R. Selten (Eds.), Bounded Rational Behavior in Experimental Games and Markets. VI, 368 pages. 1988.

Vol. 315: I. Orishimo, G.J.D. Hewings, P. Nijkamp (Eds), Information Technology: Social and Spatial Perspectives. Proceedings 1986. VI, 268 pages. 1988.

Vol. 316: R.L. Basmann, D.J. Slottje, K. Hayes, J.D. Johnson, D.J. Molina, The Generalized Fechner-Thurstone Direct Utility Function and Some of its Uses. VIII, 159 pages. 1988.

Vol. 317: L. Bianco, A. La Bella (Eds.), Freight Transport Planning and Logistics. Proceedings, 1987. X, 568 pages. 1988.

Vol. 318: T. Doup, Simplicial Algorithms on the Simplotope. VIII, 262 pages. 1988.

Vol. 319: D.T. Luc, Theory of Vector Optimization. VIII, 173 pages. 1989.

Vol. 320: D. van der Wijst, Financial Structure in Small Business. VII, 181 pages. 1989.

Vol. 321: M. Di Matteo, R.M. Goodwin, A. Vercelli (Eds.), Technological and Social Factors in Long Term Fluctuations. Proceedings. IX, 442 pages. 1989.

Vol. 322: T. Kollintzas (Ed.), The Rational Expectations Equilibrium Inventory Model. XI, 269 pages. 1989.

Vol. 323: M.B.M. de Koster, Capacity Oriented Analysis and Design of Production Systems. XII, 245 pages. 1989.

Vol. 324: I.M. Bomze, B.M. Pötscher, Game Theoretical Foundations of Evolutionary Stability. VI, 145 pages. 1989.

Vol. 325: P. Ferri, E. Greenberg, The Labor Market and Business Cycle Theories. X, 183 pages. 1989.

Vol. 326: Ch. Sauer, Alternative Theories of Output, Unemployment, and Inflation in Germany: 1960–1985. XIII, 206 pages. 1989.

Vol. 327: M. Tawada, Production Structure and International Trade. V, 132 pages. 1989.

Vol. 328: W. Güth, B. Kalkofen, Unique Solutions for Strategic Games. VII, 200 pages. 1989.

Vol. 329: G. Tillmann, Equity, Incentives, and Taxation. VI, 132 pages. 1989.

Vol. 330: P.M. Kort, Optimal Dynamic Investment Policies of a Value Maximizing Firm. VII, 185 pages. 1989.

Vol. 331: A. Lewandowski, A.P. Wierzbicki (Eds.), Aspiration Based Decision Support Systems. X, 400 pages. 1989.

Vol. 332: T.R. Gulledge, Jr., L.A. Litteral (Eds.), Cost Analysis Applications of Economics and Operations Research. Proceedings. VII, 422 pages. 1989.

Vol. 333: N. Dellaert, Production to Order. VII, 158 pages. 1989.

Vol. 334: H.-W. Lorenz, Nonlinear Dynamical Economics and Chaotic Motion. XI, 248 pages. 1989.

Vol. 335: A.G. Lockett, G. Islei (Eds.), Improving Decision Making in Organisations. Proceedings. IX, 606 pages. 1989.

Vol. 336: T. Puu, Nonlinear Economic Dynamics. VII, 119 pages. 1989.

Vol. 337: A. Lewandowski, I. Stanchev (Eds.), Methodology and Software for Interactive Decision Support. VIII, 309 pages. 1989.

Vol. 338: J.K. Ho, R.P. Sundarraj, DECOMP: An Implementation of Dantzig-Wolfe Decomposition for Linear Programming. VI, 206 pages.

Vol. 339: J. Terceiro Lomba, Estimation of Dynamic Econometric Models with Errors in Variables. VIII, 116 pages. 1990.

Vol. 340: T. Vasko, R. Ayres, L. Fontvieille (Eds.), Life Cycles and Long Waves. XIV, 293 pages. 1990.

Vol. 341: G.R. Uhlich, Descriptive Theories of Bargaining. IX, 165 pages. 1990.

Vol. 342: K. Okuguchi, F. Szidarovszky, The Theory of Oligopoly with Multi-Product Firms. V, 167 pages. 1990.

Vol. 343: C. Chiarella, The Elements of a Nonlinear Theory of Economic Dynamics. IX, 149 pages. 1990.

Vol. 344: K. Neumann, Stochastic Project Networks. XI, 237 pages. 1990.

Vol. 345: A. Cambini, E. Castagnoli, L. Martein, P Mazzoleni, S. Schaible (Eds.), Generalized Convexity and Fractional Programming with Economic Applications. Proceedings, 1988. VII, 361 pages. 1990.

Vol. 346: R. von Randow (Ed.), Integer Programming and Related Areas. A Classified Bibliography 1984–1987. XIII, 514 pages. 1990.

Vol. 347: D. Ríos Insua, Sensitivity Analysis in Multiobjective Decision Making. XI, 193 pages. 1990.

Vol. 348: H. Störmer, Binary Functions and their Applications. VIII, 151 pages. 1990.

Vol. 349: G.A. Pfann, Dynamic Modelling of Stochastic Demand for Manufacturing Employment. VI, 158 pages. 1990.

Vol. 350: W.-B. Zhang, Economic Dynamics. X, 232 pages. 1990.

Vol. 351: A. Lewandowski, V. Volkovich (Eds.), Multiobjective Problems of Mathematical Programming. Proceedings, 1988. VII, 315 pages. 1991.

Vol. 352: O. van Hilten, Optimal Firm Behaviour in the Context of Technological Progress and a Business Cycle. XII, 229 pages. 1991.

Vol. 353: G. Ricci (Ed.), Decision Processes in Economics. Proceedings, 1989. III, 209 pages 1991.

Vol. 354: M. Ivaldi, A Structural Analysis of Expectation Formation. XII, 230 pages. 1991.

Vol. 355: M. Salomon. Deterministic Lotsizing Models for Production Planning. VII, 158 pages. 1991.

Vol. 356: P. Korhonen, A. Lewandowski, J . Wallenius (Eds.), Multiple Criteria Decision Support. Proceedings, 1989. XII, 393 pages. 1991.

Vol. 357: P. Zörnig, Degeneracy Graphs and Simplex Cycling. XV, 194 pages. 1991.

Vol. 358: P. Knottnerus, Linear Models with Correlated Disturbances. VIII, 196 pages. 1991.

Vol. 359: E. de Jong, Exchange Rate Determination and Optimal Economic Policy Under Various Exchange Rate Regimes. VII, 270 pages. 1991.

Vol. 360: P. Stalder, Regime Translations, Spillovers and Buffer Stocks. VI, 193 pages . 1991.

Vol. 361: C. F. Daganzo, Logistics Systems Analysis. X, 321 pages. 1991.

Vol. 362: F. Gehrels, Essays in Macroeconomics of an Open Economy. VII, 183 pages. 1991.

Vol. 363: C. Puppe, Distorted Probabilities and Choice under Risk. VIII, 100 pages . 1991

Vol. 364: B. Horvath, Are Policy Variables Exogenous? XII, 162 pages. 1991.

Vol. 365: G. A. Heuer, U. Leopold-Wildburger. Balanced Silverman Games on General Discrete Sets. V, 140 pages. 1991.

Vol. 366: J. Gruber (Ed.), Econometric Decision Models. Proceedings, 1989. VIII, 636 pages. 1991.

Vol. 367: M. Grauer, D. B. Pressmar (Eds.), Parallel Computing and Mathematical Optimization. Proceedings. V, 208 pages. 1991.

Vol. 368: M. Fedrizzi, J. Kacprzyk, M. Roubens (Eds.), Interactive Fuzzy Optimization. VII, 216 pages. 1991.

Vol. 369: R. Koblo, The Visible Hand. VIII, 131 pages.1991.

Vol. 370: M. J. Beckmann, M. N. Gopalan, R. Subramanian (Eds.), Stochastic Processes and their Applications. Proceedings, 1990. XLI, 292 pages. 1991.

Vol. 371: A. Schmutzler, Flexibility and Adjustment to Information in Sequential Decision Problems. VIII, 198 pages. 1991.

Vol. 372: J. Esteban, The Social Viability of Money. X, 202 pages. 1991.

Vol. 373: A. Billot, Economic Theory of Fuzzy Equilibria. XIII, 164 pages. 1992.

Vol. 374: G. Pflug, U. Dieter (Eds.), Simulation and Optimization. Proceedings, 1990. X, 162 pages. 1992.

Vol. 375: S.-J. Chen, Ch.-L. Hwang, Fuzzy Multiple Attribute Decision Making. XII, 536 pages. 1992.

Vol. 376: K.-H. Jöckel, G. Rothe, W. Sendler (Eds.), Bootstrapping and Related Techniques. Proceedings, 1990. VIII, 247 pages. 1992.

Vol. 377: A. Villar, Operator Theorems with Applications to Distributive Problems and Equilibrium Models. XVI, 160 pages. 1992.

Vol. 378: W. Krabs, J. Zowe (Eds.), Modern Methods of Optimization. Proceedings, 1990. VIII, 348 pages. 1992.

Vol. 379: K. Marti (Ed.), Stochastic Optimization. Proceedings, 1990. VII, 182 pages. 1992.

Vol. 380: J. Odelstad, Invariance and Structural Dependence. XII, 245 pages. 1992.

Vol. 381: C. Giannini, Topics in Structural VAR Econometrics. XI, 131 pages. 1992.

Vol. 382: W. Oettli, D. Pallaschke (Eds.), Advances in Optimization. Proceedings, 1991. X, 527 pages. 1992.

Vol. 383: J. Vartiainen, Capital Accumulation in a Corporatist Economy. VII, 177 pages. 1992.

Vol. 384: A. Martina, Lectures on the Economic Theory of Taxation. XII, 313 pages. 1992.

Vol. 385: J. Gardeazabal, M. Regúlez, The Monetary Model of Exchange Rates and Cointegration. X, 194 pages. 1992.

Vol. 386: M. Desrochers, J.-M. Rousseau (Eds.), Computer-Aided Transit Scheduling. Proceedings, 1990. XIII, 432 pages. 1992.

Vol. 387: W. Gaertner, M. Klemisch-Ahlert, Social Choice and Bargaining Perspectives on Distributive Justice. VIII, 131 pages. 1992.

Vol. 388: D. Bartmann, M. J. Beckmann, Inventory Control. XV, 252 pages. 1992.

Vol. 389: B. Dutta, D. Mookherjee, T. Parthasarathy, T. Raghavan, D. Ray, S. Tijs (Eds.), Game Theory and Economic Applications. Proceedings, 1990. IX, 454 pages. 1992.

Vol. 390: G. Sorger, Minimum Impatience Theorem for Recursive Economic Models. X, 162 pages. 1992.

Vol. 391: C. Keser, Experimental Duopoly Markets with Demand Inertia. X, 150 pages. 1992.

Vol. 392: K. Frauendorfer, Stochastic Two-Stage Programming. VIII, 228 pages. 1992.

Vol. 393: B. Lucke, Price Stabilization on World Agricultural Markets. XI, 274 pages. 1992.

Vol. 394: Y.-J. Lai, C.-L. Hwang, Fuzzy Mathematical Programming. XIII, 301 pages. 1992.

Vol. 395: G. Haag, U. Mueller, K. G. Troitzsch (Eds.), Economic Evolution and Demographic Change. XVI, 409 pages. 1992.

Vol. 396: R. V. V. Vidal (Ed.), Applied Simulated Annealing. VIII, 358 pages. 1992.

Vol. 397: J. Wessels, A. P. Wierzbicki (Eds.), User-Oriented Methodology and Techniques of Decision Analysis and Support. Proceedings, 1991. XII, 295 pages. 1993.

Vol. 398: J.-P. Urbain, Exogeneity in Error Correction Models. XI, 189 pages. 1993.

Vol. 399: F. Gori, L. Geronazzo, M. Galeotti (Eds.), Nonlinear Dynamics in Economics and Social Sciences. Proceedings, 1991. VIII, 367 pages. 1993.

Vol. 400: H. Tanizaki, Nonlinear Filters. XII, 203 pages. 1993.

Vol. 401: K. Mosler, M. Scarsini, Stochastic Orders and Applications. V, 379 pages. 1993.

Vol. 402: A. van den Elzen, Adjustment Processes for Exchange Economies and Noncooperative Games. VII, 146 pages. 1993.

Vol. 403: G. Brennscheidt, Predictive Behavior. VI, 227 pages. 1993.

Vol. 404: Y.-J. Lai, Ch.-L. Hwang, Fuzzy Multiple Objective Decision Making. XIV, 475 pages. 1994.

Vol. 405: S. Komlósi, T. Rapcsák, S. Schaible (Eds.), Generalized Convexity. Proceedings, 1992. VIII, 404 pages. 1994.

Vol. 406: N. M. Hung, N. V. Quyen, Dynamic Timing Decisions Under Uncertainty. X, 194 pages. 1994.

Vol. 407: M. Ooms, Empirical Vector Autoregressive Modeling. XIII, 380 pages. 1994.

Vol. 408: K. Haase, Lotsizing and Scheduling for Production Planning. VIII, 118 pages. 1994.

Vol. 409: A. Sprecher, Resource-Constrained Project Scheduling. XII, 142 pages. 1994.

Vol. 410: R. Winkelmann, Count Data Models. XI, 213 pages. 1994.

Vol. 411: S. Dauzère-Péres, J.-B. Lasserre, An Integrated Approach in Production Planning and Scheduling. XVI, 137 pages. 1994.

Vol. 412: B. Kuon, Two-Person Bargaining Experiments with Incomplete Information. IX, 293 pages. 1994.

Vol. 413: R. Fiorito (Ed.), Inventory, Business Cycles and Monetary Transmission. VI, 287 pages. 1994.

Vol. 414: Y. Crama, A. Oerlemans, F. Spieksma, Production Planning in Automated Manufacturing. X, 210 pages. 1994.

Vol. 415: P. C. Nicola, Imperfect General Equilibrium. XI, 167 pages. 1994.

Vol. 416: H. S. J. Cesar, Control and Game Models of the Greenhouse Effect. XI, 225 pages. 1994.

Vol. 417: B. Ran, D. E. Boyce, Dynamic Urban Transportation Network Models. XV, 391 pages. 1994.

Vol. 418: P. Bogetoft, Non-Cooperative Planning Theory. XI, 309 pages. 1994.

Vol. 419: T. Maruyama, W. Takahashi (Eds.), Nonlinear and Convex Analysis in Economic Theory. VIII, 306 pages. 1995.

Vol. 420: M. Peeters, Time-To-Build. Interrelated Investment and Labour Demand Modelling. With Applications to Six OECD Countries. IX, 204 pages. 1995.

Vol. 421: C. Dang, Triangulations and Simplicial Methods. IX, 196 pages. 1995.

Vol. 422: D. S. Bridges, G. B. Mehta, Representations of Preference Orderings. X, 165 pages. 1995.

Vol. 423: K. Marti, P. Kall (Eds.), Stochastic Programming. Numerical Techniques and Engineering Applications. VIII, 351 pages. 1995.

Vol. 424: G. A. Heuer, U. Leopold-Wildburger, Silverman's Game. X, 283 pages. 1995.

Vol. 425: J. Kohlas, P.-A. Monney, A Mathematical Theory of Hints. XIII, 419 pages, 1995.

Vol. 426: B. Finkenstädt, Nonlinear Dynamics in Economics. IX, 156 pages. 1995.

Vol. 427: F. W. van Tongeren, Microsimulation Modelling of the Corporate Firm. XVII, 275 pages. 1995.

Vol. 428: A. A. Powell, Ch. W. Murphy, Inside a Modern Macroeconometric Model. XVIII, 424 pages. 1995.

Vol. 429: R. Durier, C. Michelot, Recent Developments in Optimization. VIII, 356 pages. 1995.

Vol. 430: J. R. Daduna, I. Branco, J. M. Pinto Paixão (Eds.), Computer-Aided Transit Scheduling. XIV, 374 pages. 1995.

Vol. 431: A. Aulin, Causal and Stochastic Elements in Business Cycles. XI, 116 pages. 1996.

Vol. 432: M. Tamiz (Ed.), Multi-Objective Programming and Goal Programming. VI, 359 pages. 1996.

Vol. 433: J. Menon, Exchange Rates and Prices. XIV, 313 pages. 1996.

Vol. 434: M. W. J. Blok, Dynamic Models of the Firm. VII, 193 pages. 1996.

Vol. 435: L. Chen, Interest Rate Dynamics, Derivatives Pricing, and Risk Management. XII, 149 pages. 1996.

Vol. 436: M. Klemisch-Ahlert, Bargaining in Economic and Ethical Environments. IX, 155 pages. 1996.

Vol. 437: C. Jordan, Batching and Scheduling. IX, 178 pages. 1996.

Vol. 438: A. Villar, General Equilibrium with Increasing Returns. XIII, 164 pages. 1996.

Vol. 439: M. Zenner, Learning to Become Rational. VII, 201 pages. 1996.

Vol. 440: W. Ryll, Litigation and Settlement in a Game with Incomplete Information. VIII, 174 pages. 1996.

Vol. 441: H. Dawid, Adaptive Learning by Genetic Algorithms. IX, 166 pages.1996.

Vol. 442: L. Corchón, Theories of Imperfectly Competitive Markets. XIII, 163 pages. 1996.

Vol. 443: G. Lang, On Overlapping Generations Models with Productive Capital. X, 98 pages. 1996.

Vol. 444: S. Jørgensen, G. Zaccour (Eds.), Dynamic Competitive Analysis in Marketing. X, 285 pages. 1996.

Vol. 445: A. H. Christer, S. Osaki, L. C. Thomas (Eds.), Stochastic Modelling in Innovative Manufactoring. X, 361 pages. 1997.

Vol. 446: G. Dhaene, Encompassing. X, 160 pages. 1997.

Vol. 447: A. Artale, Rings in Auctions. X, 172 pages. 1997.

Vol. 448: G. Fandel, T. Gal (Eds.), Multiple Criteria Decision Making. XII, 678 pages. 1997.

Vol. 449: F. Fang, M. Sanglier (Eds.), Complexity and Self-Organization in Social and Economic Systems. IX, 317 pages, 1997.

Vol. 450: P. M. Pardalos, D.W. Hearn, W.W. Hager, (Eds.), Network Optimization. VIII, 485 pages, 1997.

Vol. 451: M. Salge, Rational Bubbles. Theoretical Basis, Economic Relevance, and Empirical Evidence with a Special Emphasis on the German Stock Market.IX, 265 pages. 1997.

Vol. 452: P. Gritzmann, R. Horst, E. Sachs, R. Tichatschke (Eds.), Recent Advances in Optimization. VIII, 379 pages. 1997.

Vol. 453: A. S. Tangian, J. Gruber (Eds.), Constructing Scalar-Valued Objective Functions. VIII, 298 pages. 1997.

Vol. 454: H.-M. Krolzig, Markov-Switching Vector Autoregressions. XIV, 358 pages. 1997.

Vol. 455: R. Caballero, F. Ruiz, R. E. Steuer (Eds.), Advances in Multiple Objective and Goal Programming. VIII, 391 pages. 1997.

Vol. 456: R. Conte, R. Hegselmann, P. Terna (Eds.), Simulating Social Phenomena. VIII, 536 pages. 1997.

Vol. 457: C. Hsu, Volume and the Nonlinear Dynamics of Stock Returns. VIII, 133 pages. 1998.

Vol. 458: K. Marti, P. Kall (Eds.), Stochastic Programming Methods and Technical Applications. X, 437 pages. 1998.

Vol. 459: H. K. Ryu, D. J. Slottje, Measuring Trends in U.S. Income Inequality. XI, 195 pages. 1998.

Vol. 460: B. Fleischmann, J. A. E. E. van Nunen, M. G. Speranza, P. Stähly, Advances in Distribution Logistic. XI, 535 pages. 1998.

Vol. 461: U. Schmidt, Axiomatic Utility Theory under Risk. XV, 201 pages. 1998.

Vol. 462: L. von Auer, Dynamic Preferences, Choice Mechanisms, and Welfare. XII, 226 pages. 1998.

Vol. 463: G. Abraham-Frois (Ed.), Non-Linear Dynamics and Endogenous Cycles. VI, 204 pages. 1998.

Vol. 464: A. Aulin, The Impact of Science on Economic Growth and its Cycles. IX, 204 pages. 1998.

Vol. 465: T. J. Stewart, R. C. van den Honert (Eds.), Trends in Multicriteria Decision Making. X, 448 pages. 1998.

Vol. 466: A. Sadrieh, The Alternating Double Auction Market. VII, 350 pages. 1998.

Vol. 467: H. Hennig-Schmidt, Bargaining in a Video Experiment. Determinants of Boundedly Rational Behavior. XII, 221 pages. 1999.

Vol. 468: A. Ziegler, A Game Theory Analysis of Options. XIV, 145 pages. 1999.

Vol. 469: M. P. Vogel, Environmental Kuznets Curves. XIII, 197 pages. 1999.

Vol. 470: M. Ammann, Pricing Derivative Credit Risk. XII, 228 pages. 1999.

Vol. 471: N. H. M. Wilson (Ed.), Computer-Aided Transit Scheduling. XI, 444 pages. 1999.

Vol. 472: J.-R. Tyran, Money Illusion and Strategic Complementarity as Causes of Monetary Non-Neutrality. X, 228 pages. 1999.

Vol. 473: S. Helber, Performance Analysis of Flow Lines with Non-Linear Flow of Material. IX, 280 pages. 1999.

Vol. 474: U. Schwalbe, The Core of Economies with Asymmetric Information. IX, 141 pages. 1999.

Vol. 475: L. Kaas, Dynamic Macroelectronics with Imperfect Competition. XI, 155 pages. 1999.

Vol. 476: R. Demel, Fiscal Policy, Public Debt and the Term Structure of Interest Rates. X, 279 pages. 1999.

Vol. 477: M. Théra, R. Tichatschke (Eds.), Ill-posed Variational Problems and Regularization Techniques. VIII, 274 pages. 1999.